BANANA MAN

Peeling and Revealing

Life in the Tropics

SPECIAL EDITION

**Complete &
Unabridged**

BANANA MAN

Peeling and Revealing

Life in the Tropics

SPECIAL EDITION

Complete &
Unabridged

Clyde S. Stephens

Edited by
Dr. Ken Huff

Cecropia Press

Edited by Dr. Ken Huff.
Front Cover design: Cecropia Press
Back cover photograph: Clyde S. Stephens.

972.87
St43 Stephens, Clyde S.
Banana Man : Peeling and Revealing, Life in the Tropics / Clyde S. Stephens. -- Panama : Cecropia Press, 2023.
758 pages ; 23 cm.

ISBN 978-9962-715-45-0 (Complete work)

ISBN 978-9962-71 5-46-7 (Book 1)
ISBN 978-9962-715-47-4 (Book 2)
ISBN 978-9962-715-48-1 (Book 3)

1. PANAMA – HISTORY I. Title.
2. Bananas 3. United Fruit Company 4. Tropical America
5. Tropical History 6. Central America 7. Natural History
8. History 9. Germans

Cecropia Press
cecropiapress.com

Dedicated to Sid Banack, Sue Trafton Hamer, Walter Hamer,
and Phyllis, my wife (with her arm locked in mine).
They kept the stories alive for me to write.
Photograph: Cheryl Banack Roseland

Table of Contents

Map of Central America 13

Foreword by Dr. Stanley Heckadon-Moreno 15

Book 1 19

1. Bananas and the Twin Towers 21
2. Three Trips to Cuba 39
3. Clyde the Guide 47
4. Saving Locomotive No. 84, by Billy Saravanja 57
5. Jose Antonio Velasquez: Primitive Artist 61
6. The Yard, by Gordon Kidd 87
7. Tela Teenager, by Gordon Kidd 93
8. Turning over Old Stones of the Past 99
9. Banana People under a Dictatorship 123
10. A Crooked Path to Citizenship, by Walter Hamer 147
11. When Truth Ends and Fiction Begins 153
12. From Bananas to Cacao, Abaca, Rubber and Back to Bananas 161
13. From Rainforest to Bananas to Abaca to Oil Palm in Honduras 179
14. An Abaca Family in Guatemala, by Millicent Crichton Empedocles 185
15. The End of Abaca in Guatemala, by Mario Mena 195
16. Life on an Abaca Farm in Costa Rica, by Harold Brent and Billy Saravanja 197
17. United Fruit Company in Cuba 201
18. Cooks Tour Continued 221

Book 2 237

1. Sea Level Canal Study 239
2. Science Versus Tradition 245
3. Canals in Bocas del Toro 253
4. Taming the San San River 267
5. Hospital Point: A Lost Story of Medical History 271
6. With C. H. Ellis Through Central America and Panama, by Ed Loyd 291
7. Unforgettable Memories at Hospital Point 303
8. Natural Entertainment at Hospital Point 311
9. Tick and Spider Bites 331
10. Black Physicians in early 20th-century Bocas del Toro, by Ariel Rene Perez Price 333
11. A Boy from Bocas, by Luis "Ibu" Alvarado 345
12. From Italy to the Western Hemisphere: A Brief History of the Messina Family, by Paul Messina 365
13. Sharks, by John Munch, Dr. Erin Dillon and Clyde Stephens 383
14. Peace Corps Volunteers in Bocas del Toro, by Joseph Looby 393
15. Family Planning and the Wart 409
16. Reservations, Please 415
17. Bluefields – Panama: Going Back in Time 419
18. Striking It Rich at An Oil Well 437
19. The Zipper and La Solucion 451
20. James Norman Morris Maas: One of a Kind 455
21. Earthquake on Earth Day 473

Book 3 485
1. Close Encounters with Germans 487
2. German Prisoners of War 503
3. The Kandler Family of Bocas del Toro 553
4. Alwyn Hasso von Wedel 591
5. The Probst Family 603
6. The Stargardters 641
7. Friedrich "Fritz" Stargardter 663
Notes 687
Appendix I 723
Appendix II 729
Appendix III 733
Appendix IV 735
Appendix V 737
Appendix VI 739
Appendix VII 741
About this Book 745
Acknowledgments 747
About the Author 755
Books by Clyde S. Stephens 757

Map of Central America and the Caribbean

Foreword

IT IS A PLEASURE and an honor to write this foreword for the most recent book by my friend Clyde Stephens. Very few have researched and published so much on the banana industry of Panama, Central America, northern South America, and the Caribbean. By training, he is an entomologist but given his many interests, he is like a man of the renaissance. Besides rescuing many unknown chapters of one of the most important agricultural industries in the tropics of the New World, he gradually became keenly passionate about the natural history of these countries, their people, customs and beliefs. Although born in the United States, he is also a card-carrying "Bocasman," residing half of the year in Bocas del Toro, on the rainy Caribbean side of the Isthmus of Panama, in a solar-powered house he built at Hospital Point in 1991. This is the spectacular site where the United Fruit Company built its first hospital in 1899. He lives the rest of the year in Tavares, Florida.

I must digress some decades ago when we first met and became friends. In 1980, I was a young anthropologist for Panama's Ministry of Economic Policy and Planning and had just published my first bilingual book on the oral history of the life and times of *don* Carlos Reid, a Creole from Bocas del Toro. One day, a security guard at the Ministry called and said an American entomologist from the Chiriqui Land Company in Bocas del Toro wanted to see me. I told him to let him in. My office was at the back of the Palacio de las Garzas, the Palace of the Herons, Panama's presidential palace, near the Parque de la Catedral, the heart of this old port city. In minutes, we became friends. He told me he had worked in Puerto Armuelles where I was born and we had many friends in common. Clyde told me about his banana work in Central America with the United Fruit Company and about living and exploring in Bocas del Toro. However, he felt frustrated that there was hardly any information in Spanish about the notable history of the banana industry that had played such an important economic and social role in Bocas del Toro and Chiriqui. School children and teachers hardly had any knowledge of the vital role of the banana industry in Panama. Since banana growing for export would soon be a century old in Bocas, he wanted to write a short monograph on this fascinating history and wondered if someone would publish his article in Spanish. He had hundreds of fascinating, but little-known photographs of the birth of the banana industry and its pioneers in Panama.

At the time, I was a member of Panama's Anthropological Association and offered to translate and publish his manuscript in Spanish for Panamanian readers. Our members were delighted because many of them had done research in Bocas del Toro. In 1987, the booklet *Bosquejo histórico del cultivo del banano en la provincia de Bocas del Toro* (1880-1980) became the first special publication of the *Revista Panameña de Antropología.*

Soon, Clyde's research and publications about the banana industry bloomed. He published *Bananeros in Central America: True Stories of the Tropics* (1989). *La historia de Punta Hospital, centro médico pionero 1899-1920* (1997).

Banana People: True Stories of the Tropics (2002). *Bosquejo histórico de la provincia de Bocas del Toro, Panama* (2008). In these publications he rescues the visionary barons of the industry, the plantation managers, the engineers who built the railroads, the ports, and the towns. Then there were the *mandadores* responsible for the farms, the diverse peasantry who gradually became banana field workers in the modern agro-export industry. These workers included those who kept the rail lines, the farms or the stevedores who loaded the banana ships. Also, there was the fascinating history of banana scientists and the doctors and nurses that staffed the hospitals; they were considered the best on the Caribbean side of Central America.

The banana industry also gave rise to a particular vocabulary, in Spanish and English, that Clyde handles masterfully. Geographically, his work takes place where bananas grow along the great tropical rivers with their rich alluvial soils, such as the Sixaola, Changuinola, Reventazon, Terraba, Ulua and the Motagua.

In *Banana Man*, his most recent publication, Clyde delights us with little-known stories on the contributions of the United Fruit Company to the development of Central America, on the development of the manila hemp and rubber plantation industries during World War II, on the history of tropical medicine and on the building of banana canals. One chapter is devoted to living under a dictatorship.

One of the most interesting sections of his book is the role the Germans played in the development of the banana industry and the tragedies they suffered during the first and second world wars when they were expropriated, sent to concentration camps, and expatriated to Germany, where many fell in the hands of the advancing Russian Red Army. Some managed to return, like the Probst family from Chiriqui.

Don Clyde, as people from Bocas call him, is a down-to-earth person, concerned about the welfare of others, who makes visitors feel good at his home, and is very hospitable. He is a great host, with a great sense of humor in English and Spanish, and who likes to poke fun at himself. His interests are wide and range from the economic and social history of regions to their natural history, the people and their customs and beliefs. He is also intellectually very generous, a researcher who likes to share his findings with others.

I am sure that Central Americans, Panamanians, and Colombians will find Clyde's latest publication on the history of Chiquita bananas fascinating.

Stanley Heckadon-Moreno
Panama, March 2023

Book 1

1. Bananas and the Twin Towers

WE ALL REMEMBER WHERE we were on that horrible day of September 11, 2001. It was a day of infamy that the world will never forget. Phyllis and I were visiting Scott and Joyce Dinsmore at their jungle home near Portalón, south of Quepos, Costa Rica. We had just come back from a pleasant morning hike where we saw monkeys and parrots, and we were cooling off in the pool. When Joyce switched on their small black and white TV, she suddenly started making alarming and incoherent sounds of disbelief and we rushed from the pool to the TV.

From this remote tropical paradise on the Pacific coast of Central America, we watched in shock as the Twin Towers collapsed after Islamic terrorists crashed jet planes into the buildings. At 5:20 that afternoon, World Trade Center, Tower 1,[1] collapsed. The rest is history that changed the world.

Years before, in September 1960, I had arrived at one of the United Fruit Company's banana docks on the Hudson River in Lower Manhattan not far from where the World Trade Center with its Twin Towers would be built.

I was taking my first vacation since starting my career in banana research with United Fruit in 1959, having been on the job for 18 months. Employees were given a week of travel time on a banana ship to the United States, plus thirty days of vacation, and then a week of travel time back to the Tropics, a total leave of six weeks. Passage and meals on the banana ship were courtesy of the Company.

At the beginning of my trip, I boarded the SS *Quisqueya* as it steamed through the Panama Canal. A small launch took me to the moving ship at Balboa, and I jumped to a ladder that had been lowered alongside the ship. The Panamanian baggage handler helped me with my suitcase and a heavy box with a camphor wood chest that Dr. Pat Butler, Assistant Director of Research, had asked me to take to New York.

During earlier times, some ships of United Fruit Company's Great White Fleet had luxury accommodations for 95 tourists and Company employees. Eventually, tourist cruises conflicted with the banana trade and accommodations were reduced to take only 12 passengers who were mostly employees. We had a free run of the ship, ate with the captain and officers and enjoyed the cruise. Making ports of call along the way made the trip even more interesting.

The SS *Quisqueya* stopped for two days in Kingston, Jamaica, where I enjoyed the amenities at the stately old Myrtle Bank Hotel owned by the Company. On the second day, I hired a car and driver to see Dunn's River Falls on the north shore. The trip across the island on the narrow mountain road was frightening but the excursion was worth it. Back in Kingston, night life helped calm the nerves of a 26-year-old bachelor.

A few days later, as the ship sailed along the New Jersey coast after midnight, I heard a disturbing commotion on the main deck. The ship had stopped to pick up an elderly man in a small fishing boat because his 10-HP engine had run out of gas. He was drifting on the Atlantic Ocean and was saved only because he yelled very loudly as the ship passed close to him in the dark. This old fisherman got a free ride to the banana dock at North River, as the Hudson River is called in Lower Manhattan.

We approached New York City at dawn, and I stood by one of the officers on the upper deck. The officer told me to watch for North River Trout. I had no idea what this fish was but soon he started exclaiming, "There goes one!" Then another and another. It didn't take me long to see that he was referring to hundreds of condoms that had been flushed into the river the night before. After we had docked, thousands of green Gros Michel banana bunches were unloaded. I cleared immigration and arranged for the transfer of Pat Butler's camphor chest to Boston. Soon, I was on a bus to Ithaca to deliver some insects to Dr. Donald Strong at Cornell University. He and I had been exchanging information about *Oiketicus kirbyii*, the West Indian bagworm that had suddenly appeared by the millions in plantations and was defoliating entire banana plants. That is why I was hired. My assignment was to learn the name of the pest, find out what caused the abrupt outbreak, study the life cycle of the bagworm, collect and identify its natural enemies, and solve the problem.

THE PIERS

When the banana trade began in the late 1800s, banana ships tied up to "Banana Docks" at old slips on the East River at the end of Wall Street. Later, the United Fruit Company used Piers 2, 3, 7 and 9 along the waterfront of southwest Manhattan Island. These piers accommodated operations of the Great White Fleet that had nearly a hundred ships at one time. Ships hauled cargo to the Tropics, cruise passengers on tours, United Fruit Company employees and their families, sugar from

Company refineries in Cuba and cargo from the Tropics to North America and Europe. But bananas by the billions were the main reason for the existence of the Great White Fleet.

Each pier had its own story but Pier 3 was the most noteworthy and had lots of character. Today, there is no sign of these once-prominent docks because the Battery Park City landfill of the 1960s covered them up. Pier 3 was near the site of the Downtown Athletic Club at 20 West Sttreet, which was along the waterfront before the landfill was made.

Pier 3 is where a memorable trip began when Dr. George Wilson and his wife, Claudia, boarded a banana ship on October 3, 1963. He had accepted a job as a Plant Physiologist with the Tropical Research Department in La Lima, Honduras, and the young couple from upstate New York began their adventure in the Tropics. George describes their maiden voyage as follows:

> After leaving on the English flagship, SS *Sinaloa*, from Pier 3 in New York City, our first port of call was Kingston, Jamaica to discharge dry cargo. Captain Norman Thomas informed us that instead of proceeding directly to Puerto Cortes, Honduras, we were going to Mobile, Alabama, to repair damage caused by Hurricane Hattie en route to Jamaica. As we rounded the western tip of Cuba, the captain advised us that our ship had been detoured to New Orleans where Claudia and I could board another ship bound for Puerto Cortes. After enjoying the weekend at the Saint Charles Hotel in New Orleans, we boarded a Dutch flagship for an uneventful trip to our final destination—more than two weeks after our departure from Pier 3 in New York City.

Pier 3 stood out above the other docks because it was where tens of thousands of tourists departed for tropical cruises on banana ships. Tours typically stopped in Havana, Cuba, for a couple of days of sightseeing while the ship discharged cargo. The next stop might have been in Guatemala, Honduras, Costa Rica or Panama, where tourists boarded trains to visit banana plantations. Other tours might have included trips to the cool, scenic highlands, coffee farms, ancient Mayan cities and the Panama Canal. Accommodations, food and ambience were the best and clients frequently returned year after year. In 1930, such a tour cost $240 and included almost everything during the 12 to 17 days on the ship.

To handle so many tourists, the immense pier had a grand reception room with elegant panels of Honduras mahogany, fine terrazzo floors, and walls with United Fruit Company logos and also colorful tropical posters. Refreshments were served while passengers were waiting and getting to know each other and the tour guide. There was a lot of excitement before the boarding of the Great White banana ships. Security consisted of a single uniformed guard at the front door. The wharves on Pier 3 were able to accommodate several ships at the same time. Combined with Piers 2, 7 and 9, the Company could load or unload eight ships. Four or five ships would be tied up on any given day. It took a fleet of tug boats and barges to handle so many vessels.

THE SHIPS AND THEIR CARGO

The ships of the Great White Fleet were constantly being repainted white, a practice that helped lower the temperature of the banana cargo and reduce premature ripening. Most ships had American registry but some flew Honduran, Panamanian, Dutch or British flags. Banana ships came in daily and got priority because of their delicate, perishable cargo.

Some vessels unloaded other tropical fruits, cacao for chocolate, coffee, and Manila hemp fibers for making rope. Citronella grass was processed into an extract for insect repellent. African palm oil was sold to Colgate Palmolive to make Palmolive soap. Big stiff rawhides would be turned into leather. All of these products had their own unique aromas, and it took hundreds of trips by trucks to unload and move the products to market.

Because bananas were grown in very remote parts of Third World countries, the Great White Fleet ships were filled with freight for the return voyage to the Tropics. Sometimes, new cars were tied down as deck cargo; protective solutions had to be applied to reduce salt spray corrosion. Items loaded aboard were manufactured merchandise, canned goods and raw materials for construction. Hundreds of other supplies were needed to operate the thousands of acres of banana plantations and their ports, railroads, trains, vehicles, hospitals, schools, machine shops, power plants, offices, warehouses, airports, bridges and Company stores. It was common to dispatch one or two ships a day from New York to the Caribbean, Central and South America.

Ships also hauled building materials and supplies for thousands of comfortable Company houses that were provided free to all employees regardless of position. Homes of higher category employees were furnished and ready for occupancy. Such Company towns had to be constantly supplied and maintained by the United Fruit Company for thousands of employees.

Some ships went beyond the call of duty. The SS *Cape Ann*

of the Great White Fleet was the first vessel to arrive at the site where the MS *Stockholm* rammed the SS *Andrea Doria* on July 25, 1956, with 1,600 passengers aboard. After the ordeal of rescuing some of the victims, the SS *Cape Ann* continued to New York with 129 survivors. As the ship entered New York Harbor, all vessels sounded their horns in recognition for the valiant rescue. Captain Joseph A. Boyd and his crew received medals for their heroism.

It was a coincidence that I was in the middle of the Atlantic Ocean, returning to New York on a large troopship after serving two years in the U.S. Army in Germany. Our ship's newspaper reported the tragedy the next day when the SS *Andrea Doria* sank. When our troopship arrived in New York, we passed by the MS *Stockholm*, and I snapped a few pictures of its mangled bow, a memorable and gruesome sight.

OFFICES AND DEPARTMENTS

Although Boston was the corporate headquarters of the United Fruit Company since its founding in 1899, New York was the most important location in the Company. It was the leading port for bananas, plus there were numerous offices and department headquarters located on all the piers. These included the following groups:

> Steamship operations, freight operations, sales, Fruit Dispatch Company, passenger operations, medical department, advertising, public relations, ship stores, Maritrop Trading Corporation, ship engineering, ship building, test kitchen, port purser, research laboratory and other support operations.

In all, there were about 600 salaried employees on the

four piers. This number did not include the stevedores and longshoremen who were highly unionized.

Maritrop Trading Corporation was notable because it purchased and shipped millions of items required anywhere in the Company. This covered anything from *machetes* to airplanes and required a colossal effort in ordering, purchasing and shipping.

Another special group of executives were in the Freight and Passenger Department. They were known for efficiency and being on schedule. This was because they came from the railroad tradition where trains had to be on time. All were men, took Gregg shorthand, were excellent typists and were humorless. Their wages were relatively low but they got the job done.

Neither the piers nor the offices were air conditioned. To keep cool, every desk had a small fan. During the summer, when temperatures and humidity were in the 90s, employees could leave an hour early. One executive had a lung condition and was the only employee with an air conditioner installed in his window. There was one elevator that went to the third floor of Pier 3 where the executives' offices were located.

EXECUTIVE VISITS AND TRAVEL

Although Boston was the corporate headquarters, New York was where much of the activities were. Every week, several executives would take the train to New York on Wednesday and return to Boston on the Thursday night train. For that reason, the Company kept a suite in the Barkley Hotel and in the Chatham Hotel in lower Manhattan. A young employee like Tom McCann, who later became a senior executive, had the opportunity to meet and travel with top corporate executives like Sam Zemurray, Kenneth Redmond, Thomas Jefferson Coolidge, George Gardner and Almyr Bump.

Before good international air travel was developed, Company executives making trips to the Tropics came from Boston to New York and boarded one of the Great White Fleet ships. Plantation managers were alerted beforehand and prepared for these executive visits. Such visitors always got the red-carpet treatment.

THE MEDICAL DEPARTMENT

Dr. Edward I. Salisbury was based at Pier 3 and was in charge of the medical department that included eight medical doctors, a dozen nurses and medical technicians. The group treated crews and passengers on ships, performed minor surgery and gave required vaccinations to travelers before they left for the Tropics.

Dr. Salisbury was also the head administrator for all the medical departments in all of the countries where the United Fruit Company operated, including the many hospitals and medical staffs in the Tropics. It was his staff in New York that made sure medical supplies and medicines were properly selected and shipped to the hospitals. When birth control pills became available in 1960, the medical department supplied millions of pills that were dispensed in the various banana zones. The medical staff also had a "snake doctor," Dr. Sackett, who had a laboratory on Pier 3 that produced anti-venom vaccines against snake bites for victims on the banana plantations.

MOVIE SCENES ON THE PIERS

Vice President Hector Harris Robson, a stern Scotsman, liked movies and permitted major motion picture companies to film on the piers. This delighted the labor force because the films featured famous movie stars. In 1954, Humphrey Bogart, starring in the movie *Sabrina* was filmed dashing across Pier 3. He quickly boarded a tugboat in a rush to catch up with his love interest, co-star Audrey Hepburn. She was on a ship leaving for Paris. The movie received five nominations and one Academy Award. The following year, Bogart starred in *The Harder They Fall*, and one scene showed an actor arriving from Argentina on Pier 2 and exiting to the street via Pier 3. Bogart's co-star in this film was Jan Sterling. It was his last movie before he died of cancer the following year.

Making movies was very distracting to longshoremen and stevedores and caused substantial delays in loading and unloading ships. In 1954, movie producers asked Mr. Robson for permission to make *On the Waterfront* and he approved it. However, after reading the manuscript, it was decided that the union might be offended and the deal was cancelled.

CATS AND RATS

It is no secret that wharves attract rats. When edible cargo was spilled on the floor, multitudes of rats came for the feast. The Company had a budget of $50,000 for rat extermination and natural control by cats. A man in charge fed an estimated 150 to 200 cats. Rats were a perennial problem.

RESEARCH DEPARTMENT

The research staff had chemists who monitored banana handling, residues, ripening, disease, quality and shipping problems. Post-harvest studies were coordinated between scientists based in the Tropics and those at Pier 3.

THE END OF AN ERA

In the late 1950s, New York City told the United Fruit Company to vacate everything by 1962 in preparation for the World Trade Center and Battery Park City. This former blue-chip giant was already in financial trouble because of a devastating banana disease, but after the eviction notice, the Company plunged even deeper into a financial tailspin. Many employees lost their jobs. Others went to Company ports on the East and West coasts. Banana operations were moved across the Hudson River to Weehawken, New Jersey.

Among Pier 3 employees who remained with the company was young Tom McCann who was transferred to Corporate headquarters in Boston. At age 18 in 1952, this young man from Brooklyn had started his career at Pier 3 as an office boy. The job paid him $32.50 per week for $5^{1/2}$ days of work. He rapidly climbed the ladder to become senior vice president of Public Relations. Tom also wrote a bestseller called *An American Company: The Tragedy of United Fruit.* The book reveals with candor what only an insider can write.

BANANA MAN

It was Tom McCann who provided incredible details that made much of this story possible. In 2019, he sent me a tape recording with his stories about the banana piers and how the construction of the World Trade Center forced the banana company docks to disappear from lower Manhattan.

It took many months to move out and vacate the United Fruit Company piers. Offices were furnished with nice, solid oak desks, chairs and filing cabinets. None of these furnishings were transferred anywhere—except to the ends of the Company piers where thousands of items were dumped into the depths of North River with the help of longshoremen. Seasoned oak does not rot and all that office furniture is still there today.

BATTERY PARK CITY AND THE WORLD TRADE CENTER[1]

After the United Fruit Company and other tenants left, the piers and pilings were removed by the mid-1960s. Excavations were then made at the site of the World Trade Center, providing landfill where the old banana piers once existed. The 1,200,000 cubic yards of excavated material, plus sand dredged from the harbor, were used to expand the Manhattan shoreline across from West Street to form Battery Park City. The landfill was completed by 1976. This 92-acre addition was then converted into parks, marinas and more buildings.

This episode in Lower Manhattan's history is mostly a forgotten memory except by a few people like Tom McCann who sometimes returns to those nostalgic spots where he started his career in 1952 at Pier 3. Although Battery Park City is on top of the old pier sites, Tom knows that Piers 2 and 3 were in front of The Downtown Athletic Club, now known as The Downtown Club, at 20 West Street. Nearby were Piers 7 and 9, close to the end of Rector Street. Rector Park marks the approximate location where Pier 9 existed before the landfill converted that part of the Hudson River into Battery Park City, something that drastically changed the face of Lower Manhattan Island.

It's hard to believe that this is where the piers of the United Fruit Company once stood and where untold billions of bananas arrived from the Tropics.

World Trade Center, Tower 2, September 11, 2001.

The landfill, outlined on the left, covered the piers and created Battery Park City.

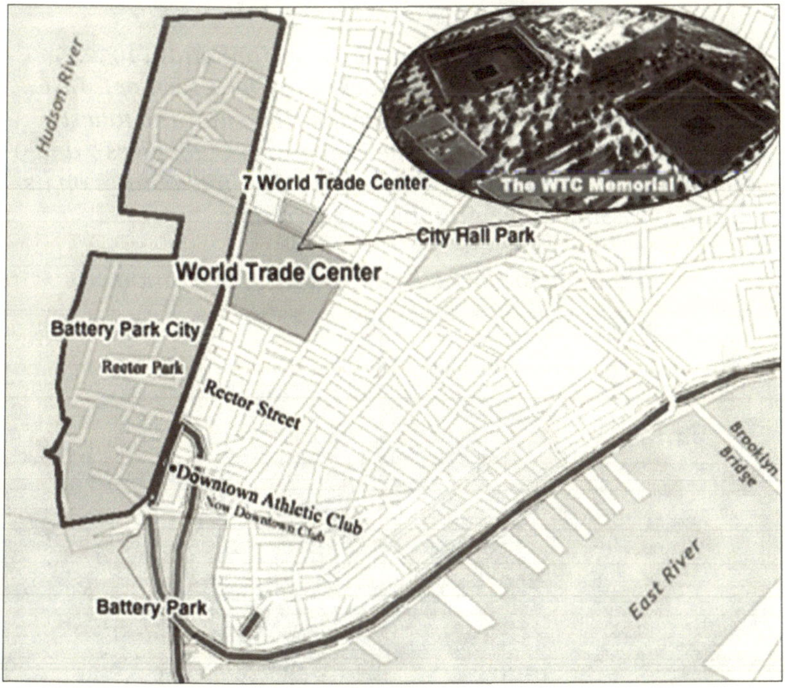

United Fruit Company Magazine, Lower Manhattan, 1952. A Great White Fleet ship is shown docked at Pier 2 (center). Just above the words Pier 3, you can see the UFC sign. The tallest building became what is now The Downtown Club. Piers 7 and 9 came next. The ship trailing smoke is the SS United States on its way to a crossing of the Atlantic in record time.

Old Banana Docks, New York City, circa 1900.

The Piers, Manhattan, 1930. The white asterisk near Pier 3 is where the Downtown Athletic Club once was located. It is now The Downtown Club.

The UFC mailing address in 1954:
Pier 3, North River, New York 6, New York.

Company promotions, circa 1919 - 1929.

The SS Quirigua, part of the Great White Fleet.

The MS Stockholm, July 26, 1956.

Humphrey Bogart starred in Sabrina, 1954, with Ellen Corby, who later won three Emmy Awards as Grandma Walton in the TV show The Waltons. Note the words FLEET and FRUIT in the background. These words were part of the UFC sign perched on Pier 3.

Bringing out and stacking bananas for shipment.

An American Company: The Tragedy of United Fruit by Thomas McCann.

Former UFC Hospital, Almirante, Panama. This is where my wife, Phyllis, gave birth to our two children in 1965 and 1966.

2. Three Trips to Cuba

AS A YOUNG TEENAGER growing up on a remote farm in Hardee County, Florida—long before I became involved in Central America and the banana business—I was fascinated by strange sounds emitting irregularly from a radio station in Cuba. Static frequently drowned out Latin rhythms with bongo drums, maracas, guitars, flutes and a cow bell. Songs were sung in a foreign language called Spanish. I was concerned that members of the family or community might discover my weird interest in a Cuban station. After all, my ancestors had settled in out-back Castalia in the 1860s, and they might be rolling over in their graves just down the road if Clyde was listening to Cuban music.

When I was 16 in 1950, the family moved 20 miles east to the small town of Wauchula to be near schools and have better opportunities. As an example, I was a delegate to the Kiwanis Key Club convention in Miami. Jim Smith was a fellow delegate and we stayed in the McAllister Hotel on Biscayne Boulevard.

After the convention was over, Jim and I noticed a flashy, erotic poster in the window of a travel agency by the hotel. We did not know what a travel agency was but we went inside and asked about the three-day trip to Cuba for $32. Jim and I counted our money and each of us had just about that amount, but not much more. The tour started in Key West, not Miami. The next day, we boarded a Trailways bus to Key West where we flew on a C-47 of Aerovías Q to Havana, a 45-minute flight. Neither of us had passports but we submitted some kind of tourist card. The package tour included two nights at the Hotel Inglaterra.

The city tour took us to the Bacardí distillery where free samples were served to each person. It did not take long for Jim and me to feel the effects of Bacardí rum and banana liqueur, something we had never tasted in our lives. We soon recovered and that afternoon, we did a lot of walking to see Havana and took a ferry boat across the harbor to visit the historic El Morro Fort.

Just before going to the airport on the third day to fly back to Key West, a man from a side street suddenly flashed a portfolio of pornographic pictures in our faces. Our eyes bugged out in disbelief. Before we knew it, he was counting the last few coins we had and then put the pictures in my pocket. Then the porno peddler disappeared in a flash and we were left penniless. Luckily, our tour included bus transportation to the Havana airport.

We had a pre-paid air ticket from Cuba to Key West, but there was no money left to buy a bus ticket from Key West to Miami. By this time, both of us were getting hungry so I stole a five-cent Baby Ruth chocolate bar in a gas station and split it with Jim. From that moment, I knew I was going to hell for stealing, and I have regretted that criminal act ever since—and still feel guilty at age 88.

We walked to U.S. Highway 1 that went to Miami and started to hitchhike. Suddenly, an old rattletrap 1936 Chevrolet stopped and the driver asked: "You boys got gas money?"

The answer was a quick "Yes," another sin committed within the same hour. As we arrived somewhere in Miami, I told the driver to stop without knowing where we were. Jim and I quickly got out of the car and said thanks for the ride. The driver demanded some gas money. We both dug deep into our pockets and gave him all the Cuban coins we had. He was obviously annoyed as we quickly disappeared down the sidewalk.

We had to find the train station because Jim and I had a round-trip ticket to and from Miami. Finally, someone in the street told us how to find the train station and that it was several miles away.

Along the way, we found a mature coconut that had dropped to the sidewalk. Like all boys and young men that I grew up with, we all carried a good pocket knife—even to school and even on the flights to and from Cuba. I opened the "eye" of the nut with the small knife blade, and we shared the refreshing coconut milk that drained into our thirsty mouths. Next, I forcefully cut off strips of the outside fibers in order to get to the hard seed where the mature copra was. That was a major challenge using a pocket knife but it worked. The last step was to crack the hard inner shell and remove the delicious, white copra. That served as our late afternoon lunch and it gave us enough energy to walk to the far-off train station.

At the train station, we had to figure out how to contact my parents in Wauchula so they could drive to Avon Park and pick us up that night. After much pondering and hesitation, we asked a lady in the train station to lend us a dime to make a collect telephone call to my parents. When my parents accepted the collect call, the telephone returned the ten-cent piece, which I gave back to the lady. The strategy worked!

The lady looked at us with compassion and asked if we were hungry. We insisted we were not but she knew better. Soon, a bowl of vegetable soup and crackers were served to us at the lunch counter. She offered more food but we modestly declined her generous offer.

My parents picked us up at the Avon Park train station at 11:00 PM and asked us about the trip to Miami and the convention. For some reason, Jim and I never told our parents that we were going to Cuba. When they found out, there was no alarm or concern. My mother and father knew I had been on a learning experience with the Key Club of Hardee County High School.[1]

We learned a lot about life on the trip. That included hearing live Cuban music in the street—one of my childhood fantasies that was finally fulfilled.

FLYING WITH RAUL CASTRO

When the United Fruit Company hired me as an entomologist, they sent me an airplane ticket from Miami to San Pedro Sula, Honduras, to start my career. On April 7, 1959, I flew on a C 47 of the TAN Airlines (*Transportes Aéreos Nacionales*). The airplane made a stop in Havana, Cuba, where the passengers got off for a break. After the break, Raul Castro came aboard. He was dressed in full guerrilla regalia, wearing a heavy beard and camouflage fatigues, looking like he just came out of the Sierra Maestra Mountains. In fact, it had been only three months earlier that he and his brother, Fidel, triumphantly took over La Habana on January 1, 1959.

Raul Castro was making the revolutionary government's first diplomatic mission to Central America—something that caused much celebration by many but serious concern by others. Accompanying him were four well-dressed men in suits carrying attaché cases. Some press agents and photographers also came aboard. I soon realized that I was on a very historic flight.

When we landed on the grass field outside of San Pedro Sula, a cloud of dust appeared. Living fenceposts[2] with barbed wire bordered each side of the airfield to keep cattle off of the landing strip.

I will never forget the blast of hot air as I got off the airplane. But the heat was normal in April because it was the dry season. There were photographers taking pictures of Castro, and I felt apprehensive in the push and shove in the small open-air, tin roof terminal. Soon, Bob Bullock spotted the new *gringo* entomologist in the crowd and started me on my career. I had planned to stay only a year or so, but before I knew it, several decades had flashed by.

1999 IN CUBA

Encouraged by friends, Phyllis and I organized and led eight nature tours in Central America after our retirement in 1991. We were not travel agents and never made money on those trips. But we enjoyed showing friends, members of tropical fruit clubs and nature societies the best of Panama, Costa Rica, Honduras and Guatemala. Each tour always included a trip to a Chiquita Banana plantation.

In 1999, after our tour group of 24 had left Panama and returned to Orlando, Phyllis and I flew directly to Havana from Panama City. A Cuban lady in Panama made all of the arrangements for seven days for $700 each. That included airfare, airport pick-up and drop-off, hotels, some meals, lots of drinks, a city tour of Havana and Varadero and a dinner club show. Our *gringo* passports were never stamped or looked at during that trip. We traveled on a special identification document. Hospitality and friendliness were incomparable. It helped that we were fluent in Spanish. During the week, we never saw other Americans.

La Habana Vieja, Old Havana, had been restored after years of neglect, and we walked on our own for hours down the narrow streets with many ancient buildings. A personal pleasure for me was listening to Cuban musicians playing on street corners and in restaurants. They were wonderful; then we stumbled upon the huge library of Dr. Carlos Juan Finlay, who was credited with discovering that mosquitoes transmit yellow fever.

As a final gesture, I took Phyllis to the lobby of the Hotel Inglaterra where Jim Smith and I had stayed a half century earlier as teenagers—two teenage boys who forgot to tell their parents that they were going to Cuba.

Clyde Stephens and Jim Smith in front of El Morro, Havana, Cuba, 1950.

Opposite Page: Phyllis Stephens in Old Havana, Cuba, 1999.

3. Clyde the Guide

FROM THE BEGINNING OF my career in bananas, I was occasionally chosen by the manager's office to be a guide for visiting dignitaries or anyone else that management didn't know what to do with. There were a bunch of them. As it turned out, I was the one who learned more than the visitors. These vignettes describe some of the most memorable people that I guided.

THE HERPETOLOGISTS

In 1959, after my orientation at the main Tropical Research Laboratories in La Lima, Honduras, I was assigned to the Coto 47 Research Department in southwest Costa Rica. An impressive staff of scientists welcomed and oriented the new entomologist who knew nothing about banana problems or the business. Names that come to mind are numerous, including Doctors Luis Sequeira and Ivan Buddenhagen (plant pathologists), James O. Harrison and Roy Roig (entomologists), Quinton Holderman (nematologist), Chris Millensted and Hermann Fabrega Proesch (agronomists).

To initiate me, the director asked me to accompany and drive three visiting herpetologists. Their leader was Dr. Arnold G. Kluge, who told me they wanted to collect poisonous snakes. I drove them in a canvas-top Jeep through banana farms that took us to a new, slummy frontier town called Villa Neily outside of the plantations. Pan-American Highway construction had just started, and there were no bridges over the wild rivers that were gushing out of the steep, forested mountains behind Villa Neily.

The herpetologists were excited to see the untouched virgin rainforests and that is where they started their searches. For hours, they turned over logs and rocks and looked at the bases of giant trees that had buttresses where snakes could be waiting. At the end of the day, and when torrential downpours started, it was time to go back to Coto 47. The three herpetologists had not seen nor collected a single snake of any kind. It was disappointing for them but a learning experience for me. It dispelled the myth that dangerous reptiles were lurking everywhere in the jungle. That experience gave me the confidence to explore and enjoy the tropical forests, always with caution, but without fear of snakes.

HOTEL ARENAS DEL DESIERTO

In 1961, Warren Breck was the assistant manager under Harlow von Wald in Puerto Libertador, Dominican Republic. Mr. Breck asked me to review disease problems with Mr. Tucker, a technical representative who was helping to develop a control for Sigatoka, a banana leaf fungus.

The Grenada Company (Chiquita) grew bananas along the Yaque del Norte River that runs through a desert valley. Fruit was also purchased from Dominican-owned farms along the same arid landscape. Although cactus, mesquite and other xerophytic plants grew in this desert, fertile alluvial loam soils produced good fruit as long as the plantations were irrigated. Otherwise, banana plants desiccated and died.

When it was time to eat lunch, I took Mr. Tucker to the only restaurant in the region. It was open-air with a thick palm thatch roof and had a view of the chocolate-colored river. Swarms of flies were about normal. Stray cats and dogs waited on the floor for handouts from diners. Customers were mostly banana people from all echelons of the nearby plantations. A very noticeable sign at the entrance read: "*Prohibida la entrada de mujeres con reputaciones dudosas,*" (Women with dubious reputations are prohibited).

As we waited for our lunch to be served, a couple of young ladies approached our table and asked if they could have a drink and join us. I politely declined their offer. After lunch, we left and once again read the sign at the entrance. This time, the connotation of the sign was more meaningful and caused us both to laugh.

After a long, dusty day of reviewing banana farms and Sigatoka disease, I drove my Company Ford truck to Hotel Arenas del Desierto, which was in Santiago de los Treinta Caballeros. This was a clean, quiet hotel where the main door was locked in order to filter out undesirables from the street. There were no flies inside and *merengue* music on the Rock-Ola was not blasting at a high volume. The hotel had a quiet bar and a nice dance floor. Guests could eat near the bar or be served in their room. Mr. Tucker liked the place.

The next morning, I put Mr. Tucker in a small "public car" with four other passengers and their baggage. It took five hours to ride across the island to the capital, Ciudad Trujillo (later Santo Domingo).

Several days later, Warren Breck called me into his office to discuss my expense account. He was surprised to see that we had stayed at Hotel Arenas del Desierto and raised his eyebrows. I assured him that we had paid by the night, not by the hour. "Ah, that was okay," he said. Later, Mr. Tucker told Warren Breck that he hoped to stay there again. I felt vindicated.

FISHING WITH A BANANA BROKER

Friedel Gellerich was the son of the largest buyer of Chiquita bananas in Germany. When he visited Panama once a year, he would always ask for me to accompany him. After we saw the harvest at a banana farm and reviewed a packing station, Herr Gellerich would say, "That's enough. Now let's go catch some tropical fish."

Each time he visited, we hiked on remote trails in the rainforested hills behind Changuinola. This is where Friedel quested to find new species of tropical fish in cold mountain streams. He told me that he had 22 fresh and saltwater aquariums in his *keller* back in Hamburg. I was impressed.

We became good friends and Phyllis invited him several times for dinner at our house at Farm 8. One weekend, he went with us in our seaworthy *cayuco* through the old banana canal to Boca del Drago for a beach party with Chiquita people.

Bananeros from Changuinola were enjoying a picnic and socializing under a metal roof with lots of picnic tables on the beach sand. In reality, the large structure was a banana packing station that the general manager, Bobby Walker, approved of. It was built by the crew of the chief engineer, Don Harris. I don't think the auditors ever found it to audit.

In mid-afternoon, an ugly weather front brought torrential rains, strong winds and huge waves that were breaking in the channel of Boca del Drago. All the bosses and their families were forced to stay overnight in the bunkhouse because they had boats that were built in the Company shop in Empalme.

Those small boats with 18-horsepower Evinrude engines were nice but not seaworthy enough to cross the turbulent channel, especially with children.

Friedel Gellerich let it be known that he had to fly early Monday morning on the Company airplane to Panama City in order to connect to New York for a flight on Lufthansa to Germany. My *cayuco* was 106 centimeters wide, 7.3 meters long and was powered with a 25-horsepower Evinrude; it was the only boat that was seaworthy, and I volunteered to get Gellerich back to the mainland. Phyllis, Linda, Carl and Friedel hunkered down and hung on as I motored our "hollowed-out log" across the wild Boca del Drago channel.

That was the last time I saw this interesting banana broker from Germany. He liked me as a banana guide because I showed him rare tropical fish in remote streams beyond the plantations. *Auf Wiedersehen, Herr Gellerich.*

THE FEAST OF DEATH

I am not proud to be writing this story because it tells of the tragedy of a man who left a confused legacy in the banana world. Yet, it happened and I will write about my brief encounter with him.

Elihu Menashe Blachowitz was born in Poland in 1921 and immigrated to the United States as a child. His name was changed to Eli M. Black, and he was an 11th generation rabbi in Woodmere, New York. Later, business became his main interest, and he acquired control of the United Fruit Company in 1968 by methods that surprised financial and banana people.

The banana company went from near bankruptcy in 1960 to a blue chip company a few short years later. This was because the company converted from Gros Michel to Cavendish varieties, shipped fruit in boxes and popularized the blue Chiquita sticker on bananas. President Jack Fox and Herbert Cornuelle managed the Company so that profits soared and that is what attracted Eli Black to get control.

Living in Changuinola for 11 years during the time the new variety and boxing were developed, I felt the affinity of this success by Chiquita. Then, an outsider cleverly worked his way in to take away the profits. Employees were confused, resentful and worried about what direction the Company would take next.

Eli Black made his debut at Farm 8 in Changuinola in 1970. Bobby Walker was the general manager and gave him the usual courtesies. This included the customary visit to my research laboratory to observe live insect pests eating banana leaves and fruit. Mr. Black got to peek into the microscope to see tiny nematodes that caused millions of dollars of damage to banana roots. I showed him specimens of the nectar-feeding bat that caused "bat scratch" on banana flowers.

It was hard to tell if Mr. Black was interested in these pest problems; he was a man of few words at that time. At one point, he asked for a piece of notebook paper and I handed him a sheet of white typing paper. He looked at me sort of strangely and then took a ruler and tore the page in half. He took one half of the paper and handed me the other half. Maybe it was an example of frugality.

Afterwards, a selected group gathered in the conference room with Mr. Black. After that, I never saw him again.

In 1973, I was transferred to Golfito, Costa Rica, to replace Russ Caid as director of the research department. A couple of years later, the general manager, Victor Heyl, asked me to accompany him and give a special banana tour to two executives who were on the board of directors and based in New York. They were Jay Wells and Robert Lehman, accompanied by the latter's wife. They knew little or nothing about banana cultivation so I gave them the "Banana Kindergarten" tour. The three of them were attentive and appreciative.

Suddenly, the tour was disrupted when a messenger found us in the plantation. He spoke discreetly with Victor Heyl who then told Jay Wells to call New York immediately. The international call had to be made via an obsolete line phone in the Coto 47 agricultural superintendent's office by using some kind of "monkey patch" connection in Golfito.

After the phone call, Jay Wells returned to our tour van. He was not the same man we had known moments before. He said, "Gentlemen, we must return to New York immediately." Their executive airplane was parked at the airport in Golfito, and the trip on the rough gravel road to the banana port would take at least an hour.

I got out of the van and exchanged inquisitive glances with Victor Heyl. He came closer and in a whispering voice said, "The old man jumped out the window." I was puzzled and asked, "What old man?" His answer was shocking. "Eli Black." It was February 3, 1975.[1]

The tour van took off in a cloud of dust to Golfito with the uneasy three visitors, Victor Heyl and the driver. I stayed at Coto 47 and joined a small group of key employees,[a] who had been invited to a gourmet lunch for the visiting dignitaries. Costa Ricans normally make jokes and smart remarks about anything at hand but this group remained quiet and somber while trying to process what had happened.

The Coto 47 guest house was a small, well-kept timekeeper dwelling on stilts. To begin with, we gathered downstairs, which was open-air and quite pleasant. No one was in a mood to go back to work, especially knowing that an incredible feast had been prepared for the dignitaries and us. Soon, the group went upstairs where a well-stocked bar was accessible. Despite the selections to choose from, several in the group drank the traditional Dewar's White Label Scotch in water, something that they had learned long ago from their *gringo* bosses. It wasn't long before conversation became more animated. Finally, the cook and her helper announced that lunch was ready.

We sat down at a beautiful table with food not normally served in Coto 47: shrimp, filet mignon, and ham with all the trimmings, plus some tasty tropical side dishes for the guests to savor and ask questions about. Fine wines were also available. It was a feast we were not accustomed to.

I always subscribed to *Time* magazine during my years in the tropics and the edition that came after Eli Black's death reported the tragic story. A leak by an insider told about anger, shouting and discord in the locked meeting room of the board of directors. The article told how Jay Wells yanked out the phone line from the wall when a member defiantly tried to make a telephone call. I thought, "My God, I was giving a tour to that man in Costa Rica just a few days before."

Mr. Black struggled with financial deficiencies, a bribery scandal, and perhaps his conscience. The rest is history. Without doubt, these were some of the darkest days that the banana empire had ever known.

TOURING THE AMBASSADOR

One of the most memorable groups that I guided was Terence Alfonso Todman and his family. He was born in Saint Thomas, U.S. Virgin Islands in 1926, and became a career diplomat.

Mr. Todman was the American ambassador to six countries, including Costa Rica from 1974-77. He had other assignments, including Japan, India and Panama.

Throughout his career, he ran into prejudice around the world because of the color of his skin, but contended it was from within—from the Americans—not the host countries nor their people. President Jimmy Carter, from Plains, Georgia, saw through all that nonsense and supported Terence Todman to the fullest. He was highly respected wherever he went.

In July 1975, the general manager in Golfito, Costa Rica, Victor Heyl, asked me to give the ambassador and his family the banana tour in Coto. I was honored and could see that the entire family, including their very attentive four children, were listening and learning about the banana world. The teenage children spoke Spanish, French, one or more languages from Africa, possibly Danish, and, of course, English. The ambassador was fluent in Japanese, Arabic, Russian, French and Spanish.

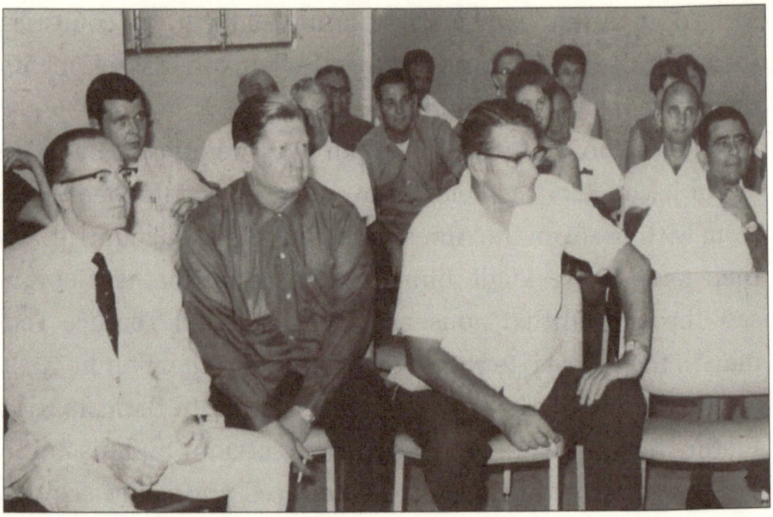

Eli Black (left); Bobby Walker (right front); Clyde Stephens (right rear) beside wife Phyllis.

L-R: Terence Todman, Victor Heyl, Mike Heyl, Todman's son, Clyde Stephens. Coto, Costa Rica, July 1975.

4. Saving Locomotive No. 84
by Billy Saravanja

AS UNITED FRUIT COMPANY developed banana plantations in tropical countries, the principle means of transportation was the railroad using steam locomotive engines. In the 1960s and 70s, all banana divisions started converting from steam to diesel electric locomotives. This resulted in cutting up steam locomotives and using the steel pieces as rip-rap to protect wharfs and buildings from seawater erosion. Such was almost the case for the Golfito, Costa Rica, banana port where Baldwin Class 80 2-4-0 steam locomotive engines Numbers 82, 83 and 84 were decommissioned by the *Compañía Bananera de Costa Rica* (Costa Rican Banana Company), the subsidiary name. Locomotive No. 82 was donated to the municipality of Golfito and parked in the civil town (non-banana company sector of town) as a permanent monument. Locomotive No. 83 was worn out and cut into pieces; No. 84 was still in good working condition and was fired up and put into service during several emergency situations. For example, when several ships collided into the Golfito wharf, severely damaged steel components and steel pilings had to be replaced. The work was accomplished with the help of steam locomotive No. 84, which was the only available steam source to activate the steel pile-driving hammer.

This old pile-driving machine had been used in the original construction of the Golfito wharf in the late 1930s, and it was activated again in the 1970s to drive replacement steel piles. The absence of such a steam boiler would render the wharf inoperative for several months and prevent the export of several million boxes of bananas. I was tasked with steel and bridge repairs and relied on Locomotive No. 84 to accomplish these challenges. No. 84 became my best friend.

In 1977, Golfito's general manager, Victor Heyl, gave an order to scrap Locomotive No. 84 in spite of my explanations and requests to keep it for emergencies. A large supply of oxygen and acetylene was ordered to cut up No. 84 into pieces of scrap metal. As fate would have it, Mr. Heyl was transferred from Golfito. During the short interim period awaiting his replacement, I was assigned to Palmar Sur as a district engineer and proposed to move No. 84 to Palmar Sur with the approval of chief engineer Ramon Cabezas.

To assure that No. 84 would be transferred from Golfito to Palmar Sur, I asked the trains operation superintendent to have the diesel-electric locomotive pull the retired old steam engine along with empty banana railcars for the trip of nearly 90 kilometers to Palmar Sur. The superintendent ignored my request and did nothing. In desperation, I filled several empty banana boxes with mangoes, plantains, pejibayes, and mamones chinos and presented the fruit to the train's engineer, oiler, brake men, and also to the train dispatcher on the promise that they would receive more rewards if they pulled No. 84 to Palmar Sur.

Locomotive No. 84 arrived in Palmar Sur as promised by the train crew and was rolled to a side track in a small park by the airport terminal building. I had the railway repair gang remove 100 meters of railroad track behind the steam engine so that nobody could move it from its final resting place. After 47 years, old No. 84 is still there and is a major attraction for the public to enjoy, take pictures and climb aboard for fun—all because of lots of persistence and a few boxes of tasty tropical fruits.

Locomotive No. 84, Palmar Sur, Costa Rica.

5. Jose Antonio Velasquez: Primitive Artist

IN THE EARLY 1960s, I was invited into the homes of several United Fruit Company people who showed me their paintings by Jose Antonio Velasquez. It wasn't long before I became interested in his art and the unusual life of this handicapped Honduran *campesino*. He sculpted a granite boulder, worked for a banana company, became a telegrapher and a barber and rose to unimaginable international fame as a painter.

THE UNITED FRUIT COMPANY CONNECTION

I was new in the Company, and three old-timers and their wives befriended me with dinner invitations. These senior employees had been assigned early in their careers to help build *La Escuela Agrícola Panamericana* (EAP) at El Zamorano, about 30 kilometers east of Tegucigalpa, the capital of Honduras. Their stories were my introduction to the personal and corporate philanthropy of Samuel Zemurray who believed that Latin Americans could benefit from a well-endowed tropical agriculture college. Zemurray pushed the idea and plans were made. Three million dollars were appropriated to build the Zamorano college despite the fact that World War II had just started in 1941.

Dr. Frederick Wilson Popenoe, who developed and directed the famed United Fruit Company Lancetilla Botanical

Gardens in the 1920s and 1930s, became the founding director of the new Zamorano school and was an icon of education, science, horticulture and achievements in Latin America.[1, 2]

The venerable Zamorano story was best told in detail by Dr. Simón E. Malo in his monumental and comprehensive book, *El Zamorano: Meeting the Challenge of Tropical America.*[3] Dr. Malo was a graduate and the director of the prestigious school. He describes how United Fruit Company engineer Harlo von Wald was selected by Dr. Popenoe in January 1942 to initiate and construct the campus in the remote Yeguare River Valley of Honduras. Harlo von Wald was my boss two decades later in the Dominican Republic, and he and his Colombian wife, *doña* Anita, related how he started building the school on raw land with no materials and no infrastructure. Work began in early 1942 during World War II when materials were scarce or non-existent, but he always found solutions. For example, the engineer from Wisconsin solved one problem by mass-producing hundreds of thousands of red clay roofing tiles on site. When no pipes were available to bring water from the mountain, he used bamboo tubes as a substitute for the moment. During the visits with the von Walds, I also learned more about Jose Antonio Velasquez and admired his paintings in their home. Alfred P. Butler was my boss in Changuinola, Panama, in 1960-61 and he and his wife, Iolene, told me about the primitive artist when Alfred organized and taught horticulture and chemistry for the new college from 1942 to 1944.

The first school year began in 1943 with 72 students from seven countries. There were strict rules and all wore uniforms designed for the classroom and for working in the field—they learned by doing. Neat haircuts were required and a barber was hired by Dr. Popenoe. The barber was Jose Antonio Velasquez who would become the foremost primitivist artist in all of Central America. He went from being a local handicapped *campesino* to an artist with international fame—truly a Cinderella story.

BIOGRAPHY OF THE ARTIST

This biographical summary was sourced from six books and a United Fruit Company magazine, *Unifruitco*.[4] Some references had conflicting data or different opinions. When there were doubts, an authoritative biography was used: *Velasquez, The Man And His Art—A Biography*, by Guillermo Yuscaran, a pen name. The author, William Lewis, a long-time resident of Honduras, interviewed the artist, his wife and children and some of the grandchildren from 1978 to 1982. The book has many photographs from private family albums, a sculpture and 18 paintings in color.[5]

Jose Antonio Velasquez was born in 1906 in Caridad in the *Departamento de Valle*, a poor, arid region in Southwest Honduras on the east side of the Goascorán River. El Salvador is on the west side of the river. Antonio was born with a cleft palate and choked when breastfeeding. He survived only because his mother learned to spoon-feed him with her milk, which circumvented choking. As a child and throughout his life, he had difficulty speaking. Children mocked and ridiculed him because of this handicap. As a result, he became somewhat withdrawn in order to escape the humiliation. Nevertheless, the family had hope. Just after birth, a Catholic priest pronounced that the infant had a talent that would be revealed later in life.

When Antonio was seven, his brothers and his father, Valentin Velasquez,[6] worked hard on the family plot to grow food for the family and their few head of livestock—pigs, chickens, oxen, a donkey and a couple of milk cows. When there was no work on their small farm, a neighbor hired Antonio to milk cows. He had to get up at four in the morning to walk to the farm. The boy rather liked the job because he did not have to talk. The solitude pleased him. Antonio's first pencil sketch—a bird in a tree—was given to his grandmother who proudly displayed the drawing on the wall. Later, in primary school, or in the plaza or up on the mountain, he drew sketches of the highlands, the church, dogs, oxen pulling heavy oxcarts, houses and people. The child expressed his own personal feelings in his art and that gave him an escape from his affliction.

When Antonio was about ten, a male school teacher met with him after class and made a proposal that might give him some relief. The teacher urged him to go up the river to the bottom of a high waterfall and shout as loud as he could into the deafening noise of the cascade. He was to continue shouting and talking to himself until he was exhausted. The teacher told him that water had a healing power. The boy complied and liked the experience. In fact, he returned regularly for several weeks. He was feeling better about himself and began to lose his fear of talking.

Antonio talked to himself, tried to sing and sometimes cried very hard. He started talking more to certain friends and even to his tormentors for the first time. Once he lost most of his fear of speaking, other children stopped bothering him; finally, he felt normal. The teacher had set Antonio on a new path by visiting the waterfall.

Another male teacher and several other compassionate people made a difference in Antonio's life and gave him more confidence and direction. The new teacher, who was also an artist, recognized his ability to sketch and encouraged him. Soon, the student got better than the teacher, and Antonio was asked to draw on the blackboard so that other students could copy his work.

When Antonio was 15, he visited a respected, self-taught educator, Purificacion Euceda, and told him that he wanted to sculpt the great Honduran statesman, Jose Cecilio del Valle, for whom their Departamento de Valle had been named. But the young man said he had no stone or tools to work with. The professor told him about the power of the mind and that nothing was impossible. That advice motivated Antonio to think differently.

The next year, his brother Fausto died from dengue fever. This sudden loss traumatized Antonio. As an escape from the pain, he became obsessed with starting the sculpture. A few weeks later, he strapped a shovel and pickax on a donkey and for two days he searched the rugged slopes of El Bejuco Mountain looking for stone. As a model to guide him in selecting the right stone, he tore out a page with the picture of Jose Cecilio del Valle from a book he had won in an art contest.

A few days later, Antonio cajoled his brother Maximilio to help him dig out a large boulder of granite from the slope of El Bejuco, high above Caridad. The following day, a couple of friends came to help them load the giant monolith on a cart pulled by a pair of oxen—a feat of engineering for four teenagers. It was mostly downhill all the way and very dangerous getting the heavy granite rock back to Caridad. Valentin Velasquez told his son to talk with the local blacksmith about making some tools to work the stone. It took a long time to get the tools but finally the young artist started hammering and chiseling the granite. Curious villagers came to see what he was doing and walked away amused. Several months later, the blacksmith delivered a package of finely tooled chisels, items to sculpt the details of the face. Many months later, around March 1924, the sculpture was finished. Antonio was 18 and his dream had come true. He gained stature in Caridad by honoring a great Honduran statesman in stone. Sculpting also relieved some of the pain after Fausto died.

About that time, 200 American Marines from a ship anchored in the Gulf of Fonseca near Caridad occupied Tegucigalpa in the name of protecting American citizens. Violence and chaos erupted throughout the region and people were killed in vicious political party rivalries after the elections. Bloody fighting went on in the mountains above Caridad.

Ernesto, one of Antonio's friends, had just returned to Caridad with his earnings from working with the *gringos* in Tela on the North Coast. He told exciting stories about his adventures and urged Antonio to go with him when he returned. They were not aware that it was a dangerous time to travel through Tegucigalpa. Nevertheless, the two young men packed their hammocks, food and *machetes* on a mule and started walking to the banana company operations on the Caribbean. They walked for over a week, sleeping under the stars or swinging their hammocks in humble homes or *bodegas* along the way. Locals told them about political violence near Lake Yojoa. People had been killed in the streets at Santa Cruz on their trail to the coast.

When they arrived in El Progreso along the Ulua River, Ernesto found friends in a boisterous *cantina* full of banana and sugarcane workers. They were told there was less malaria in La Ceiba where they could work for the Standard Fruit Company.

After they left El Progreso, it took three days to arrive at the new United Fruit Company port of Tela. They camped on the beach under coconut palm groves with other migrant workers. It was exciting to see banana ships being loaded at the long wharf in the distance. They bought *pan de coco* (coconut bread) from Black Caribs, known as *garífunas*. Antonio was thoroughly enjoying all these new experiences. After two long days of walking east on the beach with their mule, crossing streams and rivers, they arrived at the other banana port, La Ceiba. The Standard Fruit Company signed them on as employees, gave them sleeping quarters and a pass to the workers' mess hall. An interview revealed that Antonio had the aptitude for being a timekeeper. While he waited for approval, he stacked and loaded banana bunches on the wharf.

One of his assignments was on a remote banana farm bordered by a lush rainforest at the base of Pico Bonito Mountain. His job was to count the number of bunches of bananas being hauled by mules from the farm to the railcar. From his mule, he saw tropical plants and animals that he had never seen back home in arid Caridad. During slack time, he sketched pictures while sitting on his mule. The best sketch was of an egret in a pond.

Just like hundreds of other banana workers after payday, Antonio drank in *cantinas* and caroused with women. But he preferred to be a loner and walked the beaches in La Ceiba. Black Caribs who lived in villages along the beach fascinated him. He especially enjoyed listening to stories from an old *curandero*, a bush medicine man who cured snake bites. Many years later, Antonio visited La Ceiba with his son and hoped to find the old Carib but it was too late.

Ernesto and Antonio had been working for five months

when a friend from Santa Rosa de Copan got a high fever and delirium. A nurse in the La Ceiba hospital advised them to take their friend to Tela where the United Fruit Company hospital might be able to treat him better. After they got to Tela, their friend died on the third day. Antonio and Ernesto decided not to return to their jobs in La Ceiba.

In Tela, they worked together as bartenders in a *cantina* near the banana wharf on the Old Tela side of the river and quickly learned how to mix drinks and understand the basic lingo of the foreigners and the workers. Everyone was associated in one way or another with the United Fruit Company and that is what made the port such a vibrant place. Antonio liked Tela.

While he was a bartender, he taught himself how to be a barber. He had won a set of barber supplies in a card game and prospered by cutting hair in a room behind the bar. After some time, the two friends tried to work aboard a ship but that failed. In January 1927, they signed up with the United Fruit Company to sail east to La Mosquitia to work in mahogany timbering. When they were not accepted, the two decided it was time to walk back home to Caridad. Antonio had just turned 21 on February 8th.

He did not know that his future patron and benefactor, Dr. Wilson Popenoe, had already arrived in Tela in late 1925 to select, develop and direct the Lancetilla Agricultural Experiment Station in the Lancetilla Valley just south of Tela. This forested valley was only a few kilometers from the bar and barbershop where the two friends worked. Destiny delayed their meeting for 17 years until 1942 at the new agricultural college at Zamorano.

The two young men were carrying money that they had earned in La Ceiba and Tela and that made them targets for savvy bandits waiting to ambush travelers along the trails. When they passed Lake Yojoa, they heard shooting nearby and hid. Helpful residents warned and guided them to safety several times on the trail. After more than a week of walking, they arrived home at Caridad without incident.

Antonio was glad to see his family and friends once again but it took some time adjusting to the dry, xerophytic landscape that contrasted with the green, wet, tropical environment of the North Coast.

In 1928, Antonio produced his first oil painting. It had a religious theme—*La Sagrada Familia* (The Sacred Family). He gave it to his mother who proudly hung his work on the wall. Soon, the Catholic priest came to admire the work of holy art and asked Antonio to paint a mural on the wall of the church, which he did.

Soon after that, political violence disturbed the peace in Caridad when the ruling *Nacionalistas* imposed their will on the town. The Velasquez house was expropriated and the family was forced to move to their mother-in-law's house down the street. This was an outrage but there was nothing they could do.

Valentin Velasquez knew that his son should go to Tegucigalpa to find better opportunities. Reluctantly, he left. He felt the pain of leaving his aging father and the rest of the family, but after walking several hours over the rugged mountain toward the capital, Antonio felt free. It was 1928 and he was 22 years old—a man leaving home.

It took three days to walk to Tegucigalpa. Everyone in the capital wore sweaters and jackets to keep warm but Antonio was cold because he wore only a shirt. He found his uncle, Gerardo Maldonado, a lawyer, who gave him a room and a free run of his house.

Antonio applied for several different jobs and waited. Meanwhile, he started sketching scenes in the capital. An impressed art teacher arranged for him to personally present one of his freelance portraits of a former president to the Presidential Palace. In September 1929, he was accepted at the Escuela de Telegrafistas where he learned Morse code and all the aspects of the profession. He could decode and transmit without error and the professor put him at the top of the class. When the professor went on a vacation for two weeks, Antonio was chosen to be in charge of the central telegraph office in Tegucigalpa. At the end of the course, he was one of eight students out of 30 who graduated.

A day after graduation, he received a telegraph message from Caridad that his father had been murdered by the *Nacionalistas* in Caridad. Valentin Velasquez had been forced from his house at night, tortured and murdered in cold blood. Others in the village suffered the same fate. Antonio went to the funeral and vowed to get revenge. His mother told her vengeful son that "violence begets violence" and that his father wanted him to study and work in a profession. Antonio complied and returned to Tegucigalpa after a couple of weeks of contemplation.

The new graduate worked hard for seven months as a telegraph operator in several towns scattered around Honduras. He became a jack of all trades repairing equipment, setting up new offices and replacing sick telegraph operators. Later that year, when the respected professor of the telegraph school died, the position was offered to Velasquez, but he refused the important job for personal reasons.

In January 1930, the new director of the government telegraph office assigned him to San Antonio de Oriente to replace the telegraph operator until she returned from a trip. It took him three days to walk across the mountains with his backpack. He arrived at night by following a boy and his donkey with cargo. When Antonio approached the village in a valley below, he heard three sisters giggling at the window of a small store. Two of them retreated and left Maria Raquel Maradiaga, age 16, to sell cigars to the new man in their village. The girl was forbidden to talk to strangers but, when he asked, she told him where the telegraph office was located. He left and found his new quarters.

In time, Antonio and Raquel were attracted to each other but their courtship was very difficult and precarious because of her father. Finally, the intransigent father agreed and the couple was married on November 8, 1931, in the ancient Catholic church, followed by a big reception in the family home. The village was happy and everyone celebrated. Their wedding night was spent in the small telegraph office where he lived. In time, seven babies were born but two died young.

San Antonio de Oriente was formerly a mining town where the Spaniards had extracted silver and other metals since 1673. After the silver mines were depleted, the once thriving town became a quiet village with just a few people. This is the way Jose Antonio Velasquez found his new home and workplace. The landscape was picture-perfect for painting. And that is what he did—Antonio sketched and painted the village. He also operated the telegraph office.

In the late 1930s, Velasquez still made a living as a telegraph operator. But sometimes, he also walked the 40 kilometers westward through the mountains and valleys with his donkey to the capital city of Tegucigalpa to sell his paintings for just a few lempiras.

In 1942, his life took a major turn. He learned that hundreds of employees and many *gringos* were building a new agricultural college in the Yeguare River valley on the El Zamorano farm that had been bought by the United Fruit Company in 1941. Suddenly, there were jobs in the valley below and he decided to investigate how he could earn some money at El Zamorano. Cutting hair of the workers might be a possibility.

The barber walked down the mountain to El Zamorano, found a tree stump and started cutting hair with the basic tools of his trade. As workers came, sat on the stump and got their hair cut, a tall *gringo* saw what was happening and approached the barber. He sat down on the stump and asked for a haircut. Velasquez was a short man, 4.5 feet tall, and contrasted in height with the tall *gringo*. That customer was the founding director of the school, Dr. Wilson Popenoe.

Dr. Popenoe approved of the haircut on the stump and offered the barber a job to cut the hair of the first class of students in 1943. The director told him to return later. When the barber came back, he was surprised that a barbershop had been built. Even better, Dr. Popenoe had acquired a barber chair that had been left behind at the abandoned banana Division of Puerto Castilla on the North Coast. Dr. Popenoe was the first person to get a haircut in the new barbershop. After the barber gave haircuts to several students, Dr. Popenoe approved of his work and formally offered him the job for $25 per month, plus meals in the school dining room. The director also gave him a horse named Melvin to ride up and down the mountain. With this new job, he probably stopped operating the telegraph service in his village.

At the school, Jose Antonio passed the time by drawing and painting in the barbershop while waiting for the next head of hair to cut. Certain people took note of his primitivist art form. Helen Popenoe, who had been a buyer of art for a museum in Chicago, was the first to recognize the talent of the barber as a primitive artist. She suggested to her husband that this talented man would be good public relations for the new agricultural college. Wilson Popenoe agreed. They brought him new art supplies purchased in the United States and gave Antonio plenty of encouragement.

His fame grew as an artist and many people wanted his paintings. As a promotion in 1954, Helen and Wilson Popenoe arranged to exhibit 25 of his oil paintings at the Pan American Building of the Organization of American States in Washington, D.C. The paintings were shipped in advance. When the day came for the first flight in Antonio's life, Wilson Popenoe drove him to Tegucigalpa and helped him aboard a DC-3 at the Toncontin airport. To calm his nerves, the artist hid a flask of rum in his pocket. When he arrived in the United States, the flask was empty and the artist was not so afraid of customs and immigration officers.

He was met at the Washington, D.C., airport and housed at the Honduran ambassador's home in luxury quarters like he had never dreamed of. The artist said later that after a lifetime in houses with dirt floors, outdoor toilets and no electricity, the surroundings seemed unnatural. This would be his home for the next three months.

Antonio loved walking by himself and visited the great sights on the Mall. He especially liked the Lincoln Memorial. The ambassador and his wife took him to events like ice skating, art exhibitions, concerts, movies and even visited the White House. During his free time, he painted in his room. The opening exhibit at the OAS gallery of the Pan American Building was a big success, drawing over 300 guests. He met the famous American primitive artist, Grandma Moses (Anna Mary Robertson Moses) and they talked through an interpreter. She was in her 90s—and still painting. A group of barbers from Washington, D.C., presented him with a barber's trophy.

When he returned to his village in Honduras, casual attitudes about him suddenly changed and he became very popular. Everybody wanted to meet him and buy his paintings. Velasquez enjoyed the new prestige—and finer cigars.

Like many men in Honduras, he drank guaro, a spirited distillation from sugarcane. One of his drinking buddies was Dr. Wilson Popenoe. Antonio's wife, Raquel, told him sternly that he had a drinking habit. If he did not stop drinking, he would not be able to cut hair or paint. He promised to stop several times but did not. He just kept on drinking, cutting hair and painting. Drinking *guaro* started in the morning on his horse while riding on the trail down the mountain to cut hair at El Zamorano. Finally, he kept his promise. Dr. Popenoe offered him a drink to celebrate a special occasion, but Antonio told him that he had stopped drinking.

Antonio concentrated on painting scenes at San Antonio de Oriente. The Honduran government capitalized on his fame and promoted exhibitions in Panama, Spain, Belgium, Germany, Portugal, France, Cuba, Mexico, Nicaragua, Dominican Republic and other countries. In 1955, the artist received a big honor when he was awarded the most prestigious art award of Honduras, the Pablo Zelaya Sierra National Prize for Art. Other prestigious awards followed. Velasquez was also a hero to the barbers in Honduras and they presented him with more honors. In 1956, he was appointed as the mayor of San Antonio de Oriente. To add to his fame, American actress Shirley Temple narrated a film about him, *The World of a Primitive Painter*.[7]

With all this fame and prestige, his attitude toward students at El Zamorano was not like before. The new director, William C. Paddock, did not speak favorably of him nor did the students. The congenial feeling was declining. Even Wilson and Helen felt the effects although Wilson Popenoe and Antonio remained friends for life. When Dr. Popenoe told Antonio that he was retiring from Zamorano as director, Antonio felt it was also time for him to leave the barber shop where he had been cutting hair and painting since 1942. In 1956, Jose Antonio Velasquez left Zamorano College.

Others were disgruntled with the artist, too. Some of his old friends, patrons and mentors became irritated, especially when the price of his paintings jumped from $5 to $50 and $100. Regardless of those prices, he sold his art. Later, Antonio's eyesight deteriorated, and a son helped him complete his paintings. This led to accusations of dishonesty by some of his detractors and caused the family bad publicity. Later, two sons and a grandson capitalized on the Velasquez name and fame and became good artists themselves.

In 1961, after the father of Raquel Maradiaga Velasquez died, she announced that she was moving to Tegucigalpa so their daughters could go to high school. Besides, she no longer wanted to live in San Antonio de Oriente. In preparation, she had already bought a small house in 1955 in Guanacaste, a poor neighborhood in Tegucigalpa. Antonio stayed in San Antonio de Oriente alone to paint and be the mayor. Their two older sons had already moved out and were siring children, some out of wedlock. Both sons had jobs in Zamorano.

In 1964, Jose Antonio Velasquez reluctantly left his beloved village after 34 years and moved to the small, over-crowded house in Tegucigalpa with his wife. Raquel worked hard from early morning to late at night cooking, cleaning, running a *pulperia* (small store), and raising six of her grandchildren. She also answered the telephone and talked with people who wanted to meet *El Maestro* and buy his paintings.

There was no privacy for Antonio to paint. To help alleviate the problem, Raquel added another room to the house. With the high demand for his art, his two sons and a grandson expedited the mass production of paintings of their village, San Antonio de Oriente. Sometimes, all four Velasquez artists were in the room at the same time—three Velasquez generations—all painting. Each painter signed his name as Velasquez and this confused admirers and buyers. But Antonio and Raquel saw no conflict and were very proud that their two sons and a grandson turned out beautiful art with the Velasquez name. They were all good artists. Invitations were frequent for Antonio and his wife. They traveled to New Orleans where there was a large colony of Hondurans waiting to honor them. The couple also visited Washington, D.C., plus several countries for showings of his paintings. They both enjoyed the fame, honor and financial benefits but age and poor health took its toll.

News came from Antonio's birthplace at Caridad that his long-lost granite sculpture of Valle had been found in an abandoned rock pile. Restoration of the damaged sculpture was made at La Escuela de Bellas Artes in Tegucigalpa. Velasquez gave the mayor of Caridad permission to mount the work in the village and hold a grand ceremony on December 19, 1979. The large extended Velasquez family traveled from Tegucigalpa for the event in a pick-up truck carrying three generations.

The whole village celebrated the occasion all day and night with ceremonies and gala festivities. Antonio was a 16-year-old teenager when he sculpted the stone image of their great hero, Jose Cecilio del Valle. Now it finally had a place to rest in public view.

THE PATRON SAINT OF TURTLES AND THE PRIMITIVE PAINTER

Dr. Archie Carr was my professor of biology at the University of Florida in 1953. At the time, I knew nothing about his fame as a renowned naturalist and writer who was making a monumental difference in saving sea turtles and tropical habitats from extinction. He was well-known in the United Fruit Company in Honduras because he was a charismatic professor of chemistry and biology at *La Escuela Agrícola Panamericana*, starting in 1945 at $200 a month. Archie Carr, his wife Marjorie and their four children adapted well in that remote land. All of them made significant contributions in tropical science, conservation and environmental efforts later on.

Henry Hogaboom was an associate director of the college and his wife Margaret was an accomplished photographer. She frequently accompanied Archie Carr to document his work in the field. She reportedly took the first known pictures of sea turtles laying eggs at night on a lonely Honduras beach with the biologist. From her family, I acquired 126 of her prized photographs that I later bequeathed to the Zamorano College Library in 2019 where they are now available to the public. Many of these historical photographs were of high-profile scientists, philanthropists, professors and students on the campus and in the agricultural fields of the college. Archie Carr and his wife Marjorie Harris Carr and their children were among them.

One of Archie Carr's books, *High Jungles and Low*,[8] relates an experience with Jose Antonio Velasquez while the professor was getting a haircut at Zamorano. The following is a summary that he wrote:

> Not long ago, I stopped by the tiny room to get a haircut. I was carrying a quart fruit jar which contained a small, poisonous snake in the genus Bothrops. I placed the jar on the floor and sat down in the barber chair. Velasquez walked over to the jar, looked in, and turned away. "Terrible animal," he muttered as he picked up his comb and scissors.
>
> Among Velasquez's talents is a fine, if somewhat disapproving, feeling for natural history. I settled myself in the chair and prepared to learn something extraordinary. "What's so bad about the snake?" I asked. "It looks to me as if it's too small to be very dangerous." This is exactly what the barber wanted.
>
> "It's little, all right, but in my estimation, it is the worst of all the snakes we have. It is called *casera* because it enters houses. Because of its small size it can get into any house, crawling through cracks or under the door. But it lives mostly in houses of nursing mothers."
>
> This was enough to quicken anyone's interest and it did mine, but I didn't interrupt. He went on.
>
> "The *casera* comes out of the cracks during the night. It climbs the bedpost and gets under the bedclothes with the nursing woman and noses about until it finds her bosom. Then takes a nipple in its mouth."
>
> I could not restrain myself. "And bites," I suggested. "No, sucks," said Velasquez. "It nurses. It likes mother's milk better than anything else. Sometimes it just nurses and then crawls away. This is disgusting but not tragic. The really bad part comes if the baby awakes hungry while the snake is there. If the child seeks comfort at the occupied breast, the viper coils

at once into a tight ball. The domed infant, fumbling for the nipple, takes instead the coiled snake into its mouth. Thereupon the *casera* shows its true character. It crawls down into the child's stomach and strikes repeatedly at all vitals within reach. The infant shrieks, but it is too late for help."

"Does it swell up and turn black and die?" I ventured. "Why, yes." Velasquez was surprised and a little hurt with me for anticipating him. I looked down at the jar on the floor. Through the glass I could see the bright, unblinking eyes of the monstrous little viper.

"It's the worst snake we have," said the barber.

MEETING VELASQUEZ IN SAN ANTONIO DE ORIENTE

In July 1963, Phyllis and I were married in Progreso in the Departamento de Yoro where we were legally documented in a civil ceremony by the mayor of the town. The next day, a more traditional wedding was celebrated in the little chapel by the Company Golf Course in La Lima in the Departamento de Cortes. Our honeymoon was on the three Bay Islands of Utila, Roatan and Guanaja.

During that same month, we drove a Volkswagen Beetle to Tegucigalpa, then continued to the college at El Zamorano. I found Juan Leiva whom I had known at the University of Florida. The next day, Juan drove us to the beginning of the foot trail that went up the mountain to the sleepy village where Jose Antonio Velasquez lived, painted and was the mayor. It was a long, tiring walk up that rocky, clay trail through mixed hardwoods and Honduran pines, known as *ocote*. The path had been worn down by horses and beasts of burden for so many centuries that it had eroded into deep ruts and ravines. As we climbed higher, the air was delightfully fresher because of the cooler climate at 1,174 meters above sea level.

Upon arrival at the entrance of the village, we stopped at the historic bridge to view the fairytale panorama of white stucco houses with red tile roofs. On the far side of the village, the majestic old Catholic church dominated the scene just like in the Velasquez paintings.

We found the home of the artist and were told by a neighbor that he was attending a funeral in the church. Meanwhile, Phyllis and I walked up a steep cobblestone trail to the last house on the slope above the village. This traditional white stucco house with a red tile roof was a country home of Dr. Wilson Popenoe, who enjoyed time with his old friend, Jose Antonio and others in the village. Views were breathtaking and the air was fresh like eternal spring. Old abandoned silver mine tunnels were nearby. We were told that Dr. Popenoe was at his main retirement home in Antigua, Guatemala.

After the funeral was over, we met the man we had come to see. He was very cordial but difficult to understand because of his cleft palate. He welcomed us into his small, adobe house and invited Phyllis to sit on a little wooden stool. There were chickens running in and out of the house. *Don* Jose Antonio introduced us to his little grandson who was playing with the chickens and a dog in the house. The child was completely naked—a sensible tradition that avoids diaper problems.

After a cordial chat, the artist knew why we were there, like so many others before us. He showed us a small painting, then a larger one, both with similar landscapes of the village. I paid 30 lempiras ($15) for the smaller one and 70 lempiras ($35) for the larger one. He apologized that he did not have more available because he had been very busy lately.

After the visit and purchase, we walked to the church and went in. It was a perennial landmark in nearly all of Velasquez's work. Somehow, it made me feel more holy that day. By the time we walked back to the house of the artist-barber-telegrapher-mayor, it was time to say goodbye and walk down the mountain to El Zamorano with our works of art. Suddenly, a deluge of heavy rain poured out of the sky. We made a dash back into the house of our host as thunder rumbled in the distance. It was not a good feeling wondering how we were going to get our two treasured paintings back to Zamorano.

It didn't take long for the artist to come to our rescue. He carefully put the two oil paintings in a waterproof bag. Then he loaned Phyllis his heavy-duty rain cape, which she put on. It was her role to protect the paintings under the rain cape and keep herself dry.

I walked in the rain and became drenched and cold within minutes after we started walking down the slippery slope toward El Zamorano. We walked slowly and finally reached our guest room at the school in about two hours. Fortunately, there was hot water in the shower—a welcome feeling.

Sixty years later, the two paintings in our house in Florida are a daily reminder of that memorable trek to visit *don* Jose Antonio Velasquez, foremost primitivist painter from San Antonio de Oriente in the cool mountains of Honduras.

Dr. Doris Zemurray Stone, Dr. Wilson Popenoe and graduate Carlos Erben. Note his haircut by Jose Antonio Velasquez. February 1957.

Escuela Agrícola Panamericana campus at El Zamorano, Honduras, eighty years after Harlo von Wald started building from nothing.

Velasquez as a barber.
San Antonio de Oriente, 1954.

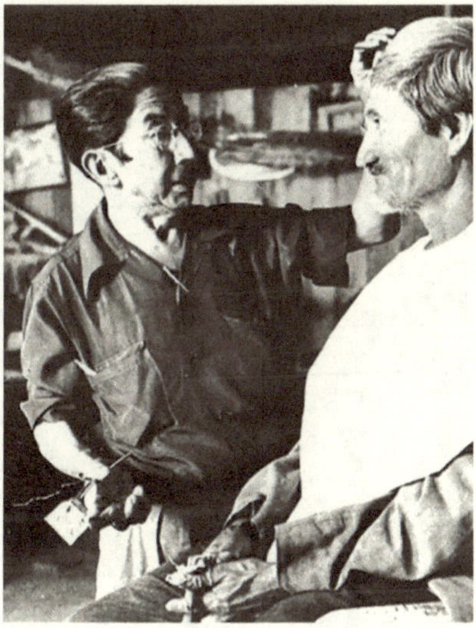

Velasquez attracting children in San Antonio.
Son Tulio Enriquez (right), 1954.

Velasquez painting San Antonio de Oriente, 1954.

Velasquez and son, Tulio Enriquez, on Melvin, 1954.

Jose Antonio Velasquez painting bought by Clyde and Phyllis Stephens in 1963.[9]

Honduras stamp issued in 1970.

6. The Yard
by Gordon Kidd[1]

SITUATED ON A SANDY lot in Puerto Cortes, Honduras, the first yard I remember was fronted by railroad switching tracks and had a jungle behind it. I was sure that the jungle abounded with tigers, jaguars, and other wild animals. There were deer, snakes, spiders, birds and swarms of mosquitoes. My father had been employed as an accountant for Cuyamel Fruit Company, a fledgling firm headed by Samuel Zemurray, "The Banana Man." I was one year old in 1921 when we moved from Austin, Texas, where I was born, to Puerto Cortes on a banana boat, the SS *Copan*. A year later, my mother and father and I left Honduras on the same *Copan* for New Orleans to await the birth of my brother, Robert. Our family lived in Puerto Cortes until 1932 when my father was transferred to Tela, Honduras, as Honduras Division accountant following the merger of the Cuyamel Fruit Company and the United Fruit Company in 1929.

The Cuyamel Fruit Company had begun development of Puerto Cortes as a major banana exporting port in 1918. The bay had been dredged to accommodate the ships and the spoil was deposited on a swamp west of the municipality and became our yard. Our house, the general manager's house and the manager's house, were among the first to be constructed on the site. A hotel for expatriate bachelor employees was erected next door. A narrow-gauge engine, No. 5, pulling a passenger car, transported the men to and from work every morning, noon and afternoon. Work was from 7:00 to 11:00 AM and 1:00 to 4:00 PM. My father, however, spent many long hours at his office, which overlooked the bay.

A boat ride I vividly remember was from Puerto Cortes to Omoa on the Omoa launch. My father tied my brother and me securely to a stern post to keep us from going overboard. Omoa was the original Cuyamel banana port on the Honduras Caribbean coast but there was nothing left except pilings from the old wharf and a few remains of buildings. But it was a great trip.

Another time, against stricct parental orders, I got in a skiff with an outboard engine and a couple of men operating it. We were moving briskly along the shoreline when the propeller hit something underwater. I remember that the motor, still running, flew through the air and fell into the boat up front, severely cutting one of the men's legs. I never learned if the wounded man survived.

Winter months were marked by continuous rain, punctuated often by thunderstorms. With the yard awash, my brother and I played under the house, which had been built on concrete stilts about eight feet above ground level. I remember seeing water covering the yard being rippled by an occasional breeze.

We relied on rain for potable water that was collected in wooden tanks constructed on top of high stilts for gravity flow. Drinking water had to be boiled and filtered. Clothing and footwear especially had to be protected from mildew by an electric light bulb burning constantly in the closet. We had to check for scorpions hiding in our shoes by vigorously shaking them.

The rainy season was followed by the dry season. The area lies about 15 degrees north of the Equator and it was hot. But the sea breeze made it livable. Insects abounded. Crews sprayed oil on stagnant water to prevent mosquito larvae from developing. Copper troughs surrounding the base of posts supporting the house were filled with tar and oil to keep insects from climbing to the dwelling above. When the *chapulines* (grasshoppers) arrived and got in the tar, a real mess ensued. Some were able to free themselves and got on my mother's white linen dress.

More dreaded was a "norther" that brought heavy rain and winds that blew down the tall Gros Michel banana plants.

A wood-slat boardwalk ran from the house to a shed out back. One day at play, running barefoot on the boardwalk, I stepped on a nail. I don't recall which was more painful: The puncture or the tetanus shot which followed at the Company hospital.

Another day, when I was about six, I took a bareback ride on our community horse, Man-O'-War. But Man-O'-War was of a different mind and he scraped me off under a curving coconut palm. The result was a dislocated right elbow, which was set in the Company hospital by Dr. Watson. During the dry season, my mother planted flowers: cosmos, zinnias and marigolds. I especially liked the cosmos for their delicate leaves and soft colors.

My mother was the social leader in our close community, having frequent bridge parties in our house and setting up a croquet court in our backyard. My brother, who was two years younger, and I learned to play the game and took great pleasure in knocking each other's ball out of the court. I remember our yard man, Mr. McCloud, a kindly Jamaican man of some years, who helped keep my brother and me out of trouble. With a *machete*, he cut the grass and trimmed the hedge along the fence and took care of the flower beds.

The only kids we had to play with were the Aycocks: Joe, Edna, Jimmy and Billy; the latter was a scamp, who later got in trouble with the law for sticking up a bank and went to jail. One day, Joe and I had a tricycle race along the macadam-like walk which had been laid along the front of our yard. Joe crowded me against the fence and caused a severe and bloody cut. I was again grateful to Dr. Watson for his kind and considerate attention.

We considered the entire community as our yard in which we roamed freely. This allowed us to become friendly with the Honduran people. We socialized with them and their children. Another notable activity was when my brother climbed the very tall Company radio tower.

I didn't know about the colorful history of Puerto Cortes until many years later. Revolutions were plotted in the Hotel Cosenza or at Mamacita's, both across the tracks from my father's office and close to the banana wharf.

General Lee Christmas had been appointed by the president of Honduras as the *comandante* of Puerto Cortes and this generated lots of incredible journalism. His colleague was Guy "Machine Gun" Molony of New Orleans. The interest of Sam Zemurray's Cuyamel Company was well protected.

Mamacita's was a well-known bordello that entertained seamen and local longshoremen. Mamacita always had good, neighborly relations with the banana company. A good example was when a revolution stopped the pay car on the railroad ,and the cashier could not pay employees in Puerto Cortes. When requested, Mamacita came forward with the cash to pay all of the employees in full and on time. There was never any lack of confidence between the two parties.

General Lee Christmas

The SS Copan was part of the Cuyamel Fruit Company fleet until 1929 when it became a United Fruit Company ship.

7. Tela Teenager
by Gordon Kidd

MY FATHER WAS AN expatriate who worked for the United Fruit Company in Tela on the north coast of the Republic of Honduras. At the end and beginning of every school year, the Company provided free transportation on banana or combination cruise/banana ships from U.S. ports to tropical divisions in Cuba, Guatemala, Honduras, Costa Rica, Panama, and Colombia, for employees' children being educated in the United States. We were generally known as "The Banana Brats."

The five-day, 1,000-mile voyage from New Orleans to Tela was an adventure filled with the excitement of being at sea, seeing friends, being on the bridge at dawn to watch the first mate "shoot the sun" with a sextant, counting off seconds on his way to the navigation table to plot the ship's position, watching the marine engineers maneuver the triple-expansion steam engine, the deck crew handling booms, hoisting engines and mooring lines, and galley crew preparing meals for the crew and passengers. Particularly on the cruise ships, fabulous food was exquisitely served. Except for cool air blowing in from open portholes, there was no air conditioning in the passenger accommodations; that was reserved for the bananas! Who cared?

Best of all, of course, was arriving "home." "Home" was the Company town of Tela, adjacent to the Honduras municipality of the same name. It was like an expensive resort, only better. At our disposal was a golf course, tennis courts, our own horses, household help, and a fabulous beach that stretched for miles in front of our Company-provided house which had running water, indoor plumbing, electric stove, refrigerator, hot-water heater, fans and a cool breeze blowing in from the sea. The Company town had electricity, ice and bottling plants, and a short-wave, dot-dash radio station for communicating with ships at sea, Company headquarters in Boston, and the outside world.

The Company also maintained a modern hospital, a first-class school for the younger kids, and a commissary where residents could purchase local meats and produce, fresh, and later, frozen fruits and vegetables imported from the States on Company steamers. Mail, newspapers and magazines arrived by banana ship. Of course, in those days, there was no TV, but families had short-wave radios, Victrolas and weekly movies and *marimba* dances for entertainment. Besides being the division comptroller, my father was an accomplished violinist and musician. It wasn't long before he had organized an orchestra which played all the hits of the 30s at weekly dances at the Masonic Lodge, a nice contrast to the regular *marimba* dances at the club.

It was something of a culture shock, coming from Arkansas, where my brother and I lived with my grandparents. We attended school in Altus and nearby Ozark. During many of those years, we drew water from a well, chopped wood for the cooking stove and broke coal for the heater in the living room. We took our weekly bath in a wash tub, went outside to the privy, had a crank telephone to call Ozark, and listened to the radio for entertainment. We walked to school in Altus and rode a school bus to Ozark.

My father's Tela house was perhaps 100 yards from the water's edge. All my brother and I had to do was pull on our suits and dash as quickly as we could across the hot sand. What a relief it was from the ambient heat of the Tropic of Cancer to dive into the cool water of the sea, paddle about in the gentle waves, and bodysurf whenever a big roller came in. There were no life guards or parents to watch over us; we were on our own. In the process, I learned to swim.

A large raft with a spring board was anchored a few hundred feet offshore. Another kid and I got sunburned on it one day—no sunscreen or sun glasses in those days—while waiting for a barracuda to move out from the raft's shade. We left, after a decent interval, when the barracuda left, presumably for lunch. After all, there were "better fish in the sea."

When we weren't swimming, we raced our horses through the surf, rode down and through a Carib village nearby, or rode up to Puerto Arturo, a ranch maintained by the Company for beef and milk cattle. I belonged to a Boy Scout troop, and learned to cook on a fire in a beach-sand pit. Every week, a cruise ship docked at the 1,000-foot-long pier that jutted into the sea. While the ship was being loaded with bananas, the passengers were enjoying the beach, visiting the famous Lancetilla Botanical Gardens, and attending a *marimba* dance at the Club. Annually, young ladies from Gulf Park College for Women in Gulfport, Mississippi, would take such a cruise, to the delight of the young bachelors employed by my father in the accounting department. A 75-foot, twin-diesel engine yacht, the Victor, built in Boston and named for a former Company president, Victor M. Cutter, and a 24-foot Chris-Craft speedboat were maintained at a landing and boathouse on the banana pier. Early every Sunday morning, my father, brother and I would make the long walk down the pier out to the boathouse and climb into the speed boat with two crewmen for a 12-mile ride out to Point Sal, and troll for Spanish mackerel, jack (tuna), an occasional red fish, and barracuda. I remember how beautiful the sun gilded the sky at dawn behind the hills. I reveled in my turn at the helm, swaying the boat left and right at speed. Fresh fish at dinner was *de rigueur*.

For my brother and me, the high point of summer was when my father had the *Victor* stocked with a week's provisions, and with a crew of five, set off running at about 4 knots-per-hour for the Bay Islands, some 50 miles east of Tela, or the Sapodilla Cayes, about 35 miles north of Tela and east of Belize, then known as British Honduras. We towed the speedboat behind, for trolling later. I loved being allowed to work the engines with the engineers.

Anchoring off one of the cayes, we would breath-hold and dive from the boat deck to the sea bottom, about 18 feet below, to bring up conch, a tropical marine mollusk with a spiral shell. Conch chowder cooked aboard was a special, acquired taste.

At nightfall, we would run the speedboat up to a flat coral reef lying about a foot below the surface, cut the engines, and walk on the reef to catch lobsters. As night fell, the crew would light up diesel-oil-fueled torches for illumination. Although this particular lobster had no claws, taste for it was quickly acquired.

Sometimes, we would go ashore on a small cay, visit with a family living there, listen to records on a wind-up Victrola, and dance with the daughter. Another beautiful Caribbean evening!

Early on, we even had our own baseball World Series, with teams from Puerto Castilla, and Progreso, two other banana-growing Honduras areas. Employees Sutton and Swofford made up a pitcher-catcher team that won the pennant for Tela.

Idyllic as it was, I was not fully appreciative. I was in love, and couldn't wait to get back to school and my high school sweetheart; the girl who many years later became my late wife. And, because I was on scholastic probation at the University of Texas in 1938, I spent that summer making up courses that I had failed. Then, in 1939, as we landed in New Orleans after summer, my passport was lifted: I was 19, war was in the offing, and I was draft bait. My 1940 summer was spent partly in rainy Louisiana with a horse cavalry troop in the Texas National Guard, participating in that year's Army maneuvers.

I didn't get back to Tela until after the war when all had changed, but never to be forgotten.

Buddies aboard the banana ship Limon, in 1949.
R to L: Gordon Kidd, Texas-born, Honduras-raised; Fletcher Ames "Buddy" Hatch Jr., born in 1920 in Santa Marta, Colombia; Jimmy Aycock, born and raised in Honduras. Kidd and Hatch were executives in the United Fruit Company.

8. Turning over Old Stones in the Past

SINCE THE BEGINNING OF my career in the banana world in 1959, I was fascinated by lost and mysterious relics left behind in remote places where *bananeros* came and went. Because of a fungus infection–Panama disease–early plantations were frequently abandoned within a decade, leaving behind only a few markers from the past. Originally, untold millions were invested in infrastructures needed to develop pestilent jungles and swamps into healthy and productive banana plantations. But when the banana people left, they took what they could and started again somewhere else. The tropical jungle quickly enveloped what remained and history of these huge investments was lost except for a few moldy photographs and remnants covered in vines and forest.

Along the way, I stumbled upon old banana relics. This chapter discloses a few of those places and artifacts left behind long ago.

BRIDGE OVER THE CRICAMOLA RIVER

The Snyder Brothers Banana Company cultivated Gros Michel bananas in the 1890s in the loam soils along the Cricamola River, whose waters led to the Chiriqui Lagoon in Bocas del Toro, a province in western Panama. I explored this river in March 1967. Bridge at Cricamola, 1921. and found a big, steel bridge that had toppled into the river because of numerous floods. Later, I found a photograph of that bridge. The bridge was built by the United Fruit Company after it bought out the Snyders in the late 1890s. Today, the bridge is a twisted mass of steel grown over by a morass of vegetation. Indigenous people live there now, but none of them can tell you anything about old banana plantations or the bridge.

DON TELEFORO

On September 28, 1968, my neighbor, Karl Steffek, an Austrian living at Farm 8 in Changuinola, and I left my boathouse at Farm 61 and motored in the Stephens' *cayuco* through the Changuinola canal to Boca del Drago to begin a weekend excursion. My large, 18-horsepower Evinrude outboard engine propelled us to Split Hill in two and a half hours. We explored some of the Chiriqui Lagoon and then tied up at a huge store and home over deep water. The owners, *don* Teleforo Serrano and his wife, were located near the mouth of the Pumpkin River, known in Spanish as *río* Uyama. The structure was built in 1934 on concrete and log pilings well above the waterline and some distance from the shore. *Don* Teleforo invited us in and showed us all the common courtesies. We ate and drank well. He provided a small room to swing our hammocks for the night. He refused to take payment for anything and seemed glad to have someone to talk to. He and his wife were especially enthusiastic about showing us recent pictures of their trip to Europe.

The store had all the basics to sell or exchange to residents along the nearby shores and rivers of the Chiriqui Lagoon. They came shopping for canned goods, rice, beans and basic supplies. A simple bar also served a few clients. Sometimes there were parties. Most customers arrived by sailing or rowing their *cayucos* but a few had small outboard engines. Amazingly, a few customers came in the dark of night with no lights on their boats and returned far across the bay the same way.

Don Teleforo traded fish, hides, rice, cacao and other agricultural products, and he earned a comfortable living. A kerosene Servel refrigerator cooled beverages and food and made a little ice. Once a week, a seaworthy launch came to re-stock his supplies and then return with cases of empty Balboa beer bottles 24 hours across the turbulent Caribbean from Colón by the Panama Canal. The boat tied up to the store in deep water. This stop was followed by other port calls at Bocas and Almirante with cargo and passengers coming and going.

The next day, after a hardy breakfast of coffee, rice, beans and plantains, *don* Teleforo recommended that we visit Mr. James Hinzie who lived along the Pumpkin River. Both of these men, Hinzie and Serrano, were children when their parents brought them to the Chiriqui Lagoon in the early 1900s seeking jobs in the booming banana business. The Serranos were Latinos from Chiriqui and the family walked several days on the cold, rugged foot trail across the mountain from Caldera on the Pacific side to Chiriqui Grande on the Caribbean side.

A VISIT WITH MR. HINZIE, A LIVING RELIC

Karl Steffek and I said our farewells to the Serranos and went up the Pumpkin River. The waterway was littered with logs and trees flushed downstream by floods. We found Mr. Hinzie living by himself in a palm thatch hut, like one of the natives. He was cooking rice and plantains over a wood fire and quietly invited us to sit down on stools.

Mr. Hinzie said he was 68 and came to that spot in 1903 with his parents. He was born in Texas, but his mother was Pennsylvania "Dutch" (meaning *Deutsch*, which is German) and his father was Irish. His father previously came to work with the Snyder brothers in the late 1890s. They developed banana plantations along the Pumpkin River before the United Fruit Company bought them out in 1899. A small railroad had been built several kilometers up the river to access the farms. Flatcars with bunches of bananas were pulled by a small steam engine to barges waiting on the Caucho River in the adjacent mangrove swamp. The diversion of the rail line from the Pumpkin River to the Caucho River was considered necessary because the Pumpkin River flooded several times every year. That made it difficult to maintain a railroad or haul fruit in launches down the Pumpkin River. The Caucho River was a calm, tidewater estuary where the rail line ended in a pestilent mangrove swamp.

Mr. Hinzie had a collection of rusty junk left over from the days when bananas grew there. It was of no use but it was there. He said that his parents followed the banana trade when the plantations died from Panama disease, but he never left.

After he explained where we could find the old railroad right of way, we walked upriver on a rustic trail until we saw a line of gravel that had once supported rails and artifacts of those times.

Upon returning downstream, we thanked Mr. Hinzie for his information and departed. I will never forget the horrible ulcers from Leishmaniasis on his face. The disease is caused by a protozoan Leishmania that is transmitted by sand flies—called *chitras* in Panama. A few years later, he died and was forgotten except for this brief footnote of history.

FINDING THE CAUCHO RIVER BANANA PORT

In October of 2010, my employee, Leovigildo Julian Lorenzo, my brother Wilson and I made an excursion to the Caucho River to find the landing where Pumpkin River bananas were brought on a rail line for transfer to a barge. I asked a couple of indigenous people on the shore if they knew anything about an old rail line, and they pointed to a man paddling his small *cayuco* in the distance. I motored at idle speed and met the man, Benito Smith. He was born in Valiente and was part indigenous. Smith was paddling with one arm and said he lost the other one in a car wreck. He had worked as a surveyor for the banana company. After a brief discussion, he and I discovered we had common friends in the Changuinola plantations.

When he told me he knew where the old rail line was in the swamp, I became hopeful and enthusiastic. Smith got in my boat and we towed his *cayuco* behind. It was a eureka moment when he pointed to a line of ballast rock elevated above the mud of the red mangrove swamp. The man said that he and neighbors sometimes walked on the old rail bed to reach higher ground for hunting. He added that there was an old steam engine abandoned in the jungle along the line. I was happy that my mission was complete, and I thanked Benito Smith with a five-dollar bill. He got in his *cayuco* and continued to paddle with one hand further up the river.

In my wild imagination, I thought of the workforce at that banana transfer port—the pestilence of mosquitos, sand flies and mud. Malaria and yellow fever were rampant in 1900. It was all driven by the law of supply and demand; suppliers hoped to make a profit, and consumers got what they demanded to satisfy their epicurean taste. They would never believe what it took to get a Pumpkin River banana onto their plate.

THE FIRST WIRELESS TRANSMISSION ON MACCA HILL

Bananas are a very perishable commodity that require expeditious attention in harvesting, loading on a ship, travel at sea and marketing before they get ripe. To get fruit harvest orders to Bocas del Toro on time in the early 1900s, a messenger paddled or sailed in a *cayuco* from Limón, Costa Rica. The precarious off-shore trip along the Caribbean coast took about two days and nights or longer, depending on the weather.

In 1903, when Panama was still part of Colombia, Mack Musgrave, a United Fruit Company electrician in Costa Rica, got support from the president of the Company, Andrew Preston, to improve communications. Wireless systems were in their infancy, but this new invention might solve lots of problems by getting fruit cuts on schedule and ready for loading when the ship arrived. In 1904, a tall tower in Bocas del Toro was ready for the first trial transmission. Bill Saravanja calculated the height of the tower to be approximately 350 feet. Mack Musgrave in Limon witnessed the first wireless transmission from Limon, Costa Rica, to Bocas del Toro. In the beginning, there were failures but problems were solved and the new technology moved forward. By 1906, two ships of the company had wireless communications—the SS *Ellis* and the SS *Preston*. Wireless installations continued until all Great White Fleet ships were equipped with ship-to-shore communications.

When the banana port of Almirante was completed by 1908, two new tall towers were constructed at nearly sea level after landfills were pumped from the bay and stabilized with tons of ballast rock from the Changuinola River.

With this pioneer success in wireless communication, a network of towers and transmission centers were built in North, Central and South America and the Caribbean.

As a public service, the United Fruit Company organized a subsidiary called Tropical Radio Telegraph Company in 1913.

With new innovations, the tall towers were no longer needed and were dismantled. In Almirante, the humongous concrete bases of those two towers can still be seen between Cedar Creek and the wharf, but no one knows why they are there.

It was different in Bocas Town, the capital of the province of Bocas del Toro. The tall tower of 1904 was built on the highest point on the southeast tip of Columbus Island, known as Isla Colon. This steep hill that rises out of Almirante Bay is called Macca Hill, a distorted name derived from Macaw Hill.

At the base of the tower, there was a large building for the staff and communications operations. Concrete steps went from the bottom near a small boat dock in the mangroves to the top of the hill, a long and steep climb. The view from Macca Hill and the base of the tower was panoramic, as photographed by J. A. Stewart in 1913.

When the tower was taken down, the historic communications center was abandoned and rapidly grew back into an impenetrable jungle. That is the way I found it a century later. After I found the bottom concrete steps, I soon discovered that it was impossible to go any further. A week later, I returned with a good *machetero* who chopped the heavy bush over the concrete steps all the way to the top of Macca Hill. Panoramic views were mostly blocked by thick foliage so we explored the flat hilltop where the tall tower had stood. The four massive concrete bases of the tower were covered in vines and bushes. Nearby were vestiges of the building of the wireless communications center. History was made here in 1904 that resulted in a huge leap into the future of tropical communications.

Just a few hundred meters to the northeast and another hill away is where, in 1993, I explored and inventoried the German Cemetery for a chapter in my book, *Banana People.*

THE BRATSI RAILROAD TUNNEL

Dr. Lloyd Berg and I went up the Sixaola River looking for a disease that is found in wild banana plants. Suddenly, the boat captain pointed to an old railroad tunnel on the right side where steep hills meet the river. Railroad builders found it impossible to go around the hills so they built the Bratsi Tunnel. The plaque on the tunnel reads 1914.

That was the time when the United Fruit Company was developing virgin banana lands in the vast Talamanca Valley on the Costa Rican side of the upper Sixaola tributaries. Just a short distance upriver from the tunnel is where the Zhorquin River, also called the Yorkin, joins the Sixaola from the Panama side.

There are five major rivers in the Talamanca Valley that all come together to make the mighty Sixaola River. These rivers constantly flood and deposit the fertile, silty loam soils that are best for growing commercial bananas. But flooding also destroys banana farms, railroads, bridges, houses and other infrastructure. Added to the forces of nature was Panama disease, which decimated the last banana plantations by 1930. That is when the United Fruit Company left the remote Talamanca Valley. Today, there are hardly any vestiges left of the thousands of hectares of plantations in the Talamanca except for the tunnel.

Dr. Bill McLarney works in environmental affairs and riverine projects with indigenous people in the Talamanca Valley, and I asked him about the tunnel. He said it is still there just like in 1914, and that it is used by vehicles going to and from the Talamanca region. Bill lauded the engineers who built this tunnel and noted that not even the major earthquake on Earth Day in 1991 caused any damage. That is certainly a proud legacy to have more than a century after the tunnel was built.

THE SAFE AT PORT MORANT

At a time when I was on assignment in eastern Jamaica in the 1980s, Ricardo Flores drove me by an abandoned United Fruit Company wharf where millions of bunches of bananas had been loaded long ago. We were at Port Morant on the southeast coast of the island where Captain Lorenzo Dow Baker had loaded his two-masked fishing schooner with a few bunches of rare fruit—bananas—in 1870. Three weeks later, he sold them at a profit in the United States. This was the beginning of an empire.

The old dock for banana ships was in complete disrepair and covered with bushes. There was no one in sight and no activity of any kind. Ricardo showed me a huge safe that was lying in the brush at an angle with the door open. There was a pool of stagnant water in the bottom that was writhing with mosquito larvae. On the side of the safe was written: *United Fruit Company*. And I was astounded at what I was seeing.

Back in Panama in Bocas del Toro in the old general office building of the same banana company, I am reminded of that discarded safe I saw in Jamaica. The former office building in Bocas Town is now El Gran Hotel Bahia owned by the Jose Thomas family. Their ancient safe is the same size as the one I saw in Jamaica and is proudly displayed in their lobby. The safe is too large to move. At one time, *United Fruit Company* was also written on this safe in the hotel but a new coat of paint blocked out that memory. It is probable that the safe was installed in the former office building not long after the headquarters were built in 1905. It is located at the south end of Calle Tercera, which is Main Street in Bocas Town.

FINDING PUERTO CASTILLA: A COMPANY TOWN GROWN UP IN JUNGLE

When we were based in La Lima, Honduras, Phyllis and I picked a long weekend in 1989 and drove east some 350 kilometers along the Caribbean coast. It was a long drive with certain risks. Our goal was to explore Puerto Castilla, a former port town and headquarters of the Truxillo Railroad Company where the United Fruit Company developed its giant banana enterprise beginning in 1912. Plantations were on the fertile loam soils of the extensive Aguan River Valley. Millions of bunches of fruit were exported via hundreds of kilometers of railroads and across many bridges to the port of Puerto Castilla.

The Company built an autonomous "Company town" on a swampy, pestilent peninsula that jutted off of the mainland. This near-impossible feat was done by dredging mud and sand to make a landfill, followed by sanitizing the quagmire with drains and mosquito control. Ballast rocks and gravel from the rivers made elevated right of ways for rail lines to access the mammoth project.

Architects and engineers designed and built a beautiful town at Puerto Castilla, providing everything that hundreds of employees needed, especially for many foreigners, mostly Americans. Depending on the job level, homes were comfortable, furnished and ready to move in. There was a well-stocked Company store, an American and Spanish school, the Puerto Castilla Hotel and restaurant for travelers coming and going, and a social club with a golf course, tennis court, baseball field and swimming pool. Some employees enjoyed diving from a high platform built in deep water off the beach. Being off-shore probably gave them some reprieve from the constant torment of sand flies.

The huge office complex accommodated top management, accounting, engineering, materials and supplies, agriculture, transportation, exports and the wharf. Public and labor relations and other departments were included. Plants and Grounds kept the town tidy and beautifully landscaped with parks and walkways. Last, but certainly not least, was the medical department and a huge hospital staffed by U.S., Latin American and European doctors and nurses. Tens of millions of banana bunches were shipped from Puerto Castilla, right up until Panama disease reduced production to a point that the Company closed and left in the late 1930s. Rolling stock from railroads, mahogany furniture, machinery and all kinds of materials were transferred to new Company banana developments on the Pacific side of Panama and Costa Rica—Puerto Armuelles, Golfito and Quepos. Employees were also transferred or released.

After the demise of Puerto Castilla, the island-like Company town literally grew back to jungle. And that is the way Phyllis and I found this former banana town. We followed the only railroad right-of-way that entered the town site and stopped at a bridge over a tidewater estuary where we encountered a guard with a machine gun. He asked why we were there and told us it was prohibited to enter. This was because the Honduran navy had a port for its coast guard there. I showed him our legal documents and fully explained that I was a *bananero* and wanted to see the former Company town. He was amazed that I knew so much about the history of the place and finally told me to proceed with caution. The guard warned me that the old roads disappeared and some bridges had rotten boards. I thanked him and drove forward.

He was right, and it was very precarious crossing ditches and canals. Bushes of all kinds scratched the vehicle because the road was not used very much. There was no one in sight until we reached the far end near the beach. Amazingly, a few *campesinos* appeared. I asked about a large two-story building and was told it was the hospital.

One insistent man volunteered to give us a tour. When we entered the main entrance, we saw entire families living in rooms, sleeping on the floor or in hammocks. The rooms and halls smelled of smoke because they were cooking on open-hearth wood fires. Our guide proudly ushered us to the surgery room. It was overgrown with vines coming through open windows. White tiles still lined the walls and floors. Giant termite nests were seen here and there. The scene was surreal and also smelled of urine along the way. After seeing some of the second floor in complete disrepair with debris everywhere, we tipped our guide a few Lempiras and left. He gave us a big smile and waved as we drove away from this monumental medical center that had been abandoned half a century ago.

I dreaded the precarious drive over what was once a railroad right-of-way but we made it without incident. That night, we stayed across the bay in a hotel in historic old Trujillo and felt so relieved to be there safe and sound. After a couple of Salva Vidas (Honduran "Life Saver" beer), Phyllis and I discussed our adventure. We agreed that we had accomplished our mission, not by just uncovering another old historic stone, but by discovering the ruins of what was once an all-inclusive, United Fruit Company banana town. That was both surreal and real.

SOSUA, D.R.: ANOTHER ABANDONED BANANA PORT

I joined the United Fruit Company in March 1959 and was moved to four places in less than two years. The fourth place was in the Dominican Republic with the Grenada Company in May 1961. My housing was the bachelors' quarters on a hill of the Company-built banana port then called Puerto Libertador, a name that honored the brutal dictator, Generalissimo Rafael Leonidas Trujillo Molina. After he was assassinated, the name of the port was changed to Manzanillo. Now it is known as Pepillo Salcedo and is right on the border with Haiti. The area is arid, dusty and windy, unlike previous environments that I had known in Central America, but employees liked their assignments there. Neighbors were socially bonded and organized excursions for recreation.

I was invited to enjoy a weekend at Sosua, some 200 kilometers to the east on the Atlantic coast of Hispañola. At the time, I did not know anything about Sosua, but I learned some interesting history from others during our trip. We stayed at the Garden City Hotel, a very accommodating wooden structure.

Sosua had produced bananas in the 1890s under previous owners. Fruit came to Sosua on ox carts and wagons pulled by mules. Later, a rail line was built to the port. Captain Lorenzo Dow Baker arranged for the acquisition of banana lands for the Boston Fruit Company. Baker and his partner, Andrew Preston, built a large wharf along the cliff on the protected side of Sosua Point.

Railcars with fruit pulled alongside the ship for loading. By 1899, 1.5 million bunches of bananas were being exported.

Valid history becomes hazy at this point, but some of the known facts are that banana tycoon Minor C. Keith had a sudden financial default and reluctantly merged with his rival, the Boston Fruit Company in 1899. Andrew Preston became the president of the new enterprise–United Fruit Company. Fruit production and shipping from Sosua continued a few more years, based on a photograph showing a banana ship there in 1902. Banana production in Sosua declined during winter months because the tropical banana plant stops growing when it gets too cold. As a result, Preston turned his attention to more perfect growing conditions in Central America where Minor C. Keith, the vice president, had extensive assets. A decision was made to shut down Sosua completely and expand production in Central America. Thus, Sosua died a sudden death, leaving behind a relatively new wharf, railroads, bridges, housing on the high ground above the Atlantic Ocean shore, a schoolhouse, a Company store, offices and the clubhouse. Eventually, the place grew back to cactus, *cambron* (mesquite) and other desert plants on the windy, xerophytic north shore. After several decades of abandonment, the structures fell into disrepair.

After lunch at the hotel, I wanted to explore the steep shore line where the wharf had existed. Getting down to the shore across the abandoned rail line was rocky and rugged. Once I got to the beach, the waters were crystal clear and snorkeling was magical. I was awed at seeing masses of sponges, corals and other forms of marine life clinging to old pilings that once supported the abandoned banana wharf and railroad 60 years before. I was floating right where banana ships had tied up to load banana bunches of the Gros Michel variety since the late 1890s. Soon, I was to learn much more.

After I changed back to dry street clothes, Dr. Otto Reif and his wife, Hansi, invited me to visit some of their Jewish friends, who lived in old United Fruit Company houses that had been abandoned for so long. This was an incredible, unexpected experience.

Generalissimo Trujillo needed to improve his depraved public image in the world and agreed to settle Jewish refugees in Sosua. Indeed, this was a noble and humane gesture. In 1940, the first European Jews came to Sosua, and some lived in the abandoned banana company houses right where I was visiting. I was told that some of these settlers had come from the huge Jewish refugee camps in Singapore and China. Others crossed the Atlantic. They brought culture, business skills and trades that were implemented in Sosua. Besides farming crops and livestock, one of their successful endeavors was dairy farming, and the production of cheese and other dairy products that were sold throughout the country. One woman was quoted that her father grew hogs, but they were kosher hogs [sic]! During the cordial visit on the front porch with a group of the Reif's friends, I heard German, Yiddish, a Yugoslav dialiect (Hansi Reif was a Yugoslav Jew), English and sometimes, Spanish. I was so impressed with this colony of former refugees that I wrote the story in *Banana People*. Today, numerous documents about this Jewish refugee settlement can be found on the internet. They are worth reading.

Top: Bridge at Cricamola, 1921. Clyde Stephens collection.

Center: Moving bananas on the Uyama River, 1901. Leslie's Weekly, January 16, 1902. Clyde Stephens collection.

Bottom: Uyama River bananas exported at the Caucho River landing, 1901. Leslie's Weekly, January 16, 1902. Clyde Stephens collection.

Left: Caucho River rail port for Pumpkin River bananas, 2020. Photograph: Mirka Stephens. Right: United Fruit Company safe in Gran Hotel Bahia, Bocas del Toro, Panama. Clyde Stephens collection.

Bratsi Tunnel built in 1914, near the Sixaola River on the Costa Rica side. Clyde Stephens collection.

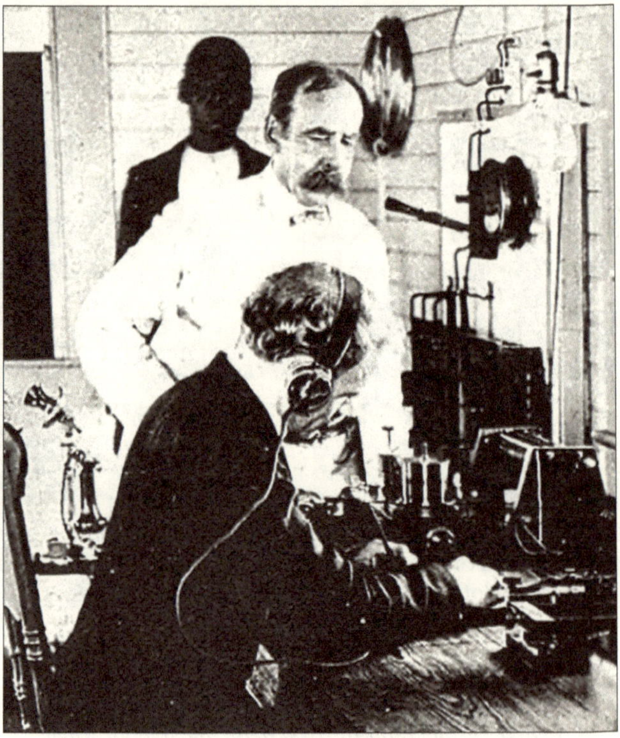

Mack Musgrave (in white) watching first historic wireless transmission from Limon, Costa Rica, to Bocas del Toro, Panama, 1904. Unifruitco, June, 1948, p. 6.

View from Macca Hill tower looking northeast, 1913. Foreground right: dry dock and boat repair. Center: Bocas Town. Photograph: J. A. Stewart.

Opposite Page: First wireless telegraph tower, 1904, Macca Hill, Bocas del Toro, Panama.

Puerto Castilla, Honduras. Hospital (foreground). Unifruitco, April 1931, p. 420.

Banana ship at Sosua, 1902. Photograph from Unifruitco.

Clyde Stephens, right; Louise and Millicent Crichton, left corner. Garden City Hotel in Sosua, Dominican Republic, 1962. Photograph: Henry Crichton.

9. Banana People under a Dictatorship

AS A NEWCOMER TO the Dominican Republic in 1961, I was told to not discuss politics. This was because the dictator, General Rafael Leonidas Trujillo, and his intelligence squad had spies all over the country. Any suspicious remarks could be reported by working colleagues, friends, neighbors or even family members. I soon observed that newspapers, radios and road signs all praised *El Jefe* (the chief), *El Benefactor de la Patria* (Benefactor of the Country), and *la Era de Trujillo* (the Trujillo Era). Criticisms were not tolerated. As with the Bible in many places, at every major hotel room in the country, a guest could review a beautiful, hardcover book illustrating the greatness of the dictator and his family and how his constituents and the military all adored him. Even the churches were targeted, all of them forced to post the following: *Trujillo en la Tierra, Dios en el Cielo* (Trujillo on Earth, God in Heaven). Many homes displayed a plaque showing their loyalty to him.

THE DICTATOR[1]

Rafael Leonidas Trujillo Molina was born into a middle class family on October 24, 1891. At the age of 16, he became a telegraph operator. Later, he turned to crime, stealing cattle, counterfeiting checks and practicing postal robbery. After several months in jail, he formed a violent gang of thieves.

Another job was weighing sugarcane and being a rural guard in sugarcane fields for about two years.

When the U.S. Marines invaded and occupied the Dominican Republic from 1916 to 1924, they formed and trained a Dominican police force to keep the peace. Trujillo was one of their trainees, and he enrolled in a military officers' training school in 1918. Upon graduation as a lieutenant, he rose rapidly up the ranks to become commanding colonel of the national police. By 1927, he was a general. He seized power in a military revolt against the president in 1930, and that was the beginning of his ruthless dictatorship that ended 31 years later in 1961. Under Trujillo, the Dominican Republic became the epitome of the so-called "banana republic"—described by one Internet source as "a country that is politically unstable, dependent on limited agriculture (e.g., bananas), and ruled by a small, self-elected wealthy and corrupt clique."[2] During his reign, General Trujillo was credited with improving the economy, sanitation, water, and health; constructing new roads, schools, hospitals and increasing the general standard of living.[3, 4] But the benefits came at a high price to the populace.

The Trujillos accrued enormous wealth by getting a cut from just about every kind of business in the country. Kickbacks and fees were normal. The dictator and his family dominated banks, the press, radio stations and the tobacco, salt, lumber, rum and sugar industries. Cerveza Presidente was the only beer in the country. He acquired cattle lands on a grand scale and went into meat and milk production. Some of these operations evolved into monopolies. Trujillo had practically the entire country's treasury at his disposal, meaning that he and his family lived royally, while much of the population languished in poverty.

At the same time, Trujillo maintained an iron grip on the entire country and its people. Estimates of 50,000 people were murdered in roundups, torture chambers and faked accidents.

The largest mass-murder that was approved by the dictator occurred in 1937 when an estimated 20,000 to 30,000 Haitians were slaughtered, mostly chopped to death by *machetes*. This included men, women and children who had illegally crossed the border from Haiti and worked in Dominican sugarcane fields for a few cents a day. During the slaughter, many escaped back to Haiti but thousands did not make it. This was known as the *Perejil* (Parsley) Massacre.

Because numerous Dominicans were dark-skinned and looked like Haitians, the police sorted them out by showing any suspect a sprig of parsley (*perejil*) and told the person to identify it in Spanish. Dominicans could say *perejil* and roll the Spanish "r" softly, but the French creole Haitians could not. They were murdered and buried in mass graves. There was an outcry from around the world, but that didn't change Trujillo's modus operandi.[5] Generalissimo Trujillo's dictatorship hit a low point in 1960 when he nearly assassinated an adversary, Venezuelan President Romulo Betancourt, with a car bomb in Caracas. Two people died but the Venezuelan president recovered from his injuries. Once again, there was international outrage that prompted the Organization of American States to dissolve diplomatic ties and impose economic sanctions on the Dominican Republic.[6]

OSTENTATIOUS WEALTH

Trujillo displayed his wealth openly: In 1955, he bought the *Sea Cloud*, one of the most luxurious sailing yachts in the world. The name was changed to *Angelita*, in honor of his daughter. When docked in San Pedro, California, the *Angelita* was used to entertain Hollywood movie stars.[7]

At the grand opening of the Dominican World's Fair in 1955, a parade featured the queen of the fair—none other than one of Trujillo's daughters, 16-year-old Angelita, who was crowned as Queen Angelita the First. Accompanied by 150 princesses, she wore a white satin gown with real pearls, rubies and diamonds that were bordered with 600 Russian ermine furs and a 75-foot train that had been commissioned for $80,000 at Fontana Gowns in Rome. Another $75,000 was spent on other regal items, like a scepter and brooch. Two very expensive hairdressers were flown in from New York.

Outside critics pointed out that the fair cost the Dominican Republic the equivalent of 30 million U.S. dollars, a third of the annual budget for the whole country. It was not a financial success, but presidents and dignitaries from other countries came to enjoy the gala event. That was important.[8, 9]

BANANAS AND EARLY YEARS OF THE Grenada COMPANY

United Fruit Company operations in the Dominican Republic preceded the arrival of Trujillo by several decades.

In the late 1890s, the Company exported Gros Michel bananas from Sosua until about 1903. In 1937 and 1938, Dr. Vining C. Dunlap evaluated alluvial soils along the Yaque del Norte River near Montecristi in the northwest. The region is wind-blown, cool in the winter months, and very arid, a desert covered with cactus and thorny *cambrón* bushes. But Dr. Dunlap approved these silty loam soils because they could be irrigated with river water and produce good bananas for export.[10] IIn 1943, during World War II, the Grenada Company and the Dominican government signed a legal contract so that the Company could grow and export bananas and other products. The original name was Granada Company, but the name was spelled incorrectly as Grenada in the legal documents and was never corrected. Walter Mosley Hamer Turnbull, an employee who was born and raised in Mexico, represented the United Fruit Company subsidiary. Soon the manager, David Cloward, and his assistants were exploring lands along the river to plant bananas.

The first headquarters of the Grenada Company were in Montecristi. From 1940 to 1947, old abandoned offices, stores and houses in the remote, rundown town were renovated and converted into Company offices and housing for banana company employees and their families.

Just after World War II, in 1946, the Company started a railroad to haul Gros Michel bunches from the United Fruit Company farms at Walterio and La Cruz to the wharf at Manzanillo near the Haitian border. There was no town; only a small military outpost and a wharf off the beach.

The banana transfer center at El Posito had 1,500 workers to clean dusty bunches of bananas before another crew stacked the fruit in a railcar. The cleaning crew rotated every eight hours, day and night. There were hundreds of trucks coming on dusty, dirt roads from the many private farms, some quite remote from the railhead. The transfer center attracted other miscellaneous workers who drifted in. This processing and transfer center was busy because sometimes there was one ship loaded every day during peak growing and harvesting seasons.

THE NEW COMPANY TOWN

To meet its burgeoning needs, the United Fruit Company decided to build a new Company town in Manzanillo just like they had done in Central America, Cuba and Colombia. Fernando Lara de Viñas describes the magic of converting a dusty pile of loose rocks in a desert into a modern Company town with all the amenities during the late 1940s and early 1950s.[11]

The new golf club was a huge palm-thatched structure with a beautiful tile floor and no walls. It was the social center for the community. The desert golf course had sandy greens and rocky fairways. Movies were shown three nights a week. Dinners and dances were popular. Music was provided by Company employees who formed a *merengue* orchestra. These musicians also played in the other club built by the Company for labor employees. Its name was Club Ramfis after the dictator's son.

As the new port developed, the Company promoted and assisted private Dominican growers who sold their fruit under contract to the Company. The fruit was transported in trucks to a railhead at El Posito across the river from Guayubin and loaded onto railcars that took fruit to ships at the new port of Manzanillo.[12] The construction of Manzanillo near the border of Haiti was promoted by Trujillo for self-serving purposes.

His strategy was that if Haitians illegally crossed the border again and invaded the Company town, then the United Fruit Company would ask the United States military to come and protect American interests in the banana port. As a result, the United Fruit Company invested millions of dollars to build Manzanillo. And typical of his obsession with inserting himself in most aspects of daily life, Trujillo gave Manzanillo the name *Puerto Libertador* (Port Liberator).

The object of such a huge investment was to increase profits by substantially boosting the export of Gros Michel bananas. Hard work and risky investments paid off when eight million bunches were exported in 1959 and everyone celebrated.

Another important export at this time was banana puree made from the Dominican variety known as Media Mata that was grown in the farms and processed at the port.

THE MANAGER AND GENERALISSIMO TRUJILLO

El Jefe Trujillo liked William Wrigley Turnbull because Bill Turnbull knew the Latino mentality and spoke fluent Spanish. He was born in 1916 in Puebla, Mexico, where he grew up. His family had both Mexican and British heritage in Mexico since the 1800s. Because of Trujillo's preference for Bill Turnbull, top management transferred him to Puerto Libertador to replace Dave Cloward. Cloward was an old-time manager who had been in Montecristi since the early 1940s to jump-start and begin the Grenada Company development, but Trujillo wanted a change. Trujillo got the man he liked although he could never pronounce the name: Turnbull always came out as Trombos.

During one visit, Trujillo came from Ciudad Trujillo to Puerto Libertador in his limousine, a long drive of six or seven hours. Before his arrival, a very forceful and armed security squadron suddenly surrounded and inspected the manager's house, inside and out. The servants were terrified. After the house passed the inspection, *El Jefe* arrived and Mildred Turnbull invited him in. He told her that he wanted to rest. She offered him a guest room, but he selected a nice chair on the front porch with a view of the ocean.

After his rest, Mildred Turnbull offered Trujillo his customary Fundador brandy, a Spanish spirit that many businesses and homes kept on hand just in case *El Benefactor* dropped by. She came out with a man-servant bringing a sealed bottle on a tray. The bottle was opened in front of El Jefe and her. The server was nervous and shaking.Trujillo came to see the great works and investments of the United Fruit Company in Puerto Libertador, but some of his family also arrived by sailing on their luxurious yacht, the *Angelita.* Bill Turnbull was not at the manager's house to receive the general because he was making special arrangements at the wharf to accommodate the vessel. When the *Angelita* arrived, the manager went to his house to greet Trujillo and then give him a tour of the Company town.

Trujillo had visited the new Company port before. In March 1951, he arrived on a converted River-class frigate of the Dominican navy, the *Presidente Trujillo*, where he hosted manager David Cloward and Assistant Manager Bert Hogge for lunch. After lunch, the managers gave the Generalissimo a grand tour of the banana farms. A dinner party was given in honor of Trujillo that night in the manager's home.[13]

FINESSING A FORMER MISTRESS

During Grenada Company banana operations from the 1940s to 1961, managers, employees and residents in Montecristi all knew about *doña* Isabel—respecting her powers and position during the Trujillo Era. At one time, *doña* Isabel was the mistress of Trujillo—one of many.[14] Walter Hamer tells it best in his mother's biography that he wrote, as told by his mother, Mildred Maxwell Hamer Turnbull.[15] The story probably took place in 1955, soon after Bill and Mildred Turnbull arrived:

> Montecristi was the seat of the provincial government and the governess was *doña* Isabel. She was a large, outgoing, light-

skinned mulatta who had been one of Trujillo's many mistresses in her younger days. As the head of the local government, she was an important factor in maintaining the Company's smooth operation and much time and effort were invested in cultivating and maintaining good relations with this unusual woman.

Not long after their arrival in Puerto Libertador, manager Bill Turnbull received an invitation to bring his family to a luncheon at *doña* Isabel's house. The invitation included Walter and Geoff Hamer, who were spending their summer vacation there. Everyone dressed up and the family was driven to the house by the banana Company chauffeur, *don* Enrique. They were greeted by their hostess wearing a great flowing colorful dress and a turban...but no shoes. They were shown to seats in the cool shade of the ground-level terrace beneath the raised house and offered a variety of fresh fruit juices and other refreshments as a lead-on to the meal.

The furnishings were gracious European pieces; the places were set with fine French chinaware and the cutlery was heavily engraved gleaming sterling silver. A gorgeous ceramic soup tureen was brought in and placed in front of *doña* Isabel. She announced that she had prepared a special treat for the occasion...the national dish of the Dominican Republic: *sancocho*. Mildred, Bill and the two boys all knew that this was the name given to the savory stew that was the mainstay of the peasantry in most of the countries of the Caribbean Basin—a conglomeration of the most plentiful and inexpensive meats and vegetables. *Doña* Isabel boasted that hers was the best *sancocho* made, as it followed a recipe handed down to her from her great grandmother. She lifted the lid of the tureen and gave the contents a stir with the big silver ladle and to her guests' great surprise, a bright yellow chicken foot spun slowly around the pot. No one said anything as she began ladling generous servings of the stew into soup bowls, which were then placed in front of each diner. To everyone's great relief, no one got the chicken foot, which, at the end of the course, *doña* Isabel removed from the pot with her fingers and fed to the big dog that had been watching the proceedings with considerable interest.

MY ASSIGNMENT IN THE DOMINICAN REPUBLIC

I was transferred from Changuinola, Panama, to the Grenada Company Division in May 1961 to replace Russ Caid in the Research and Technical Services Department. My degree was in entomology, but I was expected to deal with any kind of problem that showed up in the banana plantations, even though I was still quite naive in the business at age 27. Flower thrips and red-scale insect pests kept me busy, but I also monitored nematode pests, miscellaneous diseases and agronomy problems. In addition, there were field trials testing nutritional needs. The United Fruit Company had just converted from traditional shipping of bunches of bananas to packing hands of fruit in cardboard boxes. That conversion was progressive but it also created new problems that had to be solved.

The dry, windy and dusty environment was new to me, but I easily adapted with other employees and families and enjoyed the new assignment. Harlo von Wald was the manager and his assistant was Warren Breck. My new home was a spartan room in the bachelors' quarters on the hill with a view of the Atlantic Ocean. By looking to the west, I could see Haiti about a kilometer away. At night, Haitian drums were beating in the distance across the Masacre River that separated the two countries. As the reader will find out later, the river was quite appropriately named.

TRANQUILITY

During my assignment with the Grenada Company, my work and travels were never significantly restricted. There were always the many military checkpoints along the roads to stop at, show vehicle registration and the driver's and passengers' identitfication, but that was routine. It helped that we were in a marked Company truck. Banana company people could travel for several hours across the country on weekend trips.

AND A MOMENT OF TERROR

Not long after I arrived, out of curiosity, I asked Russ Caid to show me some of the back trails further up the Masacre River. He said that it was a sensitive and restricted area, and we would have to get permission from the military and take a guard. After the permission was granted, an armed guard jumped into the back of the pickup truck, and we drove along desert foot trails worn down by donkeys and goats. When the truck could go no further, we stopped and walked. The river split, meandering through several small channels where the landscape on the banks was xerophytic—desert with thorny *cambrón* bushes and cactus.

Suddenly, we looked up and faced a tall black man who had a string of small fish and a fishing pole. He was completely naked and was not even wearing a hat or shoes. The armed guard fidgeted with his automatic submachine gun while Russ and I stood in place for a few seconds trying to process what we were seeing. The fisherman was obviously a Haitian who had crossed the narrow channels of the Masacre River. He was probably fishing on the Dominican side but there were no international boundary markers at that point to distinctly mark the division, only a marshy wasteland where goats and donkeys browsed.

I have never seen terror on the face of any human being like I saw that day. It was because the Haitian knew the history of massacres in that very same region. He instinctively dropped his fish and pole, holding both hands in front of his body facing the guard. Both hands were open and rotated back and forth, a sort of international sign language for "No, please don't shoot." It was obvious that he was begging for his life. Soon, the Haitian was nervously saying the only word he could utter that the Spanish-speaking Dominican guard might understand: "*Non, non, non, non.*" The trembling man never turned his back to us, but slowly started walking and stumbling backwards, even across the river channels. His eyes were bulging, and his face was twisted with unimaginable terror staring at the guard with the gun. After he reached the far side of the narrow river channels, the terrified Haitian suddenly twisted around and dashed into the thorny bramble bush and disappeared into Haiti without his clothes and fish. Russ Caid and I stood there in a sort of trance with a nervous grin, trying to sort out what we had just seen. I am not sure what the guard was thinking, but he never made any gestures or threats indicating that he was going to shoot the Haitian. In fact, I don't think the guard ever said a word during the unfolding of this human drama that took place only a few paces in front of us.

MASS GRAVE

When I lived in Puerto Libertador, an elderly Company employee related an experience to me that I will never forget . He had been one of the heavy machine operators who had moved earth with bulldozers to build roads and dig canals with draglines during the 1940s.

During one job, his dragline suddenly dug up human

bodies that had been buried in a mass grave. He immediately stopped work and reported what he saw. When armed military men appeared on the scene, the operator was ordered to cover up the human remains and stop working on that project. He complied. These were remains of the Perejil Massacre.

GALLOWS HUMOR

Despite serious warnings, sometimes foreigners, especially after a few drinks, made humorous remarks but with caution. Because Trujillo was "*El Benefactor de la Patria*" (The Benefactor of the Nation), he was referred to as "Uncle Benny." From boyhood, Trujillo liked to snap bottle caps on his shirts because they looked like military medals. Bottle caps are called *chapitas* in Spanish. Later, his military uniforms were weighted down with numerous medals and awards. Thus, he was also known as "*el Chapitas.*"

Once, on the front page of a leading newspaper, a poem praising Trujillo was printed. Somehow, the editor did not see a problem with the poem. There was a single word that was printed vertically on the left side of the column by using the first letter of each line. The word was *LADRÓN* (thief). The poem was circulated widely. Readers wondered what happened to the editor.

"Gallows humor" was discretely used by some Dominicans in private conversations. There were several infamous and well-known torture chambers that had names relevant to their road markers. One was *Kilómetro Nueve* (Kilometer Nine) and the other was *La Cuarenta* (40th Street). I was amused that Dominicans would compare their *gripe* (flu) with these horrible places. One person might say, "*Ay, tengo un caso de La Cuarenta.*" (Oh, I have a case of the Forty.).

Someone else might say, "*Tengo un caso de El Nueve.*" (I have a case of the Nine.). It was macabre but descriptive.

SHORTCUT TO HAITI

Harlo von Wald was the manager of the Grenada Company and asked me to visit the small banana plantation of the West India Fruit and Steamship Company that was less than 100 kilometers west of us in Haiti. Banana companies were always curious about what their competitors were doing and politely exchanged visits if the manager approved. Exchanges of this nature were never made clandestinely or without permission. My visit was approved by Mr. Grigsby, whose office was in Cap-Haïtien on the north coast of Haiti.

From my bachelor's quarters on the hill in Manzanillo, I could see ships entering and leaving Fort Liberté, Haiti, carrying sisal products. The banana plantation in Haiti was only another 50 kilometers west of Fort Liberté. This should have been an easy drive of a few hours in my Company truck by crossing the international border at Dajabon. But that plan turned out to be a fantasy on my part. The reason was that Trujillo on the Dominican Republic side was haggling with François Duvalier, better known as Papa Doc, on the Haitian side. There was a chain across the only bridge over the Masacre River. No vehicles crossed this international boundary. Consequently, I spent four days getting to the banana plantation that was only a few hours and a short distance from my home base in Manzanillo.

Day One: I rode in a small public car (shared with other passengers) about seven hours across the Dominican Republic to the capital.

Day Two: I flew on Pan American to Port-au-Prince, the capital of Haiti.

Day Three: A government interrogation agency wanted to know what my mission was in Haiti. Fortunately, I had previously arranged with a Haitian friend, Dr. Leonce Bonfils, to accompany me. The two of us rode in a car over the hazardous roads to Cap-Haïtien where we met with Mr. Grigsby, manager of the West India Fruit and Steamship Company. He gave us a very cordial welcome and drove us to a neat little hotel overlooking the Atlantic. This is where we stayed and got acquainted with some tasty Haitian rum on the open terrace of the hotel.

Day Four: From Cap-Haïtien, Mr. Grigsby arranged for a Jeep to drive us to the plantation where Arnold Halderson, the superintendent, was waiting. He explained that he had learned his profession as a banana man with the United Fruit Company many years earlier.

The tour of their thousand hectares of bananas was routine. The banana production looked good, and the fruit was exported to Miami on small ships from Cap-Haïtien. As a courtesy, I contributed some technical advice about insect, nematode and disease control. It was an amicable exchange.

By midday, Mr. Grigsby arranged for us to explore the colossal Citadelle Laferrière, a formidable fortress south of Cap-Haïtien where Haitian King Henry I had a palace, the Sans-Souci. The king had been a slave as a young man and helped to liberate Haitian slaves after the French left the island.[16] To reach the ruins of the palace and the fort, we rode burros that knew the long, rugged mountain trail—a ride I shall never forget. At the top of the mountain, I looked northeast from the Citadelle Laferrière fortress toward where I lived in the Dominican Republic, only a few hours' drive by road. The next day, I began to return the same way that I had come, and it took the same amount of time. It would have taken only a few hours if I could have crossed the bridge at Dajabon. That was the shortcut to Haiti that I wanted but two dictators, one on each side of the border, had other ideas.

THE END OF TYRANNY

When Russ Caid and I went for a routine review of Luis Amiama Tió's banana farm, Finca Cerro Gordo, the gates were blocked. Two heavily armed guards would not let us enter. Something was in the making and we were puzzled. A few days later, Trujillo was assassinated. Luis Amiama Tió was one of the key conspirators, although he was not one of the seven conspirators in the ambush that killed the dictator.[17] On May 30, 1961, the seven conspirators, most of them with close ties to Trujillo's inner circle, assassinated the tyrant on the way to San Cristobal, his hometown, to be with a mistress. He was 70 years old. Three cars carried the assailants. A total of about 60 shots were fired during that short battle on the dark road.[18] Ramfis Trujillo flew immediately from his home in Paris to take over the dictatorship of his father. The infamous S.I.M. (Military Intelligence Service) rounded up hundreds of suspects and tortured them to death, including almost all of the prominent conspirators. The banana grower and former mayor of Ciudad Trujillo, Luis Amiama Tió, escaped the fatal roundup by hiding in a friend's house for six months.

ACTS OF FREEDOM

After the last of the Trujillo families left their country at the end of 1961, the population was still cautious with perpetual fear of the past nightmare. There was no sudden celebration of liberation in the streets until later. I remember a simple but bold declaration of independence in the manager's office in November 1961, six months after Trujillo was killed. A key employee in the office, Chery Jimenes, glanced around the room to see who was looking—everyone was looking at him. He quietly removed the traditional photograph of Generalissimo Trujillo from underneath the glass plate on his desk and daringly began his first act of liberty. The faded old portrait of Trujillo was torn into little pieces and thrown into his trash can. Nobody in the office said a word and fellow employees looked at him in disbelief. This was the beginning of freedom after 31 years of dictatorship, but many citizens still could not believe it was true.

There were other acts of freedom. During the dictator's era, popular *merengue* music was constantly played on jukeboxes in *cantinas* and clubs and broadcast on La Voz Dominicana radio, owned by Trujillo. The lyrics always extolled the machismo and greatness of *El Jefe*, his family, his military and the fearful police. But by 1962, a year after Trujillo was assassinated, song lyrics changed. A *merengue*, "*Mataron al Chivo en la Carretera*," became popular. (They Killed the Goat on the Road.) There was no doubt about who the goat was. It was a very popular *merengue* in 1962, and couples sang along as they danced. This was freedom.

FAREWELL TO THE COUNTRY AND THE BANANA INDUSTRY

In December 1962, I was transferred to the research department in La Lima, Honduras. After I left, Dominicans were arguing on how to reorganize their country after 31 years of autocratic dictatorship. Afterwards, there was disorder. True democracy was not known to most Dominicans even after the Trujillos left. The United Fruit Company was undecided about what to do with its huge investments and holdings that were being threatened by labor disputes, blockages, political disorder and chaos. In time, many Company employees were transferred to Santa Marta, Colombia, and to other banana divisions.

In July 1964, Phyllis and I went on a vacation through Santa Marta, Colombia. When we visited banana friends that I had known in the Dominican Republic, I learned firsthand about what happened to the Grenada Company after I left. It was nice to see our friends again, but sad to hear about their stressful exodus from the Dominican Republic when the United Fruit Company shut down its banana operations. After the closure of the Grenada Company, several banana companies attempted to rehabilitate the industry in the Dominican Republic, but the trade has struggled to recuperate like before.

My travels through the Dominican Republic and Haiti.

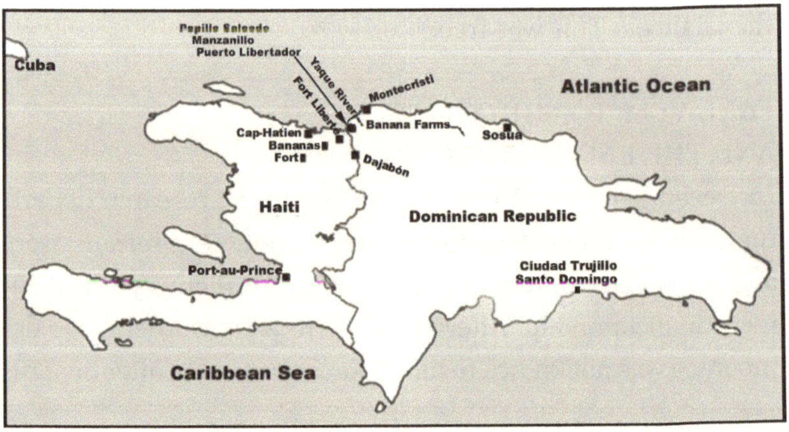

Captain Rafael Leonidas Trujillo Molina, 1922.

Trujillo poster. Translation: In this home, Trujillo is a national symbol. Rectitude, Liberty, Work and Morality. 1955, Year of the Benefactor of the Nation!!

The yacht Angelita which again sails as Sea Cloud.

Gros Michel banana bunches.

Manager David Cloward and his assistant Bert Hogge, exploring Yaque del Norte River and future banana farm sites, 1943.

Puerto Libertador in 1949.
Starting from nothing in 1945, manager David Cloward and staff were praised for building this town.

Puerto Libertador, where eight million bunches of Gros Michel bananas were exported in 1959.

Bill Turnbull and Mildred in Puerto Libertador between 1955 and 1959.

Trujillo at the manager's house, Puerto Libertador.

Bill Turnbull (right) giving Trujillo a tour of Puerto Libertador.

Trujillo and Turnbull inspecting the Company commissary.

Generalissimo Trujillo (center) visiting Puerto Libertador between 1955 and 1959.

Manager David Cloward, circa 1945, with a tarpon caught in the Masacre River at the exact spot where I caught a snook in 1961. Haiti is across the river.

Trujillo with his two daughters on either side of him and doña Isabel behind with her niece at the manager's house in Puerto Libertador in the 1950s.

Rafael Leonidas Trujillo
October 24, 1891 - May 30, 1961.

Generalissimo Trujillo, in full regalia.

Visiting the Jewish refugee colony with high school students home for vacation. L to R: Clyde Stephens, Bobby Batson, Ian Dahlgard, Joe Batson. Sosua, 1961. Clyde Stephens collection.

10. A Crooked Path to Citizenship
by Walter Hamer[1]

I WAS BORN IN the United Fruit Company's hospital in Puerto Cortes, Honduras, in 1932. My mother was American, my father British. The American citizenship regulations at that time would only grant U.S. citizenship to foreign-born children if both parents were Americans. Since this was not my case, my father registered my birth with His Britannic Majesty's Consul, and I became a "Subject of the King and Citizen of the United Kingdom of Great Britain and Northern Ireland." This created a special bond between my father and me as I grew up, and he often referred to us as "a couple of Englishmen."

Nationality was never an issue living in the melting pot that was the community of United Fruit Company employees. The company towns that we lived in were populated by people from all parts of the globe of differing colors, sizes, shapes, and languages. In my one-room-school class there were three different skin colors and four different languages, but nobody noticed.

The Company provided its families with free schooling through the 8th grade, after which most kids were sent away to school to live with family members or go to boarding school. When the time came for me to go to high school, I was enrolled in a military school in Tennessee. Six weeks after taking me there, my father was killed in an airplane accident in Costa Rica. Our common British citizenship now became a very important factor in my life, as it provided a bond back to my beloved "fellow Englishman."

By the time I graduated, I qualified for, and was granted, resident alien status and issued the coveted "green card." After finishing high school, I moved to New Orleans to live with my now-retired grandparents and to study civil engineering at Tulane University. Shortly after settling into my college life, I received a letter from the Selective Service Office informing me that I had to register for the draft. Thinking this was surely a mistake, I armed myself with my birth certificate, green card, and passport, and presented myself, per instructions, to Local Draft Board 41.

Local Draft Board 41 consisted of a standard-issue government metal desk set among many others like it on an upper floor of the downtown post office and was manned by a very large woman wearing a very large scowl. When I explained to her that I had not registered because I was a foreigner, her response was to the effect that she was fed up with foreigners coming to her country to take advantage of its wonderful educational opportunities, but who were unwilling to participate in the country's defense. I informed her that I was neither unappreciative nor unwilling, only uninformed, and that I was prepared to put things right. A short time later, I walked out with my freshly printed draft card and a simmering dislike for the opinionated bureaucrat at Local Draft Board 41. Because I was enrolled in a formal course of study, I had reluctantly been given a student deferment, and so was not in danger of being drafted for the moment. At the end of my freshman year, I spent the summer with my mother and stepfather, who were living in the Company's Honduras Division in the town of La Lima. I took a temporary job working in the field with one of the Company's engineers, and by the end of the summer was convinced that I wanted to be an agricultural engineer. I investigated school choices and decided that I wanted to go to Cornell University, the alma mater of my summer boss. Application was duly made, and I was accepted subject to obtaining a year of practical farm experience, which was a prerequisite to entering the Ag-Engineering program.

My stepfather assured me that I could get a qualifying job with the Company, so I planned to finish my current semester at Tulane, then put in a year at a Company farm. As the time drew near, I went to the draft board to report my new educational trajectory and upcoming temporary change of residence. I was happily informed by the unpleasant bureaucrat that she would not consider the year of farm work as part of my education and would immediately reclassify me as 1-A, the top draftable category. She also assured me that, at the current rate of induction, I could be assured of being called before the year was up.

In the face of this possibility, I decided that I might as well enlist in the service of my choice and get the military obligation behind me. I attempted to volunteer for the Air Force but was told that only U.S. citizens could enlist. I spoke to the ROTC Office on campus and was told that the program was for the training of officers and since a commission could only be held by a citizen, I could not participate. I was subsequently contacted by the Air Force recruiter who informed me that further study of the regulations revealed that foreigners could be enlisted if they first completed a Declaration of Intent. This involved going before a judge and swearing under oath that you intended to renounce your present national status and become a U.S. citizen. The thought of giving up my British citizenship distressed me, so I consulted His Majesty's consular office where I was comfortingly told that "once one is British, one is always British." I was also informed that the Australian Air Force was actively recruiting men for the Korean War and that that could be accomplished through his office. I saw this as a solution to my dilemma–military service where I could make use of my flying skills while conserving my citizenship status. When I told my nemesis that I was off to Australia to return in a year as a military veteran (a draft-exempt classification) she was pleased to inform me that foreign military service did not qualify for such status.

After consultations with my family, it was decided that I should go ahead and enlist in the Air Force at the end of the winter semester to get my problem behind me and get on with my life. In early February 1952, I traveled to Lackland Air Force Base in San Antonio, Texas, for basic training.

A couple of weeks into Basic, we were all tested for qualification for pilot training. I did well on the tests but told the testing officer I couldn't enter the flight program because I wasn't eligible for the required commissioned officer status. He said they were desperate for pilots, and he would look further into the case. Several days later, I was summoned by the officer and told that he had checked the regulations and consulted with the Judge Advocate's Office and that it was determined that, as the Declaration of Intent was good enough for enlistment, that it was also good enough to hold a commission. This marked the beginning of my five-year service as an officer and pilot in the Air Force.

I never met another foreign officer in the Air Force, and my situation was generally considered odd by the administrative types. At each of my numerous changes of station, I would be called up to the Adjutant's Office within a few days of arrival to "straighten out some anomalies in my records." The anomalous citizenship question was explained at length and, surprisingly, I never had anyone question my story.

I returned to Tulane in 1956 and completed my civil engineering studies. I graduated, got married, worked on several overseas building projects, and had a daughter. Forward to 1962, when I was working in San Juan, Puerto Rico. My wife and I were tired of going through different immigration lines in our many travels, and our daughter, born in Puerto Rico, was an American, so we decided to get the family all under the same umbrella. I became a naturalized U.S. citizen in the District Court of San Juan on August 20, 1962, the day following my thirtieth birthday.

11. When Truth Ends and Fiction Begins

GABRIEL GARCIA MARQUEZ WAS one of the most celebrated novelists in the 20th Century and his name is recognized around the globe. He was born in 1927 in Aracataca near the north coast of Colombia where the United Fruit Company employed thousands of workers in banana plantations. The young boy was raised by his grandparents and he bonded closely with his grandfather who told him stories of war and strife in Colombia. When he was eight, his grandfather died and his parents took him to live in Barranquilla, a large port city by the Magdalena River. The change from Aracataca was drastic and turned the child's world up-side-down. The boy never forgot his early childhood nor his influential mentor, his grandfather.

Eventually, he studied law for a while but quit and became a journalist. Assignments took him to Bogota, Paris and Spain. In New York City, he was a correspondent for the news service of Fidel Castro. Garcia Marquez moved to Mexico and dedicated more time to writing. In 1982, he won the coveted Nobel Prize for Literature for his short stories and novels. Among the titles was *One Hundred Years of Solitude* in which he described in gory, graphic details how the Colombian army slaughtered 3,000 banana workers and demonstrators in 1928 in Ciénaga, where thousands gathered to strike against the United Fruit Company. Readers were awed and mesmerized. As a result of that novel, he was quoted universally by the press, academics, writers of every genre, students and teachers who considered the novelist as the authority on what happened. The Colombian army commander was harshly criticized, and the banana company was vilified as the cause of this tragedy that killed thousands.

In 1990, Garcia Marquez gave a TV interview in London. When he was asked about the banana massacre, the famed novelist shocked the interviewer and listeners with comments that were totally unexpected. He said that only a handful of people, maybe three or five, had died during the strike, not 3,000 as he wrote in his book. He made it clear that he was a novelist and could write whatever inspired him to make a point. He was quoted as follows:

> The banana events are perhaps my earliest memory. They were so legendary that when I wrote *One Hundred Years of Solitude*, I wanted to know the real facts and the true number of deaths. There was talk of an apocalyptic massacre. Nothing is sure, but there can't have been many deaths....It was a problem for me when I discovered it wasn't a spectacular slaughter. In a book where things are magnified, like *One Hundred Years of Solitude*, I needed to fill a whole railway with corpses. I couldn't stick to historical reality. I couldn't say there were 3 or 5 or 17 deaths. They wouldn't even fill a tiny wagon. So, I decided on 3,000 dead because that filled the dimension of the book I was writing. The legend has now been adopted as history.[1]

In the interview, he admitted that he could not stick to historical reality regarding the banana strike.

This interview caught the attention of Eduardo Posada Carbó, a Colombian who earned his doctorate from Oxford University. He was inspired to follow through with a thorough investigation of all the numerous conflicting theories, concepts, reports, ideologies, charges, political speeches and beliefs about what happened. The result was a monumental publication that helped to set the record straight.[2] This scientific paper has 98 references that back up his findings. The title represents a fresh approach to find the truth about what happened between December 5 and 6, 1928, in Ciénaga: *Fiction as History: The bananeras and Gabriel García Márquez's One Hundred Years of Solitude.*

Besides Garcia Marquez, other authors have distorted history in ways that awe and convince their readers. Ideologies, not investigative journalism, dictate their viewpoints. Chilean communist poet and Nobel Prize for Literature laureate, Pablo Neruda and his poem about the United Fruit Company is a classic example. It is doubtful that the poet knew anything about banana plantations, their workers or the economics of those capitalistic enterprises. His poem flowed from an ideological mind and from fellow travelers who are compelled to discredit capitalists.

MORE IDEOLOGY

In 1982, Philippe Bourgois arrived in Sixaola and Changuinola to begin studies in the plantations of what was formerly the United Fruit Company, then United Brands Company and now Chiquita Brands International. These plantations were on both sides of the Sixaola River that divides Panama and Costa Rica. He was in his twenties and convinced top management that he wanted to study the Company and its people as an anthropologist. The young scholar gained almost unlimited access to what he wanted. This included the banana company infrastructure, farms, packing stations, labor camps and the port. Socially, he enjoyed the golf course, the golf club and engaging with employees at all levels. Mark Wenholz and I spent much time with Philippe, sharing what we knew, reviewing banana history and sorting through hundreds of my old photographs that dated from the early 1900s. We were glad to help and facilitate with whatever we could.

It was nearly three decades later, in 2011, that I sent Philippe a courtesy copy of my book, *Outline of History of the Province of Bocas del Toro, Panama* (2008). This was just after Monsignor Ganuza gave me a book by Philippe Bourgois in Spanish, *Banano, Etnia y Lucha Social en Centro America.*[3] When I read it, I felt betrayed. In a personal letter to him, I replied: "I would have guessed that the book had been written by Fidel Castro or Carlos Luis Fallas." Fallas was the renowned communist writer and banana company hater in Costa Rica. Philippe replied in a letter to me: "I also admire your energy as a genuine on-site historian. It is true that I am left-of-center in my politics and was even more left-of-center back then in my twenties."

Once again, it was ideology that dictated the direction of his book. Of the 331 pages in Spanish, 13 pages were references that backed up comments about the past that started a century before. (He did his homework.) This included every sin that the United Fruit Company ever committed. Nearly every page of the text was a diatribe of harsh criticism that made the local management and administrative employees look like conniving, repressive schemers who plotted against the labor force of about 6,000 employees. Just about anyone with a monthly salary was vilified because they had a better Company house and more benefits than labor employees. Incriminating terms, like exploitation, apartheid, racism, abuse of the masses and class struggle, set the mood. This attitude was the Marxist concept of a classless society—the opposite of a capitalist infrastructure where there are incentives and distinctions in employee benefits based on the job.

Philippe bitterly noted that the labor class was not allowed in the golf club or on the golf course. One has to imagine thousands of laborers and their big families enjoying such social and recreational benefits within the small, protected zone where executives and administrators lived.

Bourgois emphasizes overcrowded living quarters in some instances. That was because friends and relatives of labor employees, plus squatters, moved into Company-provided houses without permission—they were not employees and did not belong in a private Company house. It was a constant struggle trying to control this problem.

REALITIES AND BENEFITS

Most laborers lived on banana farms where they worked. The banana company provided good worker houses for each employee with a family. The home had a kitchen, a living room and bedrooms, potable water, screens, toilets, showers and electricity—facilities and benefits that the majority had never had in their lives or dreamed of having before they came into the privately-owned banana zone. There was typically a soccer and baseball field in the housing compound, plus a playground for the children. A school and a commissary to buy food and supplies were nearby. The employee and his family received free medical care in the Company hospital that was later transferred to the government. This is also where their babies were born and cared for.

Last, but not least, banana company workers earned higher wages and benefits far above other agricultural labor in the country, especially on coffee and vegetable farms or on cattle ranches. Banana workers got an annual paid vacation and other social benefits based on laws of the country and union agreements. Unlike many seasonal workers, banana laborers had a paycheck every month of the year. That gave them more stability and security and a much higher standard of living and education than they would have received in their traditional, isolated settlements and villages.

Changuinola and Sixaola were developed for the purpose of growing and shipping bananas with the expectations of making a profit. That is the design of capitalism. Hundreds of millions of dollars were invested in these outback areas that never had anything before. To make the entrepreneurial business function efficiently, there was never a classless society. At the top, there was management, followed by employees down the hierarchy with different job descriptions, pay scales and benefits. Incentives and rewards to move up the Company ladder of success were always there. By and large, that has occurred in the form of more material goods and better education for children and adults. With promotions, some of the labor class became monthly employees.

ACKNOWLEDGEMENTS

Philippe Bourgois never gave the Chiquita Banana Company any credit for its huge economic, social or cultural contributions in the banana zone. There was no appreciative mention for running an extensive, complicated company that generated extraordinary benefits for thousands of employees who otherwise came from places where there would have been poverty and little opportunity. Managers, superintendents, overseers and monthly foremen were never complimented on their professional jobs to plant, cultivate, harvest, pack and load delicate, perishable fruit on schedule on ships at the port. Just managing 6,000 employees scattered over 7,000 hectares was a major challenge.

So many administrative employees supported and enabled Philippe Bourgois to accomplish his anthropological study. Not one of their names was mentioned in acknowledgements. This was a sign of leftist ideology that compels devotees to distort history and propagate references that will not tell the other side of the story. Subsequently, future scholars with these leanings quote such publications, ad infinitum. It is rare to find unbiased references like the book *The United Fruit Company in Latin America* by Stacy May and Galo Plaza.[4]

Although our ideologies differ, I respect Philippe Bourgois for his renowned academic accomplishments as an anthropologist. In the letter he wrote to me in 2011, he closes with a sympathetic consolation where he quotes the astute statement by the economist Joan Robinson: "The only thing worse than being exploited by capitalism is not being exploited by capitalists."

This postcard features items made from abaca fibers grown in Changuinola from World War II until 1954. Rope, which was vital for securing warships, is illustrated in four sizes by Hasso von Wedel, the official photographer for the United Fruit Company. Thanks to Vicente Pascual, the owner of the card, we can appreciate this piece of history.

12. From Bananas to Cacao, Abaca, Rubber and back to Bananas[1]

COMMERCIAL BANANA DEVELOPMENT IN Panama started in the 1880s in plantations along rivers bordering the Chiriqui Lagoon and the Almirante Bay. But within a few years, the Gros Michel banana variety died from a fungus called Panama disease, *Fusarium oxysporum*. This disease may have come on corms dug from banana plants along the Chagres River and the Panama Railroad that crosses the isthmus.

An effort to revive the industry began after 1900, when new Gros Michel bananas were established in the virgin soils along the Changuinola and Sixaola Rivers and into the vast Talamanca Valley on the Costa Rican side of the border. Shiploads of Jamaican and various other Afro-Antilleans were brought by the United Fruit Company to the province of Bocas del Toro to work the plantations.

By the first decade of the 20th century, there were an estimated 5,000 to 6,000 Afro-Antillean employees scattered from Chiriqui Grande to Bocas Town, to the new banana port of Almirante and to all of the plantations on both sides of the Changuinola and Sixaola Rivers. Some brought their families, but most did not. Most laborers spoke Caribbean English, but the immigrants from the French West Indies spoke French patois.

But history was repeating itself: Panama disease once again began to decimate the banana plantations.

CACAO

In plantations where Panama disease was killing Gros Michel bananas, cacao was planted to keep the soil productive. By the mid-1920s, there was no banana production in Changuinola at all, and the entire area was planted with cacao. Because of this, Almirante had the world's largest cacao processing plant.

The best-known use of cacao is chocolate. Merriam-Webster notes that "Chocolate starts as the seeds in the fruits of the cacao tree (*Theobroma cacao*), a tropical plant that produces flowers and fruit not at the tips of branches but on the trunk."[2]

Large payrolls were no longer needed to sustain slower, more latent, cacao operations. The cacao payroll was not large to begin with, plus women did some of the work. Consequently, the men drifted elsewhere to make a living if they could. This situation continued up until the beginning of World War II.

World War II: ECONOMIC IMPACT ON CHANGUINOLA

When the Americans declared war on the Axis Powers after the bombing of Pearl Harbor on December 7, 1941, Japan caused an immediate impact on Bocas del Toro by invading the Philippines the next day. Suddenly, the main global supply of rope was in the hands of the enemy, Japan. Rope was made from fibers of a banana plant called Manila hemp (*Musa textilis*), also known as abaca. In Spanish, it is *cáñamo de Manila* and it is a critical, strategic war material. Navies could not tie up their war ships without rope. Thus, there was an immediate crisis for the American Navy and its allies.

ABOUT ABACA

Abaca is a banana plant that grows from a big bulb in the soil, but its agronomy, horticulture and cultivation were very different from commercial bananas that were grown for the fruit. This was the challenge that seasoned old banana men faced in order to jump-start this wartime emergency effort.

Abaca was selected long ago because the strong fibers in its trunk can be made into rope and other forms of cordage.

Unlike the edible commercial banana, abaca is genetically normal in that the species is a diploid. This means that it reproduces sexually. That is, the large female flowers are pollinated by the small male flowers of other abaca plants. With this sexual union, numerous hard, black seeds develop in the pulp. When wild mammals eat ripe abaca fruit, they distribute seeds that sprout wherever they land.

Bananas that humans eat were selected by early humankind because the fruit had no seeds—they are triploids. Abaca can also multiply vegetatively from the planting of corms, suckers or bulbs that sprout from the base of the mother plant.

When the abaca trunk with its strong fibers was fully mature, it was cut with a *machete* near ground level. Leaves were chopped off, and the heavy trunk was hauled by mules to a flat railcar that was pulled by a steam locomotive to the abaca processing plant in Base Line.

The trunk is mostly water that must be removed in order to dry and process the strong fibers to make rope. Removing the water was done by inserting abaca trunks into rotating crusher rollers similar to how sugarcane juice is extracted at sugar mills. After the fibers were dried and compacted into bales, they were shipped from the port of Almirante. The ship was at risk of being sunk by German submarines looking for prey in the Caribbean and Atlantic. This was a strategic war material that the enemy constantly targeted.

JAPAN

In the 1920s, Japanese immigrants gained control of the world's main supply of Manila hemp on the Philippine island of Mindanao. This caused concern by American cordage manufacturers. In 1925, Harry T. Edwards, a plant explorer for the U.S. Department of Agriculture, shipped about 1,000 abaca corms from the Philippines to the Panama Canal Zone Plant Introduction Gardens at Summit. Soon after, the United Fruit Company acquired and planted them in Changuinola. Dr. Vining C. Dunlap of the Changuinola Research Department, followed by Dr. Wilson Popenoe, were in charge of multiplying and studying abaca. By 1938, there were 800 hectares (2,000 acres) of abaca in Changuinola, comprising the only collection in the Western Hemisphere. This was the reason why Changuinola had the first abaca plantations after World War II began in 1941.

RAMPING UP PRODUCTION

Before the end of the war in 1945, the United Fruit Company had expanded to over 11,330 hectares (28,000 acres) of abaca production in Panama, Costa Rica, Honduras and Guatemala. All of this production was under contract with the United States government to provide rope as a strategic war material to tie up war ships of the navy and merchant marine fleet.

Money was no object to get abaca into production. The United States government and its Reconstruction Finance Corporation paid for just about everything. They paid for a monumental airlift of workers from San Pedro Sula, Honduras, direct to a grass airfield at a cacao farm, Finca Costa Rica, near the Sixaola River on the Costa Rican side of the international boundary; they paid the United Fruit Company to chop down its cacao plantations in Changuinola and plant abaca;

they paid all costs and gave the United Fruit Company 10% above its payroll expense as a commission. During the 1930s when there were no bananas, practically all of the infrastructure suffered from neglect. Railroads, locomotives, bridges, housing, machine and repair shops had to be repaired or replaced and made functional. Hundreds of new labor houses were built along the railroad tracks and in the new abaca farms.

Money was so freely available that an estimated 3,000 Salvadoran and Honduran men were flown in to work in abaca on the Panamanian side. Many laborers brought their women and children, which doubled the total to about 6,000 Central Americans. A DC-3 or C-47 had about 28 passenger seats, which suggests that some 214 flights might have been made. If flights arrived after dark, lanterns and flashlights lined the airfield so pilots could see where to land. It is probable that TACA Airlines of El Salvador made the chartered flights.

A medical team from the Company hospital met all flights at the airport and gave each person a quick medical exam. If needed, men, women and children were sent to the Almirante hospital for further exams and treatments. After this procedure, each one was sent back to Changuinola to their assigned living quarters.

With hundreds of people being assigned in so many directions, it took a major effort to make sure that no members of any family got separated and lost. Dealing with 6,000 new arrivals in a strange land to jump-start a major war effort was no easy task.

Besides the airlift of Hondurans and Salvadorans, laborers from Nicaragua and Costa Rica came to Changuinola by walking down the Caribbean coast or arriving in small coastal boats. It is not known if some of their numbers were in the estimated 6,000 newcomers.

Old banana and cacao farm numbers in Changuinola were changed to Philippine names, for example, Filipinas, Manila, Mindanao, Luzon, Davao, and Cebu. This was because The Philippines had been the world supplier of *Musa textilis*. The contract with the U.S. government expired in 1948 but the United Fruit Company continued abaca production until 1954. George Munch was the general manager of all four country operations, first based in Guatemala, then later in Almirante, Panama.

CENTRAL AMERICAN NATIVE RUBBER: ANOTHER STRATEGIC WAR MATERIAL

In the late 1890s and early 1900s, new banana plantations were developed by planters coming to seek their fortunes in Bocas del Toro. One such person was Dr. Paul Osterhout, a medical doctor and planter from Texas. He first invested heavily in turbulent Bluefields, Nicaragua, as a banana grower, druggist and medical doctor. But politics and bad times caused him and his wife to move to Bocas del Toro, then Colombia, in 1895. Dr. Osterhout and his wife, May, wrote a letter every week to their families back in Texas.[3] I have copies of numerous Osterhout letters that were donated to the Rice University Woodson Research Center.[4]

These letters described conditions in Bocas del Toro: Racism, political turmoil, debt, banana operations, yellow fever, malaria, and another crop of special interest—rubber. During the research I did for this book, I wasn't able to ascertain if rubber plantings in Bocas del Toro were *Hevea brasiliensis* from the Amazon or *Castilla elastica* found in Central America. It was about this time when the British introduced the Brazilian species of rubber to Southeast Asia and became a leading world producer of rubber. Competition in Southeast Asia caused Bocas del Toro growers to abandon their rubber trees. This is evidence that *Hevea brasiliensis* from the Amazon was the favored rubber tree species that had been planted in Bocas del Toro by Dr. Paul Osterhout, et al. After the competitive decline, rubber plantings in Bocas del Toro nearly disappeared.

But World War II changed the economics. A new world crisis suddenly arose because the Japanese invaded and controlled the foremost rubber supply in the Dutch East Indies and British Malaya, especially near Singapore. Once again, the U.S. government turned to the United Fruit Company for assistance. This time, it was to help produce rubber.

Hevea brasiliensis was never grown commercially on a large scale in Central America. However, a native species in the fig family, known as the Panama or the Mexican rubber tree, *Castilla elastica*, grows naturally in the tropical forests from Yucatan, Mexico, to the northern part of South America.

A tough, young banana man, Aaron King, who graduated from the Wharton School of Business, was given a leave of absence from the United Fruit Company to join the U.S. government's Rubber Development Corporation. His job was to organize and manage rubber collecting from thousands of wild trees in the forests from Mexico to Panama. This job kept him in the rainforests of Middle America on a mule or in a *cayuco*. King managed hundreds of rubber collectors in those seven or eight countries. He was known as a tough guy who could get the job done. The raw material was tapped from rubber trees and molded into huge, heavy bales, then floated out of the mouths of rivers, where they were loaded onto ships waiting at sea.

I knew Aaron King when he was the manager of bananas in Almirante in the early 1960s. Everyone was scared of this man because he said things like, "That goddamned research boy has burned his house down." That was when our house in Changuinola burned to the ground in half an hour on February 17, 1966. But when we were alone, Aaron King became a slightly more likable person with lots of unbelievable stories to tell—and I listened.

He told me about his experiences collecting rubber from the Yucatan to Panama. Because this product was a critical war material, German submarines sank several United Fruit Company ships off of the Central American coast. When the ships sank, bales of rubber would escape through the hatches and float on the sea. As the rubber washed ashore, vigilant natives claimed salvage rights and sold the bales back to Aaron King. This happened several times off the coast of Honduras and the Bay Islands and on other shores. After the war, Aaron King returned to the United Fruit Company as a banana man.[5]

As a measure of caution in Bocas del Toro, it was decided to interplant Panamanian rubber trees alongside traditional *madre de cacao* trees to shade the cacao or supply rubber if the need arose. However, after the war, rubber trees were no longer used for rubber. They were chopped down when Chiquita converted its cacao farms to banana plantations in the early 1970s, using a disease-resistant variety called Valery. That was half a century ago, and Panamanian rubber trees are not common anymore where they once thrived. In fact, the thousands of banana employees there today have no idea that those silty loam soils produced cacao for 40 years before Valery bananas were planted.

OTHER IMPACTS ON CHANGUINOLA: LABOR

With the outbreak of the war in 1941, men were suddenly in demand in the Panama Canal Zone, Panama City, Colón and David. Excess and unemployed workers in Bocas del Toro quickly took these new, wartime jobs, leaving their wives and children behind at the cacao farms, or in Changuinola, Almirante and Bocas Town. English-speaking, Afro-Antillean men from Bocas del Toro were especially in demand in the English-speaking Panama Canal Zone and surrounding cities. As a result, there was no labor supply left in Changuinola to initiate abaca production and its exportation during the war.

Latinos from Chiriqui and other provinces of Panama also could not be recruited to work in Changuinola because they too had good wartime jobs elsewhere in Panama. There was a reserve of indigenous labor in Bocas del Toro, but recruiting was not very successful. Some prospective Ngäbe workers from the Cricamola River went to the Company hospital in Almirante for medical exams, but many were in poor health and were rejected. The healthy men were sent to Changuinola to work in abaca. There was a problem of longevity on the job because of irregular coming and going.

The only solution for these labor shortages was to import thousands of workers from other countries.

Finding workers at the administrative level was more straightforward: Personnel like timekeepers, overseers and superintendents were transferred from other United Fruit Company operations in Honduras, Guatemala, Costa Rica, Cuba, Jamaica and Puerto Armuelles in Panama.

LANGUAGE

Since the early 1900s, Caribbean English—known in the vernacular as *guari guari*—was the dominant language at the banana and cacao farms, the port of Almirante, the islands and various scattered communities. Spanish was spoken by Latinos in the province, but was not the main language until later. English was the dominant language of the international community in the capital of the province, followed by German and French. Ngäbe, Buglé and Naso tribes spoke their own indigenous languages.[6] But with the arrival of 6,000 Spanish-speaking Central Americans, their language, customs and culture soon drastically predominated the Changuinola abaca area.

AIRPORTS

Transportation by air was not easy, and it was sometimes difficult to get a seat on any airplanes coming in and out of the Sixaola airport on the Costa Rican side of the border. From Changuinola, a passenger had to go by railcar to Guabito on the Panamanian side, then cross the "big bridge" to check into Costa Rica, and continue on a railcar to the grass airfield a few kilometers away. Priority was given to anyone involved with the war effort. United Fruit Company and Panamanian government officials also got priority. To get to Panama City from Changuinola, a passenger had to fly from the Sixaola grass field to San Jose, Costa Rica, then fly to the Panamanian city of David, and on to Panama City.

Before the war, there was an airfield in Bocas Town, the capital of the province, but hardly anybody used it. There was little need to stop there. The Bocas airfield was neglected and in bad shape. However, during the war, improvements were made so it would be safer.

Finally, in 1944, trees in the old pastures at Base Line were chopped down and removed so the Changuinola airfield could be built with gravel and grass. Flying in and out of the Sixaola grass field on the Costa Rican side was no longer necessary.

CUSTOMS AND CULTURES

In addition to the Spanish language, food, music, dancing, entertainment and customs were Central American in nature. For example, Panamanians ate thick, yellow corn tortillas, whereas, Central Americans ate thin, white corn tortillas. Central American Independence Day on the 15th of September became a major celebration. Costa Rican bands and *marimbas* crossed the border to perform on the Panamanian side for the Central Americans. This was something new for Panamanians because *marimbas* were not part of their musical heritage.

After paydays and on other special occasions, Latinos celebrated by drinking—frequently over-indulging. Then came the violence and *machete* fights.

Teofilo Alvarado's son, Carlos, commented that about a dozen students traveled from Changuinola every school day to the American School in the port of Almirante. They traveled by motorcar on the railroad for over an hour.

On school days after a festive weekend or after a payday, the first stop in Almirante was the hospital to transfer mutilated bodies to the hospital morgue. The bodies were hauled on a small, flat railcar that was pulled behind the student motorcar. When there was a chance, some of the students would lift up the sheets covering the bodies and have a peek. A year later, a new American School was built at Farm 8 in Changuinola, where the students walked to school from their homes.

THE END OF THE ABACA ERA AND MORE TRANSITIONS

After the war ended in 1945, many Central Americans returned to their countries. The Panamanian government insisted that other workers leave based on contracts that had allowed them to work in Panama. During the next few years, the United

Fruit Company paid for several flights to repatriate them. However, hundreds stayed to work in the old cacao farms. Then, in 1950, the United Fruit Company happily converted from abaca to bananas, which is where the remaining Central Americans worked. By this time, numerous Central American men had married Panamanian women and had Panamanian children; they had no intention of going back home.

INDIGENOUS PEOPLE

In the early 1950s, Guna Yala men from the San Blas Islands were contracted to work in abaca, which was being phased out by the newly-revived banana industry. The contracts were very stern and dictated by the chief of each island. A selected group of 50 Guna Yala men would come as a labor gang to Changuinola to work for three months and then return to their families in the islands. There was a strict arrangement about withholding earnings to share with the family and community when they returned. The Company paid for their transportation by boat to and from the Almirante port. The labor gang had special housing apart from other laborers and could not mix with the other races or any women.

At one time, there were about 500 Guna Yala men under contract in Changuinola and they obeyed the contract. That is, until the abaca plant closed in 1954. The United Fruit Company told the Guna Yala men they could go back to San Blas with the paid ticket they had, or they could stay and continue to work for the Company. They also could come and go as they pleased with no contract but at their own expense.

Many stayed and earned a good reputation for working in cacao, in Company kitchens to feed laborers, as servants, houseboys, hospital helpers and later as banana packers. The chiefs in San Blas were not pleased with this new arrangement because many of the men never returned to their island homes. In addition, some Guna Yala women began immigrating to Changuinola to be with their men and make a new way of life in the banana plantations.

To close this sequence of changes, in the 1950s, Ngäbe from both the Cricamola River and Valiente Peninsula increased by great numbers on the banana payroll, and to this day are the leading working group in the Changuinola plantations and the labor union. Many have become monthly employees and are climbing the ladder of success.

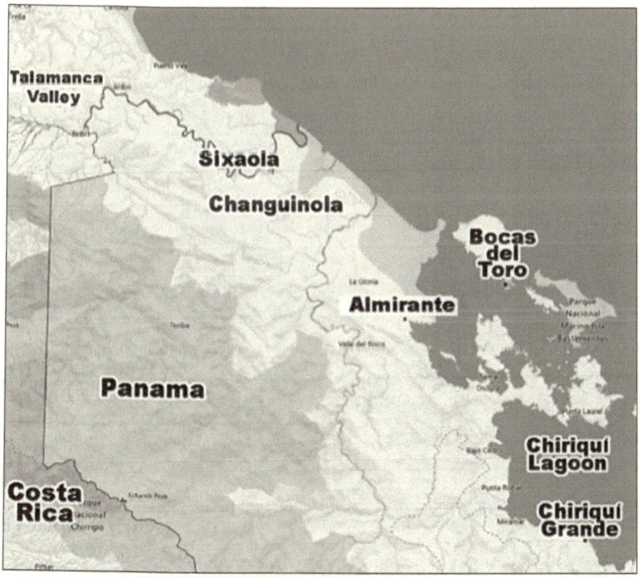

Map of Caribbean coast.

Woman husking seeds from cacao pods.

Harvesting abaca in Changuinola. Ngäbes first appeared as laborers during the abaca period.

Transferring abaca trunks from mules to a railcar.

Drying abaca fibers from the plant's trunk.

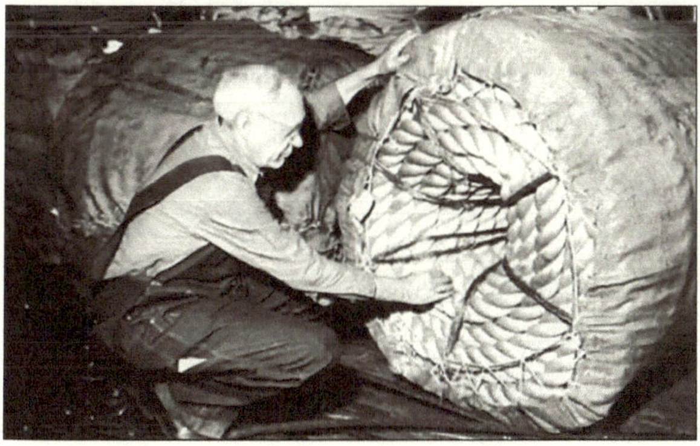

Finished abaca fiber rope.

Abaca put to use.

Teófilo Alvarado, Superintendent of Abaca, Changuinola, circa 1950.

Castilla elastica, Panama rubber tree.Photograph: Carmen Galdames.

Tapping rubber. https:// upload.wikimedia.org/wikipedia/ commons/thumb/a/ a3/ Sri_Lanka-Rubber_ plantation_(5).JPG/1200px-Sri_Lanka-Rubber_plantation_(5).JPG.

Russ Caid illustrates parts of the abaca (Manila hemp) plant. Female flower bunch (right), the small male flower bud with pollen (just above his shoe) and a mature abaca fruit bunch with hundreds of seeds (upper left). The trunk with strong abaca fibers that are used to make rope is on the ground.

Aaron King (2nd from left) transferring from Tiquisato, Guatemala, to La Lima, Honduras, in 1957 to manage bananas.

13. From Bananas to Abaca to Oil Palm in Honduras

THE 1920S WAS A notable decade for bananas and abaca in Central America, and especially in 1925. The United Fruit Company had acquired vast virgin lands on the right bank of the Ulua River in Honduras and had built railroads from the new port of Tela to the interior banana plantations near Progreso in the Department of Yoro.

One of these banana districts was called Guaymas. Harry Hinson Hutchings was the assistant superintendent of this district in 1925 and describes how he and his men converted an extensive virgin rainforest into a huge plantation with the Gros Michel banana variety. In his memoirs, he writes about the incredible challenges in surveying the land, building housing, draining the land, planting bananas, felling the forest on top of the banana corms in the soil and finally harvesting the first crop.[1] At the same time, far across the Pacific Ocean, another situation was occurring in the Philippine Islands. Japanese immigrants settled in Mindanao and gained control of the production and export of abaca.

As a result, in 1925, Harry T. Edwards, a plant explorer of the U.S. Department of Agriculture, brought 1,000 abaca corms and plants to Changuinola, Panama. These were cultivated on lands of the United Fruit Company and production had increased to 800 hectares by 1938. This was the only source of Manila hemp plants in the entire Western Hemisphere.[2] In 1928, different varieties of the abaca collection in Panama were transferred to Dr. Wilson Popenoe, director of the Lancetilla Botanical Garden of the United Fruit Company near Tela, Honduras. Selective breeding was carried out to improve the yield of the fiber and its strength.[3] Dr. Popenoe also maintained 20 hectares of abaca in Panama for research purposes. His efforts and results were important in order to select the best varieties of *Musa textilis* when the time came for expanding abaca production during World War II.[4]

On January 3, 1942, the United States government made a contract with the United Fruit Company to plant 8,000 hectares of abaca and supply the government's Defense Supplies Corporation with fibers to make rope for tying up merchant marine and war ships, plus supplying rope for the Army and the Air Force. The agreement was to sell the product without profit to fulfill urgent war needs.[6] Emergency measures were accelerated to generate abaca production in Panama, Costa Rica, Honduras and Guatemala during World War II. In Honduras, the Guaymas District was available because Panama disease, *Fusarium oxysporum f. sp. cubense*, had decimated Gros Michel bananas and the land was not being used.

The new name was Manila District. Each farm had a name from the Philippines: Bataan, Samar, Davao, Mindanao, Cebu and Luzon.[7] On the north end of these abaca plantations, Guaymas Town was shown between Bataan and Samar. The area could not be reached except by railroad.

Abaca districts in each of four Central American countries had a large fiber processing plant to receive tons of harvested abaca trunks. The machine, called a decorticator, removed water from the trunks and separated the fibers. This remarkable apparatus was developed by United Fruit Company engineers in the banana port of Almirante, Panama. Fibers were extracted, dried and compacted into a large bale, like cotton or hay, so it could be loaded onto a ship. During the war, these ships with war materials like abaca and rubber were prized targets for German submarines. By the end of the war in 1945, the United Fruit Company had 11,300 hectares of abaca in Central America. The original contract between the U.S. government and the United Fruit Company to produce abaca as a war effort without a profit terminated on December 31, 1948. However, the government agency, the Reconstruction Finance Corporation, made a new agreement with the Company to continue producing abaca with the stipulation that the Company would be paid a management fee.

In 1949, the administration of all four abaca operations in Central America was managed by George Munch, who was also the general manager in Bananera, Guatemala. He was transferred to Almirante in 1952 to manage the Chiriqui Land Company.

Because the Philippines were recuperating from World War II and gaining back their former abaca industry, and because growers were destined to establish new banana plantations again in the early 1950s, abaca operations stopped in all four countries in Central America in 1954. However, bananas were not planted in the Guaymas District again. Some years later, African oil palms were planted by private land owners. The only vestiges of abaca left are scattered plants in the wild that sprouted from seeds and a few Philippine names that dot the landscape.

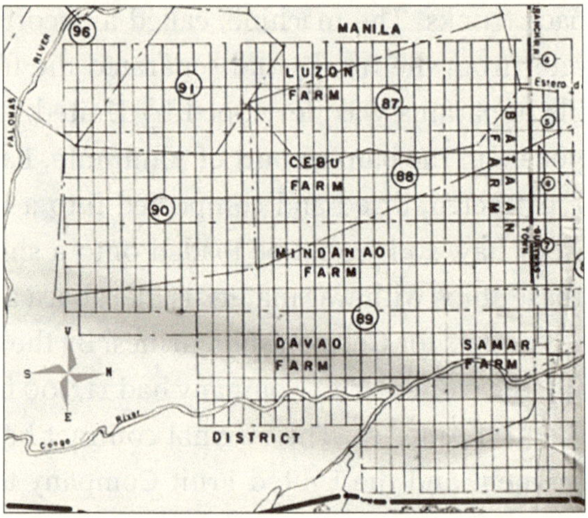

Map of abaca farms in Guaymas, Honduras, 1946, drawn by the Tela Railroad Company.

Mats of young abaca trunks. Honduras, 1946. Photograph: Hermann Wimmer.[5]

Abaca processing plant, Guaymas, 1946.

Harvesting abaca trunks transferred to railcar, 1946.
Abaca trunks transferred while the bosses watch, 1946.

Decorticator machine removing water and separating fibers that are drying (upper left). Guaymas, Honduras, 1946.

14. An Abaca Family in Guatemala[1, 2]
by Millicent Crichton Empedocles

FROM SCOTLAND TO HONDURAS

My grandfather was David Henderson Crichton who was born in Edinburgh, Scotland, in 1881. He married Ann Porter Johnson, four years his junior. During World War I, he was gassed in by the Kaiser's army. After the war, his doctor advised him to seek a milder climate in order to cope with his damaged lungs. As a result, he got a job in Tela, Honduras, with the United Fruit Company in 1919. At age 38, he began his career in the Tropics in the machine shop, the power plant and in refrigeration at the giant banana company.

Meanwhile, the rest of the family remained in Scotland. Their four children were still in school, and Ann's mother was blind and could not be left alone. David would return to Scotland as often as possible. Consequently two more daughters, Jean and Nan, were born. In 1925, when David's mother-in-law died, Ann, her three daughters and their youngest son, Joey, moved to Tela to begin a new life in the American Tropics. A bit later, Ian Crichton was born. By 1929, David was working for the Tela Railroad Company, the subsidiary name of the United Fruit Company in Honduras.

My grandmother, Ann, did not like the Tropics, especially when she came upon snakes in the wood pile under the house. When she collected firewood to cook with, she always took a *machete*. Each time she killed a snake, she declared that she was going back to Scotland. Ann died in Tela in her early fifties from cardiomyopathy.

The Crichton children continued their education at the United Fruit Company school in Tela. As adults, they scattered: Jean married Sarge Labate, who worked in abaca in Guatemala, then bananas in Quepos, Costa Rica, and Puerto Armuelles, Panama. Ann married Axel Dahlgard, a purser on United Fruit Company ships. Rena went to Brooklyn, New York, and married Manny Silva. Joey married Nancy in Edinburgh, and they briefly lived in Progreso, Honduras, but returned to Scotland. He may have worked on banana ships at one time. Ian, the only Crichton baby born in Honduras, was raised in Tela. He married Blanca, from Honduras, moved to Tegucigalpa and became a bank officer. One of their daughters still lives in La Ceiba, Honduras.

STARTING LIFE IN GUATEMALA

Henry "Scotty" Crichton, my father, born in Edinburgh in 1908. After getting a degree in electrical engineering from Heriot-Watt University in Edinburgh, he went to New York in 1929. Henry is in the family group photograph in Tela in 1931. The United Fruit Company hired him in 1932, and he became the engineer on the banana ship, SS *Tivives*. In 1943, he and Millicent Doig were married in Nyack, New York, in the house that the bride grew up in.

About the same time in 1943, Henry was drafted to serve in World War II. Lucky for Henry, his father in Tela informed his son in New York that he could be exempt from military service by going to Los Andes, Guatemala, and working for the United Fruit Company. The job involved the production of abaca. Henry quickly signed up for the job and departed for Guatemala. Jean's husband Sarge was also drafted. He became exempt by taking a similar job in the abaca processing plant in Guatemala. The two couples lived in the Los Andes abaca farm next door to each other.

My mother stayed in New York because she was pregnant with me. I was born in 1944, and when I was six months old, my mother took me to Guatemala. My father sent a picture to my mother in New York to show her the house they would live in. I can't imagine what her reaction was when she saw that barren landscape all alone in a tropical wilderness. As a young New Yorker, she and I joined my father in the Los Andes farm district, where he was running the abaca processing plant.

LIFE IN LOS ANDES

We lived near the Motagua River in the Los Andes abaca farm district for five years. The houses were built high up on stilts so flood waters could flow underneath. As a very young child, I do not remember a flood, but an earthquake caused us to run from underneath the house.

There were no special amenities and getting in and out of Los Andes was only by railroad. Our house was set back from the railroad, and we used two models of self-propelled railcars for making short trips and a motorized car for longer trips. My father would sometimes take me to the abaca processing plant and perch me on a bale.

LIVING IN THE TROPICS AND MY EDUCATION

My mother taught me the 1st grade underneath our stilt house in Los Andes. After five years there, my father was transferred an hour away to the United Fruit Company town of Bananera, now known as Morales.

The general manager and his staff lived in fine homes around the golf course. It was country-club living—a tennis court, swimming pool and a nice social club. My mother shopped in a Company store. Most important was the American School that had grades one through eight. That is where I attended the 2nd grade. We also had good medical care there.

In 1951, my father was transferred to the banana plantations of the Damas District in Quepos on the Pacific side of Costa Rica. The American School was in the banana port of Quepos, and I had to ride a train every day to go to the 3rd grade. If I missed the train, I missed school that day. My first cousins, the Jean Crichton Labate family, lived in the Parita District, and they also rode the train to school in Quepos.

The next move was to Sevilla, Colombia, where I enrolled in the fourth grade in the American School of the banana company. In 1953, my sister, Louise, was born in New York City. When she was a few months old, my mother brought her to Sevilla.

After I graduated from the 8th grade in Sevilla in 1957, I went to a boarding school: The Northfield School for Girls in Northfield, Massachusetts, followed by college at Middlebury in Vermont where I got a degree in mathematics. Each summer vacation from school, I would return home at Company expense to wherever my parents were living.

After Sevilla, my father was transferred down the same railroad tracks to the banana port of Santa Marta, Colombia. Other assignments took the family to Puerto Armuelles, Panama; followed by Manzanillo, Dominican Republic. Then, my father was sent back to Puerto Armuelles to develop his invention called the "Hanging Tractor" that pulled a hundred bunches of bananas on a cable to the packing station. This was followed by a transfer to La Lima and finally back to Tela, where he started life in the Tropics with his parents in 1925 as a 17-year-old teenager. Henry "Scotty" Crichton had gone full circle, and it was time to retire after 41 years with the United Fruit Company. In 1973, the couple settled in Fort Myers, Florida, where they wrote a chapter about an adventure in Guatemala. David Crichton also contributed a chapter about his adventures in Honduras in the same book.[3]

L-R: Joseph, Rena, David Henderson Crichton, Henry "Scotty", David, Ann (standing). Edinburgh, Scotland, circa 1914.

The picture that Henry sent to Millicent
to show her the house they would live in.
Los Andes abaca farm, 1943.

Crichton family in United Fruit Company house in Tela, Honduras, 1931.
L-R: Ann, Jean, Henry, David, (holding Ian), Joseph, Nan (Ann)
and David Henderson.

Two models of self-propelled railcars.
Hand pump cars were used for local, short trips.

Both Millicents in motorized railcar for long trips.

Crichton children in Tela, Honduras, mid 1930s. Jean (on the horse), Nan, Ian.

Los Andes abaca farm and the Motagua River. Crichton house along the railroad (arrow). Note the river (upper left) and abaca plants.

Henry "Scotty" Crichton by bale of abaca fibers.

Millicent perched on a bale at the abaca processing plant.

El Prado American School, Sevilla, Colombia, 1957.
Pat Hatch and Millicent Crichton. Louise Crichton is reading the sign.

El Prado American School, grades 5-8. Sevilla, Colombia, 1957.

15. The End of Abaca in Guatemala by Mario Mena Stricker[1, 2]

THE LOS ANDES ABACA farms and the processing plant closed in 1954. Later, Valery bananas were planted on these soils owned by Julian Presa. The farms were given new names: Chinook, Kickapoo, Pamaxan and Valle de Oro. The fruit was sold to the United Fruit Company.

In 1972, the United Fruit Company reached an agreement with the United States Anti-Trust Division and was required to leave Guatemala. Consequently, Julian Presa sold his fruit to Del Monte, the new banana company that replaced Chiquita. The new company formed BANDEGUA—Bananera de Guatemala. The next three general managers for Del Monte were former United Fruit Company employees: Eddy Taylor, Roy Wells, and me(1977).

New independent banana growers were developed by Del Monte. Meanwhile, Domingo Moreira, Julian Presa, Jack Loeb, Sammy Gordon and I founded COBIGUA at Puerto Barrios. As a result, Del Monte lost all of its independent banana growers in 1989—a big loss for Del Monte. Because I was a founding partner of COBIGUA, Del Monte replaced me with Gary Bratcher, who was transferred from their Far East Operations. With all of these complicated events occurring, Chiquita secretly showed interest in returning to Guatemala despite having left the country to comply with the anti-trust agreement. Later, in 1989, Chiquita made a deal with me and the partners of COBIGUA and bought us out. Included in this purchase were the banana farms of Julian Presa, once known collectively as Los Andes, where abaca was produced until 1954. Chiquita now owns these banana farms, and the plantation area name is Chinook, as seen on a satellite map.

When the transaction between Chiquita and COBIGUA was made, I was retained to assure an orderly transition by communicating with the Guatemalan government, labor unions and the news media. During this time, there was some opposition in the Guatemalan congress and the news media, questioning the return of the successor of the United Fruit Company. It was my job to elucidate Chiquita's new presence in Guatemala.

In 1999, I became COBIGUA's general manager and I held that title until 2013.

Mario Mena (4th from left) at a farewell for manager Fletcher "Buddy" Hatch. Changuinola, Panama, 1968. L-R: Lydia Britt, Buddy Hatch, Frances Moffitt, Mario Mena, John Calley, Don Harris. All were born and raised in Central American banana areas except John Calley.

16. Life on an Abaca Farm in Costa Rica by Harold L. Brent[1] and Billy Saravanja[2]

REMINISCENCE OF HAROLD L. BRENT

In 1943, my brother Sammy and I were small boys riding the train with our father from Limon to San Jose, Costa Rica. My father, Robert Oliver Brent, had been transferred from the United Fruit Company in Honduras to build the abaca plant in Bataan during the emergency war effort to grow abaca and export fibers for rope. When we passed through areas where abaca was planted and saw some labor houses along the tracks, my father knew that was the place for us to get off the train. But the train was not slowing down to stop. My dad instinctively reached up and pulled the emergency cord. The train stopped suddenly. The conductor came running through the railcar and asked why he had stopped the train. In fluent Spanish, my dad answered that this is where he was getting off. The train was not stopping, so he pulled the cord. My father walked down the aisle towards the rear of the train carrying two large suitcases with Sammy and me trailing behind. I remember that there was no caboose. He stepped down on the gravel along the track, put the two suitcases down and set Sammy and me down beside him. The conductor waved a red flag to the engineer, and the train disappeared down the track.

We looked around and saw lots of abaca plants and a few Company houses for the workers. Soon, it got dark and we found a place to sleep.

Lodging was in a two-room labor barracks in La Zonita while my father built the abaca plant. He also built administrative houses in La Zona, where the American School and other buildings were constructed. Administrative employees moved out of the labor barracks as soon as their houses were completed. This was a tropical wilderness that was connected to the outside world only by a railroad running through it. In 1946, after the war, the Company transferred my father to Progreso, Honduras, as construction superintendent.

REMINISCENCE OF BILLY SARAVANJA

I recall living in the Bataan abaca farm when I was a boy in the early 1950s. My father, Johnny Saravanja, was building an extension on the abaca plant and was assisted by Ramon Cabezas.

Bataan was along the main railroad line between Limón and San Jose, and it was the headquarters for all the other abaca farms. There were four other abaca farms near Bataan. Davao was to the south. Luzon, Leyte and Goshen were northeast of Bataan. Manila was isolated along the Pacuare River near La Perla and had a dining hall and social club. Mindoro was by the main rail line southeast of Pocora. Rodolfo Portocarrero stated that there was a network of some 200 kilometers of railroad tracks connecting those abaca farms. The farms were accessed with small motorized railcars that had been transformed from flat hand-pumped railcars. Bataan was the headquarters for all seven abaca farms.

The American School was in Bataan and was on the second floor above the social club next to Bill Newell's house. Elizabeth Newell did not have to walk very far to school. One year, an American woman came to teach. She lasted only a few days before returning back to Limon.

I went to the 1st grade there and my brother, Johnny, who was in the 3rd grade, when it closed in 1953. Between 1953 and 1954, my mother home-schooled us. That was when abaca operations closed, and we were transferred to Puerto Cortes, Honduras, where my father started building a new wharf.

The Company hired "Doctor" Balma to take care of medical problems in the Bataan farm dispensary. He did not graduate from the University of Costa Rica Medical School, but his good reputation qualified him for the job in Bataan. He delivered babies, fixed broken bones, stitched up cuts, gave injections, treated colds and other ailments. Dr. Balma sometimes prepared patients to be medevaced to the big United Fruit Company hospital in Limon. Patients were put on the Northern Railroad train when it passed through Bataan. He lived in the designated doctor's home in the center of the golf course. His wife, Pepita, played golf with my mother, Alva Bailey Saravanja.

When my father constructed administrative houses in Bataan, he stored leftover lumber and materials from each house. With this scrap wood, he built a golf clubhouse in the back corner of the golf course that had only seven greens. Golfers replayed greens one and two to complete the nine holes. The club had a bar and a movie hall. But there was no movie projector.

MOVIE NIGHT

Once a week, a United Fruit Company employee, a young, well-dressed Costa Rican whose name was Fairweather, came on the train from Limon to Bataan with a large Bell & Howell portable projector. He brought several reels of film with the movie of the week. The projector was set on top of the bar. When it got dark, Fairweather started the movie. When he changed to the next reel, the lights were turned on briefly.

On one occasion, Fairweather had too many highballs while he was waiting for it to get dark to show the movie. This caused him to put on the wrong reel. When he nervously fumbled around to get the right reel, it fell off the bar counter and went rolling down the aisle, unwinding the film on the floor. My brother Johnny jumped up and chased the reel almost to the screen. He gave it back to Fairweather who somehow managed to get the movie shown that night.

Such was life in a remote abaca farm in Costa Rica.

Bataan was the headquarters for all the abaca farms in Costa Rica. Map: Bernal Saborio.

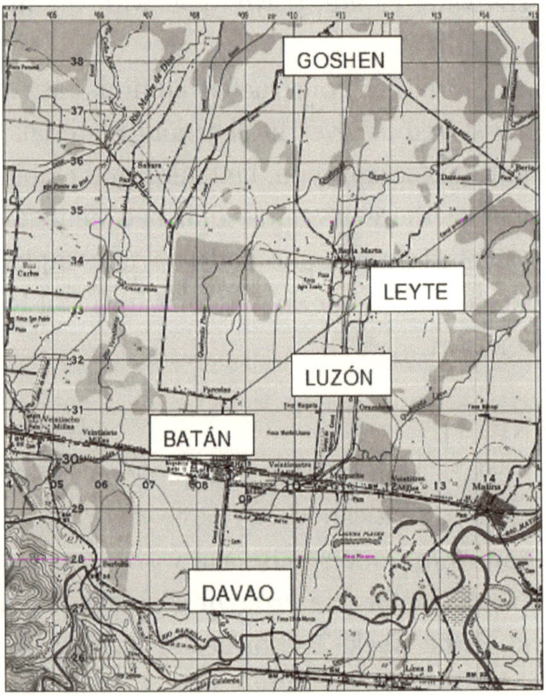

17. The United Fruit Company in Cuba

EARLY ON

In 1887, Alfredo Dumois Gessé bought land at Banes, province of Oriente, in northeastern Cuba. He was financed by his brother Hipolito and other investors and formed the Banes Fruit Company that exported bananas and other tropical fruits.

Because of the war of independence from Spain in 1896, Banes town and the Dumois businesses were ruined. The Dumois investors sold their assets to the United Fruit Company in 1899, when Andrew Preston had control of the new merger of the Boston Fruit Company and the Central American empire of Minor Keith. The new owner, United Fruit Company, continued to produce and export bananas until 1906, when the Company fully switched to sugarcane.[1, 2]

There were several reasons for the switch from bananas to sugarcane. One was because bananas stop growing during the cool winter months in Cuba. This led to a corporate decision to abandon banana production in Cuba and the Dominican Republic and relocate to adept areas in Central America where Minor Keith was well-established. Another reason was because the Company wanted to expand sugar production in Cuba. On the south side of Nipe Bay, another sugar port was developing. It was called Preston.

THE HOGGE FAMILY: HEROES IN MANZANILLO

Before the Hogge family became sugarcane and banana people a generation later, they left an historical legacy in Cuba. William Hogge was a British colonial, who was born in Lucknow, India. He married Mary Emerson, who was born in England in 1858. A son was born in Germany, a daughter in Persia, and a son and three daughters in Cuba.

In 1895, William was assigned by the British Cable Company to operate a cable and telegraph station in Manzanillo on the southeast coast of Cuba. Near the end of the Spanish-American War in 1898, American war ships bombarded Manzanillo. A large gun shell landed on the Hogge house where the cable telegraph was located but did not explode. Spanish officers disarmed the shell and saved the house, the family and the telegraph station. The empty gun shell was used as a doorstop afterwards.

William and Mary Hogge became heroes at the end of the Spanish-American War and were lauded in the Manzanillo newspaper that began with the following headline:

> *Recordad con fraternal afecto a los que salvaron de la más terrible destrucción a nuestro solar nativo. Para la distinguida familia Hogge tienen los manzanilleros un sincero y eterno agradecimiento.* (Remembering with fraternal affection those who saved our native home from the most terrible destruction. For the distinguished Hogge family, the people of Manzanillo are sincerely and eternally grateful.)

In five columns of print, the story relates details of their heroism. The following is a summary:

In 1898, the North American war ships blockaded Manzanillo, along with Cuban revolutionaries on land. After three bombardments, it seemed like the whole world was falling on their heads. People were advised by the Spanish authorities to take refuge wherever they could. People went in all directions. But the Hogge family stayed in their home and continued to operate the telegraph cable station. If they were going to die, the family would all be together and that was their destiny.

There was not much telegraph traffic that night, but after the fourth bombardment by American war ships, a sensational circumstance occurred. A telegraph cable message arrived and was taken by Mary Hogge, who knew how to receive and transmit Morse code. At that moment, the city was being destroyed. The message related that Spain, Cuba and the United States had signed a peace treaty and the war was over. But the American war ships had not gotten the message and continued to blow up Manzanillo with more destruction and deaths.

William Hogge knew it was his duty to deliver the cable message, even during the height of the bombardment. When his small boat approached the war ships, they started shooting at the light on the boat. The boat retreated and waited until daylight. Hogge displayed a British flag that enabled him to arrive and deliver the urgent message that the Spanish-American war was over. Hostilities ceased immediately and Manzanillo was spared from more demolition. The protocol to end hostilities was signed on August 12, 1898, in Washington D.C.

In recognition of her bravery, a street in Manzanillo was named Avenida Mary Hogge. She was also given a set of silverware. For William Hogge, the king of Spain, Alfonso XIII, awarded him an honorary white cross for bravery and a red cross for bravery under fire, plus a large, inscribed loving cup. Queen Victoria had to approve of these honors from the king of Spain because William and Mary Hogge were subjects of the British crown. A letter from the British Foreign Office, dated August 6, 1900, gave this approval. William Hogge continued to serve as the British Cable Company agent until he died in Manzanillo and was buried there. He was replaced by his wife Mary until she left to live in New Orleans in 1909. Mary Hogge was replaced by a son-in-law, Santiago Banning. On occasions, she returned to visit the grave of her husband and to visit her children and grandchildren in Manzanillo.

The residents of Manzanillo always paid their respects to Mary Hogge when she returned. During visits, visitors saw half a dozen blond children playing in the street, all speaking Spanish. The Hogge family legend lived on in Cuba. Later, several of them became executives with the United Fruit Company in Cuba, the Dominican Republic and Central America. This included some of the Cook family, who descended from the Hogge children of Manzanillo.

THE HOGGES AND THE COOKS IN PRESTON

Albert Levi Cook (1881-1952), was born in Parsons, Kansas, and worked for the Santa Fe Railroad. In 1905, he found his way to Preston, Cuba, and became a cashier.

Albert became friends with Frederick William Bertram "Bertie" Hogge who was born in Cuba and also worked for the United Fruit Company. Bertie Hogge invited Albert to visit his parents and siblings in Manzanillo on the southeast coast of Cuba. This is where Albert met Bertie Hogge's sister, Anita Aurea Hogge. Bertie and Anita were the son and daughter of the heroic couple, Mary and William Hogge, of British Cable Company fame. Anita was born in 1884 in Guanabacoa, just east of Havana and was bilingual because she had studied in Cuba and Scotland.

Albert and Anita fell in love and married in Manzanillo in 1908. The newlyweds made their home in the sugar port of Preston where their first son, William "Bill" Wylie Cook was born nine months later. Eighty-five years later in Gulf Breeze, Florida, Bill Cook gave me interviews and copies of his many photographs and documents from Cuba, Honduras, Guatemala and the Dominican Republic.

Bill Cook began life in Preston with several relatives. When his aunt, Lois Cook, came for a visit, she met and married an employee, Norman Moseley. Bill Cook's mother was Bertie Hogge's sister. Bertie later became the manager and lived in the opulence of the manager's mansion. He died in 1923.

Because Albert Cook grew up with horses back in Kansas, it was only natural that he was an excellent player on the Preston polo team. In 1911, while his wife and two-year-old son were visiting relatives in England, he was suddenly transferred to Puerto Cortes in Honduras.

MARK TRAFTON'S CHRONICLES IN BANES

Mark Trafton began his long career with the United Fruit Company in 1926. He described some of the Company operations in Cuba in his memoirs:[3]

> Banes was the headquarters of one of two large United Fruit Company sugar plantations in the Oriente Province of Cuba.[4] The other one, adjoining Banes, was Preston,[5] named after Andrew Preston, the co-founder of the Company. The Banes Division included some 50,000 acres of sugarcane and pastures divided into about 35 farms. The large sugar mill was located in another nearby Company town at the port of Macabi where

Company boats loaded bulk raw sugar for refining at the Revere Sugar Refinery in Boston.

The Banes Company compound included the general office, a large clubhouse-restaurant-guest quarters, commissary, hospital, meat market, about 50 first-class employee residences, tennis courts and a polo field. A nice Atlantic beach was nearby. The native town was separated from the Company compound by a small bridge over a river. It was a typical, well-ordered Spanish village with its park, theater, club, a few small family stores, several blocks of residences and "Pepy's" *cantina* conveniently located just across the bridge.

Most of the personnel at the executive level in the offices, in the sugar mill, at the machine shops, hospital and on the farms were English or Jamaican colonials. There were only a dozen Americans, all in the accounting and merchandise departments. Well-educated Cubans filled a few executive and supporting positions.

All cane-cutting laborers were Haitians brought in by the thousands under contract just for the harvest season—known as *la zafra*. I had about 150 Haitians on my farm. At night, they chanted and danced around a bonfire in a voodoo ritual, spurred on by drinking fermented cane juice.

Dressed in a white jacket and jodhpurs, black leather riding boots, and a white helmet—riding a spirited horse around the cane fields was fun. I had been transferred to Tasajeras Farm by Banes and I could ride my horse to town in 25 minutes to play polo on the Banes team, not the farm team. Mr. Harty, the manager, made sure of that.

During his vacation to Boston in 1928, Mark was transferred directly to Puerto Armuelles, Panama, to begin his career in bananas. He could not return to Cuba to say goodbye and left a broken heart behind in Banes.

CASTRO FAMILY AS NEIGHBORS

In 2023, I interviewed Pauline Howe McCaffery, age 95, who related many memories about life in Cuba, right up to the time she left in 1960. Her mother was from England and married an Englishman who was born in Calcutta, India. Pauline was born in Havana in 1927, but the family lived in Sagua la Grande on the north coast. This is where her father was based with United Railways of Havana, a British company that carried passengers and cargo. Pauline spoke Spanish from early childhood, but it was different with her father. He had a contract with the railroad company that required any new employee to speak, read, write and understand Spanish. Employees had to know the business and the ways of the Cuban people. Otherwise, they were sent home. Mr. Howe was engaged before he left England, but he could not go back to get married for three years because of a minimum time clause in his contract. Pauline and her sister were sent to school in England, but the rest of their education was through home schooling and in English-speaking schools in Cienfuegos and Habana as boarders. During vacations in Cuba, she has fond memories of excursions from Sagua la Grande to the outer islands off the shore of the Atlantic Ocean.

In Havana, Mr. and Mrs. Howe became friends with another British family by the name of McCaffery, and that is how Pauline met young Tom McCaffery. Tom's father was also employed by United Railways of Havana. Tom was not born in Cuba because his Irish mother went to New York City to have the baby, then returned to Havana with the newborn in 1925.

Tom became fluent in Spanish at an early age because he went to a Catholic school in Havana that taught in Spanish. Later, he was sent to a boarding school in Canada, then he got a degree in engineering from the University of Toronto. The United Fruit Company hired Tom as an engineer in 1951, the same year that he and Pauline were married. He was assigned to work at Preston in the province of Oriente. (This northeastern end of Cuba was renamed as the province of Holquin in 1976). Preston was the location of the United Fruit Company hospital where the McCaffery sons, James (1952) and Michael (1953), were born.

In 1954, Tom was given a challenging promotion in Macabi, the Company port town where the Banes sugar mill was located. His duty was to overhaul and restart the malfunctioning sugarcane processing plant that had been misaligned by a previous manager. His successful repair was appreciated by Company officials and local Cuban employees. They were back in business. He was promoted to assistant manager in Banes and operations were running smoothly. That is, until one of the former neighbors became a revolutionary.

FIDEL CASTRO

Fidel Alejandro Castro Ruz was born nearby in 1926 on a large estate near Biran, just 50 kilometers south of the United Fruit Company sugarcane operations of Preston. His father was Angel Maria Castro Argiz, from Galicia, Spain. He started poor, became a soldier in the Spanish-American war in Cuba and later acquired 10,000 hectares from the United Fruit Company. He was a contractor for the American company and sold his sugarcane to them for processing. Fidel's father expanded his various businesses and could afford a large, fine country home where Fidel and some of his many siblings were born. Fidel's mother was Lina Ruz Gonzalez, an unmarried servant in the Castro household.

Because of the Castro proximity to the *yanqui* company, Fidel knew *La Compañia.* In fact, his first marriage in 1948 was to Mirta Diaz Balart, the daughter of a United Fruit Company lawyer and prominent politician who had been the mayor of Banes. The couple had met at the University of Havana, where she studied philosophy. Fidel Castro got a degree in law. They had one son, known as Fidelito. The couple was divorced in 1955, reportedly because of his infidelity. Also, after the revolution began on July 26, 1953, the rebel had little time to be with his wife and son because he was on the run in the mountains of Oriente and elsewhere, attacking military barracks, killing, terrorizing, and fighting the forces of Fulgencio Batista, the dictator of Cuba.

Pauline McCaffery remembers the acts of terrorism against the United Fruit Company and its employees. Four Americans who lived in the remote Guaro sugarcane district were kidnapped on the job and held in the mountains. It was common to raid Company areas and loot materials and supplies. Burning, destroying and terrorizing Company people was routine.

Rebels controlled the roads, the urban area of Banes and the vast outlying terrain. Vehicles on the roads to or from Banes were stopped and passengers were told to get out and walk. The vehicle was then burned. As a result, transport of cargo and people came to a stop, unless the rebels gave special permission to pass. The only way to get somewhere was to walk long distances on deserted roads.

Pauline McCaffery had eyeglasses that needed fixing. The only way to get help was to find a courier who walked many hours to the optometrist in Holguin.

Near the end of 1958, sandbags were stacked around critical areas of the McCaffery house to protect against bullets. The family and community of employees were nervous knowing that rebel raids had been made in Banes. As the situation grew worse in 1959 and the early 1960s, Tom saw no hope for the future and resigned from the Company in 1960. Pauline packed many of their belongings that would be shipped later. She traveled with their two sons some 300 kilometers west to Camagüey for the last time. They only needed a one-way ticket, but the revolutionary government required international travelers to buy a round-trip ticket. Pauline still has the unused return portion of her Pan American World Airlines ticket from Miami to Camagüey.

When she departed from Cuba, she was allowed to take only five dollars in cash. Consequently, arrangements had to be made to have cash waiting for her at American Express in Miami. She proceeded to Toronto where their next baby was born in 1960.

Tom stayed behind and was able to ship some of their possessions. He was one of the few employees who prepared in time. The great majority of employees remained complacent with the optimistic belief that the United States government would protect the Company and its people in Cuba. They were wrong.

Henry Murray told me how he and others were told to pack their suitcases and fly out immediately. Tom told me the same story about others. They were obligated to leave all their household and personal effects behind, including money. Pauline McCaffery said that a few men returned to Preston during the transition and found a way to salvage and ship some personal items of a few friends who were suddenly forced out earlier. Most were not so fortunate.

A decade later, Tom McCaffery was the general manager

of *La Compañia Bananera de Costa Rica*, based in the banana port of Golfito.

THE END OF UNITED FRUIT COMPANY IN CUBA

The last audit of United Fruit Company properties in Cuba was made by Carl Stevenson and his assistants in 1958. One of his auditors took a picture of Fidel Castro in the Company area where they were auditing. Perhaps this picture was an ill omen.

Several months after Fidel Castro made his triumphant entrance to La Habana, Wolfgang Schulz and his auditing team were sent to Cuba to audit the United Fruit Company books. After arriving, they were denied entry to Cuba. Nevertheless, the auditors were required to get their passports stamped to enter. Immediately, immigration officials directed them to the departure line where their passports got an exit stamp. They were put back on the same Pan American Airlines airplane that the group had arrived on. This was foretelling the future again.

On August 6, 1960, the world learned that Fidel Castro had nationalized American companies, including everything that the United Fruit Company owned in Cuba. This included several Company towns, railroads, ports, two sugar mills and thousands of hectares of sugarcane fields and cattle pastures.

The United Fruit Company was just one of many American enterprises that were expropriated. The Company was not large compared to other sugar producers because "its mills rarely produced more than 5% of the national output, and its landholdings were in fifth place among the North American sugar monopolies [sic]."[6] This included 36 sugar mills throughout Cuba. When nationalizations began in 1960, Fidel Castro ordered that his own family estate be expropriated. It was a shock to his mother to see bearded, armed men invade her yard where she was waiting with a rifle. One of them had to explain to the elderly woman what had happened.

Today, the large Castro home is a popular tourist attraction. It is doubtful that the tour guide mentions a report from *Forbes Magazine*: "*Forbes* estimated Fidel Castro's personal net worth at $900 million."[7] This was quite remarkable compared to the wealth of his old nemesis, corrupt and brutal dictator Fulgencio Batista Zaldivar, who left with only an estimated $300 million, plus untold millions in fine art and other assets.[8] The two men were born only 51 kilometers from each other: Castro in Biran and Batista in Banes—both sugarcane regions.

The Dumois brothers, L-R: Alfredo and Hipolito.
Photograph: Unifruitco, June-July, 1949.

Cuba with cities relevant to this chapter. Map: Ken Huff.

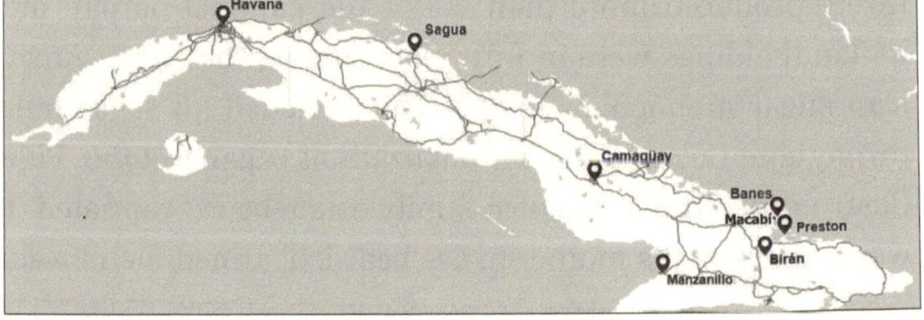

Mary Hogge, heroine of Manzanillo, Cuba, circa 1898. Photograph: Bill Cook collection.

William and Mary Emerson Hogge (rear), Frederick William Bertram "Bertie" and Anita Aurea (front right). Manzanilla, circa 1906. Photograph: Bill Cook collection.

Preston Sugar Mill (rear). When Bertie Hogge was manager, he lived in this executive house (center). Photograph: Bill Cook collection.

Albert Levi Cook Photograph: Bill Cook collection.

Albert Cook and Anita Hogge Cook, recently married, in the home of auditor S. L. Slessor. Preston, 1908. Photograph: Bill Cook collection.

Opposite Page: House in Preston, Cuba, where William Wylie Cook Hogge was born in 1908. Photograph: Bill Cook collection.

High society social gathering. Manager O. G. Sage (standing, 6th from left). Preston, circa 1908. Photograph: Bill Cook collection.

Nipe Bay Company staff, 1908. Albert Cook (2nd row, 2nd from the left). Bertie Hogge, who was Albert Cook's brother-in-law and who became manager (6th from the left). Photograph: Bill Cook collection.

Staff of the Nipe Bay Company.—Cuban.—Division
PRESTON.—Cuba.

Mark Trafton in jodhpurs.
Photograph: Mark Trafton collection.

Cashier Albert Cook paying laborers. Preston Cane Farm, 1910.
Photograph: Bill Cook collection.

McCaffery home in Preston.
Photograph: Pauline McCaffery collection.

Entrance to Hospital, in Preston.
Photograph: Pauline McCaffery collection.

Golf club in Banes, 1950s.
Photograph: Pauline McCaffery collection.

McCafferys at Easter, 1954. Photograph: Pauline McCaffery collection.

Center: American School. Banes, 1958-1959. Photograph: Pauliine McCaffery collection.

Bottom: Golf club members. Tom McCaffery, 2nd from right. Banes, Cuba, 1950s. Photograph: Pauline McCaffery collection.

Bus to Camagüey: Tom, Pauline, Jim and Michael. Banes, 1956. Photograph: Pauline McCaffery collection.

Center: Weighing sugarcane, Guaro, 1960. Photograph: Pauline McCaffery collection.

Bottom: Jim and Mike McCaffery. Cane-weighing station. Guaro, 1960. Photograph: Pauline McCaffery collection.

Sandbags for safety from bullets, Banes, Cuba, 1958.
Photograph: Pauline McCaffery collection.

The family home where Fidel Castro grew up at Birán, Cuba.
Photograph: Shutterstock.com. ID: 1741529831. https://www.shutterstock.com/g/DeBora+Rachelle.

18. Cooks Tour Continued

FROM SUGARCANE TO BANANAS

Albert Levi Cook, from Kansas, was suddenly transferred in 1911 from sugarcane operations as a cashier in Preston, Cuba, to Puerto Cortes, Honduras, where he was assigned to *El Ferrocarril Nacional de Honduras* (the Honduras National Railroad). The new job may have been a way of getting him into the country for other assignments. Cook arrived before his family came. His wife, Anita Aurea Cook Hogge, and young son, William Wylie Cook, sailed on October 21, 1911, on the maiden voyage of the SS *Sixaola* that departed from England, where they had been visiting her family.

A letter from the treasurer of the United Fruit Company in Boston announced the appointment of A. L. Cook as auditor of the Ceiba Division in La Ceiba, Spanish Honduras, effective April 1, 1912. This was the beginning of United Fruit Company developments on the north coast of Honduras, where the Vaccaro brothers and Salvador D'Antoni had been growing and exporting bananas to New Orleans for a decade.

Some 72 kilometers west of La Ceiba, the United Fruit Company acquired a large area of beach frontage west of Old Tela across the Tela River. In 1912, the Company started construction on an open sea wharf over a kilometer long. At the same time, offices, houses, a hospital, commissary, American

School, club and golf course, machine shop, power plant and other structures in the Company town were under construction by hundreds of employees. A railroad began on the wharf and extended east and west to new banana plantations. Thus, the huge enterprise was aptly named Tela Railroad Company. This giant project took several years to build and was ongoing with new developments.

In the beginning, before the Company town was built in New Tela, the first employees lived on the east side of the Tela River in Old Tela, where they had temporary offices and living quarters. A dining hall was behind the office.

Albert Cook was transferred from La Ceiba to Old Tela in 1913 as the accountant. Felix Aycock reported to Albert Cook in the beginning, but later Aycock became a manager in Tela. Years later, Aycock's career took a drastic turn when he joined United Fruti Company's nemesis, Samuel Zemurray and his Cuyamel Fruit Company at Omoa and Puerto Cortes. The Cuyamel Fruit Company extended all the way to the Guatemalan border. When Samuel Zemurray unexpectedly acquired the United Fruit Company in 1932, Felix Aycock eventually became the general manager of the Tela Railroad Company. Others also became managers, including several who joined another competitor, the Standard Fruit Company in La Ceiba in Honduras, in Nicaragua during the rebellion of the Sandinistas, and in other countries.

Life for these pioneer employees in Old Tela was austere. As an accommodation, Albert Cook arranged for his family living quarters to be directly over his accounting office. He had a special reason—his wife. Although Anita was born and raised in Cuba, she retained much of her English upbringing. At precisely three o'clock each afternoon, Mrs. Cook would tap on the floor upstairs to remind her husband below that it was time for tea and toast with strawberry preserves.

In 1913, after the first new houses were finished in Tela, the Cook family moved from Old Tela to a fine home on the beach.

Because Albert Cook's new accounting office was not ready, he continued to cross the Tela River each day to his temporary office in Old Tela. To get there, he rode his prized horse from the new house to the old office across the river. Meanwhile, an army of employees with many skills continued to build the new Company town in Tela along and behind the Caribbean beach. When ships brought in new arrivals, passengers were transferred to lifeboats that stopped near the beach. Black Carib Indians, now known as *garifunas,* were hired to physically lift up and carry elegant ladies in long dresses above the waves to the seashore. A capricious rumor spread around beforehand that these native porters sometimes took women back to their Carib village further down the beach. When one porter tried to lift a woman out of the boat, she beat him on the head with her umbrella until he dropped her. He was just trying to do his job. The incident caused lots of merriment.

Many experienced banana men came from other countries to build Tela. One was John Wilson Lawrence, who began his career in the 1890s in Bocas del Toro, Colombia, which is now a part of Panama. J. W. Lawrence sent a postcard from La Ceiba, Honduras, in October 1912. He was soon transferred to Tela and became a superintendent assessing lands for banana plantations.

Albert Cook was the accountant in Tela until 1916 when he returned to Cuba. It appears that he was associated with a railroad in the Cuban town of Guantanamo on the southeast coast. This town is located inland, north of the Guantanamo

Bay Naval Base. The Cooks' first son, William Wylie Cook, was born in Preston, in 1909, but no more children were born until 1923, when Henry Hackett Cook was born in Guantanamo. In 1925, Albert Emerson was also born in Guantanamo. The last son, Richard Robert, was born in Miami, Florida, in 1927.

WILLIAM "BILL" WYLIE COOK'S CHRONICLES

Bill Cook had hundreds of pictures and stories to share and it would take a book to include so much memorabilia. In addition to what has already been written in the previous chapter, the following is a brief outline of his notes and interviews in the 1990s in Gulf Breeze, Florida.

SCHOOLS

London, England (Kindergarten); Colegio Padre Varela in Spanish in Guantanamo, Cuba, 1st and 2nd grades; 3rd to 6th grades in Episcopal Mission Schools in Guantanamo and Camagüey, Cuba; 7th to 9th grades in Winfield, Kansas, USA, living with Cook grandparents; 10th grade in Gulfport, Mississippi, and Miami, Florida; 11th grade in Coral Gables, Florida; 12th grade in Winfield, Kansas, and New Orleans, Louisiana. He attended Louisiana State University, Tulane University (1930-32) and also studied accounting correspondence courses.

OTHER RESIDENCES AND EMPLOYMENT

Honduras: Puerto Cortes, La Ceiba, Los Planes, Tela, La Lima; Dominican Republic: Puerto Libertador; Guatemala: Retalhuleu; Ecuador: Guayaquil; United States: Gretna, Louisiana; New York City; and Gulf Breeze, Florida, to retire.

Bill Cook married Mary Louise Corbin in 1934, and they had a daughter, Billie Louise, in 1936. Their son, Wylie, was born in Tela, Honduras, in 1944. Bill Cook was employed by both banana companies: United Fruit Company and Standard Fruit Company. He was the manager for Standard Fruit Company in Guatemala. His career ended in 1974 as a vice president. The couple retired to Gulf Breeze, Florida.

VIGNETTES OF BILL COOK

FIRST ELECTRIC FAN

When I was five in 1914, we lived in Tela, Honduras. Electric power lines had not been installed but Mr. Hackett gave my father a fan. At least, it looked like a fan. The device had a container below for alcohol. When the wick was lit, the hot flame rose and propelled blades that turned the fan. This stirred the hot air. We were proud to have the first fan in Tela.

UNHAPPY COOK

When my father rode his horse from our new house on the beach in Tela to work in Old Tela, he left the horse behind the kitchen and dining hall. The cook left some pies on the kitchen windowsill to cool but they did not last long. The horse ate all the pies and the cook was not happy.

FROM CUBA TO KANSAS

My parents sent me to live with my Cook grandparents in Winfield, Kansas, from 1922 to 1925 to attend the 7th through the 9th grades. Father took me on the train from Guantanamo on the Guantanamo Railroad to San Luis where I changed to the Cuba Railroad Company train to Havana. Father would tear two twenty-dollar bills in half and give one half of each bill to the conductor. The other two halves were given to me. If I was well taken care of during the long trip, I would give my two halves to the conductor when he helped me off the train in Havana. It always worked.

From Havana, I boarded a ship to New Orleans, then a train to Fort Worth. From there, I made the final change to Winfield, Kansas, where my grandparents were waiting for me. When I arrived at their house, my grandmother took my coat and cut out the seams and concealed pockets that held extra money that my mother had hidden, just in case I had special needs.

A HONDURAN BIRTH CERTIFICATE

I was a good baseball player and in 1934 the Honduras team selected me to go to El Salvador to play in the Latin American Olympics. But I had just married and did not want to leave my bride at home alone. Also, there were concerns about my nationality. I was born in Cuba, but there were questions about my getting American citizenship. I did not want to cause any complications and politely declined the offer to play baseball. The Honduras military *comandante de armas* sent a message for me to come to his office. He asked me why I would not play baseball for Honduras and I told him. Soon, he reached into his desk drawer and pulled out a document. It was a Honduras birth certificate with signatures and seals showing that I was born in La Ceiba, Honduras, and was a Honduran citizen.

I knew I was in trouble and went to Mr. Chute, who was the Standard Fruit Company manager. He got me out of that difficult situation and that gave me peace of mind.

SMUGGLING MONEY

During the world economic depression in the 1930s, banks closed and countries were in financial chaos. In 1934, I was the cashier for the Standard Fruit Company in Los Planes, some

70 kilometers over the mountain range south of La Ceiba. This was in the Aguan Valley where the banana plantations were located and where hundreds of employees worked. The accountant called me by phone and asked me to get all the money possible and bring it to La Ceiba. It was my problem of how to get the money to La Ceiba on the north coast of Honduras.

I stuffed my shirt with $10,000 in cash, put a coat over the shirt and climbed into the cab of the steam locomotive that was pulling a banana train destined fornLa Ceiba. As we approached the bridge before La Ceiba, armed men stopped the train. We were not sure if they were bandits, rebels or military men. I removed my .45 Automatic pistol and hid it in the sand box of the locomotive, then sat on the lid of the sand box. Fortunately, the armed men were on our side and let us continue. I delivered the hidden money to the accountant as planned. It was another adventure for me as a cashier.

FIRST GOLF CLUB PRESIDENT

In 1949 and 1950, when Bill Cook worked for the United Fruit Company in Puerto Libertador, Dominican Republic, there was much construction going on. The banana company was building an arid and rocky community at the new banana port on the north coast bordering Haiti. As was customary in any United Fruit Company town, a golf course was made, but the fairways had no grass. It was on rough, dry terrain and golf balls frequently bounced off of rocks. The greens were not green because they were made of sand, not grass.

A large club house with a beautiful tile floor was built by the Company. Although there was very little rain, the roof was made of a tight palm-leaf thatch. Golfers and other members socialized at a nice bar by the kitchen. Movies were shown three nights a week. Holidays were celebrated with dances and lively *merengue* music provided by local musicians. Later, a pool was added. In a dusty environment, water was always refreshing.

For the club to get legal status and be incorporated, Bill Cook wrote the official letter in 1949 to President Trujillo, the dictator of the country. The letter was recorded in the public record. Because of this application and its approval by President Trujillo, William Wylie Cook became the first president of the *Club de Golf de Manzanillo, Puerto Libertador, R. D.* His own certificate issued in his name on November 29, 1949, was signed by him as president of the club. It is probable that William Wylie Cook also signed the membership certificate for his first cousin from Cuba, Bertie Hogge, who was the assistant manager to manager Dave Cloward. "On the founding date in 1949, 51 original members were recorded by nationality as follows: 31 Dominicans; 10 North Americans; 4 English; 1 Spaniard, 2 Hondurans; 1 Panamanian; 1 Mexican; and 1 Nicaraguan.

Seventy bonds for $25 each were issued to founding members and were supposed to pay dividends. Somehow, the dictatorship found a way to absorb the value and members lost their investment.

A THANKSGIVING TURKEY

When the Cooks lived in Guayaquil, Ecuador, Bill was the assistant Manager for the Standard Fruit Company from 1962 to 1970. One of the banana producers who sold his fruit to the Company was from China. He had been in the army in Formosa (Taiwan). The man liked Americans and their ways. Because of this endearment, he arranged for his son and daughter to deliver a Thanksgiving turkey to the Cook's apartment on the sixth floor.

The daughter knocked on the door and told Bill and Mary Lou that her father had sent them a Thanksgiving turkey. Mary Lou thanked the girl and waited for a moment. The girl said goodbye and left. When the door opened enough to receive the gift, the son threw a live turkey into the living room and departed with a nervous smile and goodbye. The unusual gift was kept on a small, enclosed porch and waited for the day of reckoning.

TRAINING FOR THE BAY OF PIGS

Bill Cook was the manager of the Guatemala Division of the Standard Fruit Company from 1960 to 1962. His office and home were on the Pacific side of the country in Retalhuleu. One morning, around 7:30, when he was meeting with a Guatemalan banana producer, the quiet was suddenly interrupted when a colonel from the Guatemalan military burst through the swinging doors into his office without a knock on the door or a salutation. A backup man with a sub-machine gun followed him and gave credibility to what was about to take place.

Bill Cook was ordered to get the Standard Fruit Company Cessna 182 and the helicopter off of the airfield by 5:00 P.M. because the military was taking over the airport. A pilot was quickly found and flew the airplane out, but the helicopter was a problem. Its pilot had just married and was on his honeymoon somewhere in the mountains. After some fast searches, a helicopter pilot was found, and he flew the aircraft out before the deadline.

Almost immediately, a big crew of workers arrived and started constructing a large wooden wall around the entire airport so the public could not see what was happening inside. Bill Cook learned that Cubans were being trained by members of the American Air Force who had come from Tennessee. They were preparing for the Bay of Pigs invasion. The airfield was normally a private landing strip used only during the day. But when airplanes came in to land at night, bright landing lights were turned on and lit up the Cook house and community outside of Retalhuleu. Life was different after that.

Temporary offices and living quarters.
Old Tela, Honduras, 1912.

Tela Hospital under construction, 1912-1913.
Bill Cook collection.

*Old Tela staff, 1912. Felix Aycock (top row, arrow).
Others in this photograph also became managers.
Many joined their competitor, Standard Fruit Company.*

*Saint Charles Hotel, aka Balderach Hotel, in Old Tela, 1913.
Bill Cook collection.*

View of New Tela, possibly taken from Tropical Radio Telegraph tower. Bill Cook collection.

Tela wharf, over a kilometer long. 1912-1913. Bill Cook collection.

The Tela River divided Old Tela and New Tela (right rear).
Bill Cook collection.

Boats and porters. Bill Cook collection.

Postcard from Lawrence in La Ceiba, Honduras, 1912. J. W. Lawrence collection.

John Wilson Lawrence on motorcar. J. W. Lawrence collection.

Nº 000013

ASOCIACION INCORPORADA DE ACUERDO CON LAS LEYES DE LA REPUBLICA DOMINICANA

CLUB DE GOLF DE MANZANILLO · FUNDADO EN EL AÑO 1949 · SELLO

Una Acción

ASOCIACION QUE NO TIENE POR OBJETO BENEFICIOS PECUNIARIOS

CLUB DE GOLF DE MANZANILLO
PUERTO LIBERTADOR, R. D.
CAPITAL AUTORIZADO Y PAGADO: $1,750 DIVIDIDO
EN 70 ACCIONES DE $25 CADA UNA

Este Certificado Comprueba que William Wylie Cook *es el dueño de* Una *Acción enteramente pagada del Capital Social del CLUB DE GOLF DE MANZANILLO, transferible solamente en los libros y con el consentimiento de la Sociedad al devolver este Certificado debidamente endosado.*

En Testimonio de lo cual se sella con el Sello de la Sociedad y lo firman los funcionarios autorizados de la misma, hoy 29 *de* Noviembre *de 19*49–

A. A. [signature] — Secretario
W. W. Cook — Presidente

Golf Club membership of Bill Cook, first club president, signed by him. Bill Cook collection.

Private banana farm records of John Wilson Lawrence, Tela, Honduras. J. W. Lawrence collection.

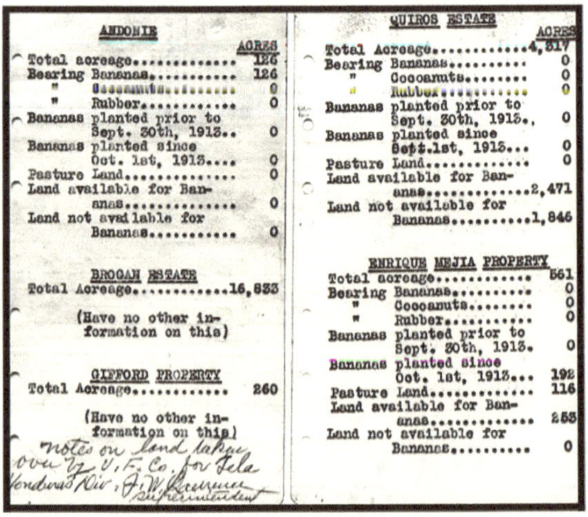

ANDONIE

	ACRES
Total acreage..............	126
Bearing Bananas...........	126
" Cocoanuts............	0
" Rubber.............	0
Bananas planted prior to Sept. 30th, 1913..	0
Bananas planted since Oct. 1st, 1913....	0
Pasture Land.............	0
Land available for Bananas.............	0
Land not available for Bananas...........	0

BROGAN ESTATE

Total Acreage.............16,833

(Have no other information on this)

GIFFORD PROPERTY

Total Acreage............ 260

(Have no other information on this)

notes on land taken over by U. F. Co. for Tela Honduras Div, J. W. Lawrence Superintendent

QUIROS ESTATE

	ACRES
Total Acreage............	4,317
Bearing Bananas..........	0
" Cocoanuts........	0
" Rubber..........	0
Bananas planted prior to Sept. 30th, 1913.,	0
Bananas planted since Sept.1st, 1913...	0
Pasture Land.............	0
Land available for Bananas.............	2,471
Land not available for Bananas...........	1,846

ENRIQUE MEJIA PROPERTY

Total acreage............	561
Bearing Bananas..........	0
" Cocoanuts........	0
" Rubber..........	0
Bananas planted prior to Sept. 30th, 1913.	0
Bananas planted since Oct. 1st, 1913...	192
Pasture Land.............	116
Land available for Bananas.............	253
Land not available for Bananas..........	0

Sons of Albert Levi Cook and Anita Hogge Cook. L-R: Richard Robert, 1927, Miami, Florida; Henry Hackett, 1923, Guantanamo, Cuba; William "Bill" Wylie, 1909, Preston, Cuba; Albert "Bertie" Emerson, 1925, Guantanamo, Cuba. Bill Cook collection.

Mary Louise and Bill Cook, Gulf Breeze, Florida, 1994.

Book 2

1. Sea Level Canal Study

I WAS THE BANANA research guy in the Changuinola banana plantations, and managers always asked me to take care of visiting scientists. One such person was Dr. James A. Duke of the Battelle Memorial Institute, a science and engineering organization. Later, he reciprocated with an invitation for me to collect and take notes on insects in Darien in eastern Panama, where extensive scientific studies were in progress. The studies were undertaken because the lock-based Panama Canal was being burdened by more and larger ships, and among the solutions whirling about was the idea of a sea level waterway slicing through the tropical forests of eastern Panama.

Under one proposal, atomic energy would be used to blow up a mountain range that was in the proposed canal's path. This was a staggering thought for most scientists and non-scientists alike. Yet it was scientists who were tasked to determine the probable ecological impact of such a massive undertaking.

Such was the situation in April 1966, when I flew from Changuinola in Bocas del Toro to Panama City and stayed overnight in James's small, high-rise apartment. Early the next morning, he and I flew in a small, single-engine airplane to the wilds of Darien and landed on a short, grass air strip somewhere near Santa Fe. From there, a native dugout, called a *piragua*, took us down the Sabana River to Boca Lara (now Puerto Lara), where a field camp was located.

I was assigned a bunk bed with a mosquito net in a large open-air, palm-thatched *bohio* (hut) where about 30 members of a scientific team were lodged. Some had set up their microscopes and lab equipment either under the dry *bohio* or outside on makeshift tables in the sunlight—until the rains came. The scene was impressive and unforgettable; I was awed by the biodiversity and wild beauty. James introduced me to several renowned scientists, and this made me feel so humble, shaking their hands. Two of them were the famous Odum brothers: Dr. Eugene P. Odum, "The father of modern ecology" from the University of Georgia, and Dr. Howard T. Odum from the University of Florida. Other scientific titles at the site were mammalogist, ichthyologist, biologist, ornithologist, herpetologist, geologist, hydrologist, meteorologist, botanist—all the kinds of -ologists. In Boca Lara, I met Horace Loftin and Ed Tyson, who later assisted me with bat studies in the Changuinola banana plantations.

The manager who kept the camp going—a major task—was Jim Berlin. Jim had been my neighbor and close friend in Changuinola, where he had been the manager of a cattle company, the Ganadera Bocas, owned by Panamanian tycoon Mario Guardia. *Don* Mario replaced Jim with a son-in-law, Julio Alcedo, another close friend of mine. When Jim and his wife, Paula, moved to Panama City, they left their two daughters, Linda and Maureen, to live with Phyllis and me so the girls could finish the school year at the American School in Farm 8.[1] I told Jim the latest news about his daughters, and in return, he explained to me lots of things going on behind the scenes at the Boca Lara camp and about the mixture of native and foreign people there.

Indigenous Choco men wearing *taparrabos* (loincloths) walked around the camp with no pants. Some were hired to help with different chores in Boca Lara. Their *piraguas* were chartered to take scientists up and down the rivers. Choco-women came to the camp grounds and caught the attention of new arrivals like me. They were painted, wore skirts, but had no upper garments, meaning that they were bare-breasted. It was even more novel that the women had brassieres slung over their shoulders. Christian missionaries had preached to them that it was not nice to go without brassieres. The women were not shy about this issue, and were laughing and giggling as they attempted to put on the contrivances. Most failed on their first attempts, but they were having fun. Some brassieres were strapped on upside down. It was all symbolic of how foreign do-gooders can mess up people's lives and cultures.[2]

Although James was a renowned botanist and ecologist, he mixed and related with the indigenous people like an anthropologist. As a result, he learned some of the Choco language and much about their lives. James pointed out a young girl reaching puberty. She had black face paint with certain designs, and this meant that she was ready for marriage. The black paint came from the fruit of *Genipa americana* and lasts about a week on the body. James illustrated this tree on the cover of his Darien Dictionary that compares and defines words in Choco, Guna, English and Spanish. I still have a copy of this book, plus other scientific publications on his work in Darien.

James told me that the Choco people in the camp gave all the *gringos* a Choco nickname. Mine was *Adíchichi*, which means grasshopper—and was very appropriate. They had seen me collecting insects, which included a large, colorful grasshopper. I still have that original collection.

After I returned home,[3] I realized that the trip had taught me much about Darien and what science and scientists can do to make good decisions that affect the world. "Gradually, as more experts began to come in, more and more questions were raised about the viability of a sea-level canal via atomic charges," said my good friend, Dr. Stanley Heckadon-Moreno, a highly respected Panamanian anthropologist. The sea-level proposal was rejected, but the decision could have gone the other way, with approval to blow up that vast wilderness. In the end, science, logic and common sense prevailed. I shudder to think of the other option.[4]

This large insect is the one that led to my nickname:
Adíchichi or grasshopper.
Clyde Stephens collection.

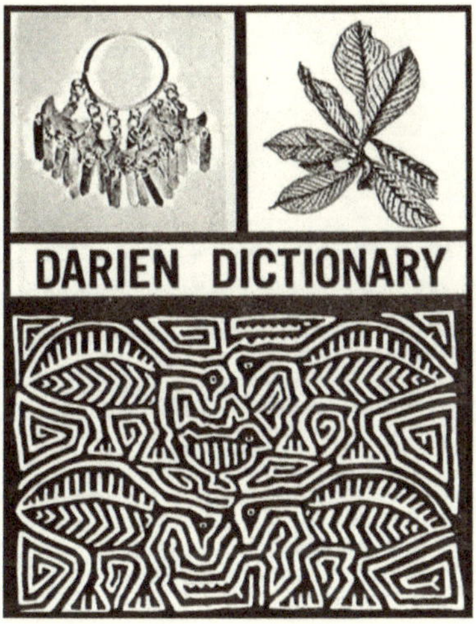

Darien Dictionary, James A. Duke. Battelle Memorial Institute, IOCS-Bioenvironmental Program, 1967. This dictionary shows a Genipa fruit used for black body paint. The book translates English, Spanish, Guna and Choco words.

Painted Choco women in Darien, Panama.
Photograph: https://en.wikipedia.org/wiki/EmberáWounaan.

2. Science Versus Tradition

IN THE 1920S, THE United Fruit Company sent Doctors Vining C. Dunlap and Otto August Reinking of the research department to the Philippines, Indonesia and Southeast Asia to collect banana varieties. These banana collections were taken to Panama and kept in quarantine at Flat Rock on Columbus Island (Isla Colon) in Bocas del Toro. Later, the plants were moved to the mainland in Changuinola at Farm 6 for holding and studying.

The two scientists probably never knew how important their collections would be a few years later. They had brought back multiple banana varieties like Valery that were resistant to Panama disease. Valery was planted throughout Changuinola and other countries in the 1960s and was a huge financial success that saved the United Fruit Company from bankruptcy.

The massive losses of banana plants before the 1920s are what motivated the Dunlap Reinking trip, but the motivation came from Company scientists, not from Management. The Company remained committed to the Gros Michel banana, even though it was highly vulnerable to Panama disease and was generating millions of dollars in losses.[1]

Dunlap, Reinking and others surmised that the Company would eventually face a doomsday scenario if solutions were not found—solutions such as: The first attempts at banana breeding, traveling around the globe collecting varieties resistant to Panama disease, and collecting different genetic types of bananas that might be useful in the breeding effort. Over time, it was the Dunlap-Reinking approach that saved the Company.

GROS MICHEL

The limits of the Gros Michel banana were reached by the 1950s; losses from disease and pestilence were dragging down profits and harming investor confidence. Regardless, Management continued to bet on Gros Michel despite its susceptibility to Panama disease. Drastic intervention was needed if the Bocas Division was to survive; the best path forward appeared to be what is known as flood-fallowing, a method of controlling disease by flooding the land while it lies fallow, meaning unused.

The first flood-fallowing began in 1950 when all the soils in Changuinola were flooded for six months for the purpose of killing Panama disease. Water was pumped from the Changuinola River at Empalme and channeled through an extensive network of aqueducts. This flooded the soils of old, diseased farms where Panama disease had killed millions of banana plants, long before cacao and abaca were planted there. Engineers built high levees based on natural contours, and these levees impounded waters from the river in lakes about two meters deep. For example, Farm 13 in Empalme had lake numbers one through five. Each lake was about 40 hectares (99 acres) in size. These lakes attracted ducks and other waterfowl and duck hunters took advantage.

After six months of flooding, the lakes were drained and Gros Michel plantations were once again established. It was not long before the new plants started dying. Flood-fallowing for Panama disease was a total failure and a multimillion-dollar fiasco. Management and shareholders were shocked at the loss. The only option left was to shut down all operations in Bocas del Toro and leave, or switch to a Panama disease-resistant variety.

Out of desperation, the Company planted one of the banana varieties that Drs. Dunlap and Reinking had brought from the Philippines in the 1920s. Called Lacatan, it had been shown to be resistant to Panama disease. Unfortunately, the decision to plant Lacatan sparked a familiar pattern: The extent to which banana managers were in conflict with the highly educated banana research scientists who made recommendations based on science.

Before the Lacatan stock was shipped from Jamaica and planted in Changuinola, the nematologists expressed alarm that the Company might bring in corms infested with nematodes (roundworms). Still, the managers and corporate officers had no concern about these microscopic worms in the roots, and completely ignored precautions and recommendations by the Company nematologists—a typical antagonistic attitude by Management against banana scientists at that time.

Lacatan corms were placed aboard a ship and sent to Changuinola for planting between 1958 and 1960. This was the first time that a banana variety was dehanded and shipped in a box—also something the old-timers said could not be done. Nonetheless, it was the way of the future for all of the banana world, and as expected by the scientists, devastating banana nematodes were scattered all over Changuinola. The nematodes took over, the crop declined and managers faced the predicament of what to do next.

The solution was to send Aaron King to Almirante to close the entire division that had been losing money. He was transferred there in 1961, and became impressed with the good, silty loam soils in Changuinola. He also noted that costly irrigation was not needed in the rainy Bocas del Toro plantations. There were other good reasons why he liked the division. So instead of closing the whole division, he did the opposite, saving the Bocas del Toro operation and the local economy from disaster. This was done in part by a tough, fiscal policy that made him unpopular. Fortunately, the nematologists eventually came up with an unlikely proposal that might solve the problem. After the flood-fallowing for Panama disease in the 1950s, the nematologists had discovered that nematodes did not survive because their plant food sources had been drowned. That is, *Radopholus similis* and *Pratylenchus coffeae*, both devastating nematode pests of banana roots, were not present after the flood-fallowing. This was good news for everyone except the financial analysts. This time, managers and corporate officers worked with the scientists and decided to invest the incredible sum of money necessary to flood-fallow the same soils again, this time for nematode control, not for Panama disease.

When the flood waters were drained after six months, a different Panama disease-resistant variety was selected to be planted throughout Changuinola and Sixaola. This variety was called Valery, a clone that was another of the varieties collected by Drs. Dunlap and Reinking in Saigon, Indochina, in 1925.

Valery was chosen because it had a shorter, more robust trunk that gave it the strength to stay upright longer. In addition, because Valery was shorter, more plants could be planted per square area.

Although the soil had been purged of damaging nematodes by the flood-fallowing, there was fear that some corms in the nursery farm, Farm 24, might be infested. To make sure that the corms were free of nematodes, each one was peeled with a sharp butcher knife by skilled indigenous workers to remove any possible nematode infestation. Afterwards, each corm was cooked in a tank of hot water for ten minutes to kill any microscopic pests that remained. After several years, over ten million cooked corms had been processed in the treatment plant near Farm 62. These clean, cooked "seed bits," as they were called then, were planted in clean, nematode-free soil.

The results were phenomenal. Production of two boxes of bananas for every bunch of Valery harvested was common. The Bocas del Toro Division returned to profitability and the Company's stock soared in the mid-1960s. Also, the scientists gained credibility with the production and management officers.

At the same time, 13 new banana farms were sold to experienced Panamanian employees, who sold their boxes of fruit to Chiquita.

Of course, the experiment with nematode control could have gone the opposite way. Certainly, credit also goes to managers like Aaron King who did much to reverse the fate of the United Fruit Company in the Bocas del Toro Division. Thus, it was the recommendation of several competent scientists in the research department who made a big difference. Some of them were Clive Loose, Quinton Holderman, Estle J. Wehunt, Dale Edwards, and Bill Mai. A new generation of scientists followed those pioneers in this field. Except for Clive Loose, I knew them all, and they taught me everything that I learned about nematodes and their control. I was fortunate to work under these giants in science and nematology.

ABACA

Another example of science versus tradition involved abaca plants, also known as Manila hemp, which the United Fruit Company had collected in Changuinola in the 1920s.

In a letter written in 1959 to Dr. Norwood C. Thornton, Director of Research, Dr. Wilson Popenoe reminded the director that in 1937, he (Popenoe) had to beg for $300 to keep the experimental planting of abaca in the Almirante Division going for another year. Five years later, in 1941, abaca became a strategic war material and urgently needed. As Simon E. Malo writes:

> Tunnel-vision *bananeros* could not see growing anything but bananas. Luckily, when the war broke out in the early 1940's and Manila hemp was badly needed by the U.S. Navy and Merchant Marine for rope to tie up war ships, Popenoe's 50-acre experimental abaca planting was the only one in existence in the western hemisphere. It eventually supplied the seed for more than 28,000 acres of abaca that the United Fruit Company planted for the war effort.[2]

One must ask these questions: What if that experimental planting had been destroyed? Where would the Americans and its allies gotten ropes to tie up their war ships?[3]

It has been a century since abaca and bananas were brought to Changuinola and 80 years since the Japanese controlled abaca production in the Philippines. Certainly, a word of thanks is due to Harry T. Edwards and Drs. Dunlap and Reinking for having the vision to introduce abaca to Changuinola in the 1920s, which helped win World War II. In addition, other varieties they brought back to Panama, like Valery, changed the fortune of the giant banana company, especially in the 1960s, once banana production had been restored. And as a testament to its quality, Valery is still grown today in some parts of Changuinola and in other countries.

Dr. Otto Reinking collecting bananas in the former Dutch East Indies, now Indonesia, 1925. Unifruitco, Nov. 1927, Vol. III, No. 4, p. 212.

Flood-fallowing (left) was a total failure in controlling Panama disease, but it was effective in controlling nematodes (right). Clyde Stephens collection.

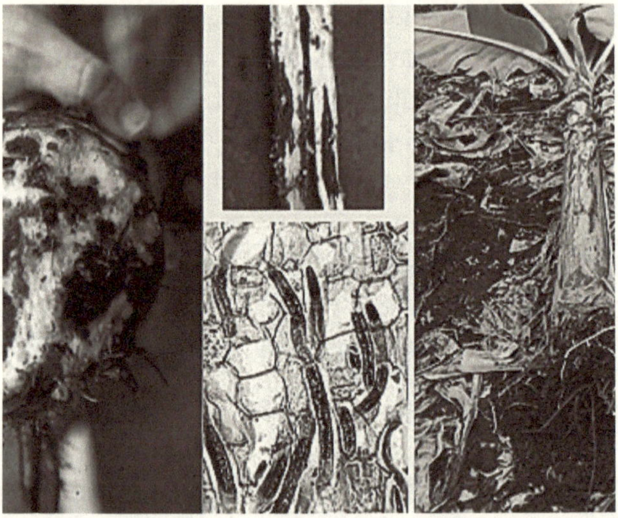

Nematodes (center bottom), and the damage they can cause, including toppling banana plants. Photograph: Burrowing nematodes: Florida Division of Plant Industry, Florida Department of Agriculture and Consumer Services, Bugwood.org. https://www.forestryimages.org/browse/detail.cfm?imgnum= 5384272#collapseseven.

Corms being inspected, peeled and cooked to kill nematodes in Changuinola in the 1960s. Clyde Stephens collection.

3. Canals in Bocas del Toro

PANAMA IS A LAND known for the Panama Canal, but there were other canals, too. I refer to canals dug by banana people for hauling bunches of Gros Michel fruit from remote plantations to ships waiting in the harbors. That was in the 1890s and early 1900s when Bocas del Toro was still part of Colombia.

Long before the age of banana canals in western Panama, there was a proposal to build a canal between Bocas del Toro and the Pacific. In 1860, an American, Lieutenant St. Clair Morton, and his assistant, Thomas Jekyll, proposed a "Canal of locks that would have to be built as an interoceanic canal cut through the Isthmus between Almirante Bay in Bocas del Toro and the Golfo Dulce in the Pacific." Anyone who has walked or driven on this proposed route between the two oceans would find such a scheme completely preposterous, but in the 1850s and 1860s, there was a rash of outlandish ideas schemed up by the Americans, British and French to connect the Pacific and Atlantic for their own imperialistic reasons. Thanks to an in-depth and scholarly study, Giselle Marin Araya of the University of Costa Rica, revealed all the implications, interactions, wild proposals and intrigues of building a railroad, tunnels for ships, a land road, a colony for freed American slaves, colonial developments and a canal from the Almirante Bay, or Robalo River or Chiriqui Grande across the high Continental Divide to the Pacific.[1] It was not until the 1980s that the historic foot trail was paved over with a road between Chiriqui and Bocas del Toro.

THE FIRST BANANA CANAL

Michael Theodore Snyder was a canal builder. He came to Bocas del Toro in 1889 and married a local woman, Seveline Mathews Shepard, in 1890. Soon, his two brothers, Charles Louis and Joseph Alfred, joined him in developing some of the first banana plantations in the Chiriqui Lagoon. One of these plantations was at Manati Creek and the Biarra River, now known as the Guariviara. Their banana railroad lines could only go so far along the banks of the rivers. So to solve this problem, a barge canal 2.5 kilometers long was built through the swamp to Manati Cove, a small protected bay to the east. From there, fruit was transferred to launches or banana ships waiting in the deep Chiriqui Lagoon.

In 1985, Larry Mahan, an authority on Snyder Brothers history, and I hired a derelict, leaky *cayuco* powered with a 15-horsepower engine in Chiriqui Grande and found that canal. It was straight with tall spoil banks on each side and was navigable to the Guariviara River, just like my old United Fruit Company map showed. As adventurers in history, we felt rewarded.

In 2015, I retraced my "footsteps" and went back to see the old Snyder Canal of the late 1800s. There was no sign of the canal that we had seen in 1985. It had completely disappeared because of floods of the Guariviara River. Not even satellite views suggest a previous canal in that area.

THE SECOND AND THIRD BANANA CANAL

Minor C. Keith, a railroad builder and banana tycoon, invested and gained control of the banana industry in Bocas del Toro, then part of Colombia, in the late 1890s. Suddenly his empire defaulted because his banana broker went broke. Minor Keith swallowed this bitter pill and united with his nemesis and competitor, the Boston Fruit Company, controlled by Andrew Preston. Thus, two big banana companies merged in 1899 and formed the United Fruit Company. In this corporate agreement, Minor Keith continued his role as the "top banana" in Bocas del Toro, but under Andrew Preston. They bought all the assets of the Snyder Brothers and other banana interests in 1899.

Just prior to the sale, the Snyders had started a barge canal from the northwestern shore of Almirante Bay, where there was a protected entrance southwest of Donato Point. Their motive was to haul bananas from the new plantations in Changuinola to Knapps Hole, near Boca del Drago. The Snyders were part of a land rush by investors, speculators and developers who were acquiring the excellent silty loam soils along the Changuinola and Sixaola Rivers. This was all virgin territory and was not infected with Panama disease. They entered a wilderness with no infrastructure of any kind. There were no railroads, no canals, no housing, no maintenance shops and no labor force. Getting in and out of the mouths of the precarious Sixaola and Changuinola Rivers was extremely hazardous. They were not reliable water routes and that was the reason for building a canal.

After Michael Snyder dug nearly three kilometers of his canal through the unstable and pestilent swamp toward Changuinola, he realized that he had chosen the wrong route and stopped digging. It was around this time, at the turn of the 20th century, that the new United Fruit Company bought out the three Snyder brothers, plus other banana growers. Charles Louis and Joseph Alfred Snyder packed up and left Bocas del Toro.

It was different with Michael Snyder. The United Fruit Company offered him half a million dollars in gold to complete a barge canal from Boca del Drago to the Changuinola River by 1903. Snyder complied, took his gold and left to pursue similar business opportunities elsewhere. His Bocas wife and many of his Bocas-born offspring were in the United Kingdom and France, where the children were being educated, as was the custom then. From that time, it appears that the parents split and never reunited. However, Snyder apparently continued to support his large family. One of his sons joined the Canadian Army during World War I and was killed in France. Some of the children married in Europe, one stayed in Europe and only one, Ella Olla, returned to Bocas del Toro with her husband, Manuel Azael Bendiburg Castillo, a Panamanian judge.

The new canal, 11 kilometers long, was constructed in more stable terrain where the wetlands meet the sandier strata behind the beach. Nevertheless, the entrance to the canal from Almirante Bay was not stable because wave action filled the mouth of the channel with sand. This made it shallow and difficult for barges and tugboats to get in and out of the canal. The other end of the canal entered the Changuinola River and changed its course after each flood. Changuinola's expanisve wilderness was ripe for development, but it required lots of capitalistic entrepreneurship and fortitude. After the completion of this new banana canal in 1903, the United Fruit Company transported numerous administrators, engineers, carpenters, agriculturalists, overseers, timekeepers, foremen and about a thousand imported farm laborers. Railroad builders, locomotive train engineers, and train conductors were needed to construct and operate railroads that moved the materials and people. Later, these men would operate trains for hauling hundreds of thousands of bunches of Gros Michel bananas from the plantations to a port along the Changuinola River at Farm 6.

With this pioneer group came materials of every imaginable kind—heavy equipment, machinery, cranes and spans of steel for building bridges over the three rivers. Unassembled steam locomotives required skilled technicians to put the components together and get them running on new rails that were laid across the swamps, creeks, rivers and future banana plantations. Box cars, flat cars, passenger cars and small motorcars for supervision and inspections were needed. They hauled materials, wood, building supplies and tin roofs for constructing hundreds of houses, bachelors' quarters, offices, stores and a social club. Medical clinics were needed because yellow fever and malaria were still prevalent. Communication was improved with a vast network of telephone lines scattered across the Changuinola flood plain and along new rail lines through the plantations.

A major task was the felling of giant trees in order to plant rhizomes of Gros Michel bananas. This work required hundreds of laborers to clear the forest and plant bananas. Experienced banana men from elsewhere came to direct the development of the new plantations. This was a skill and trade that only a few men knew. In summary, it was the 1903 banana canal that jump-started this grand agricultural enterprise that covered thousands of hectares. Men and supplies came by way of the canal because there was no other route. Without the canal, it would have taken many more years to build a railroad from the future port of Almirante to Changuinola.

SHIPPING BANANAS FROM CHANGUINOLA THROUGH THE CANAL

Trains pulled flat cars carrying bunches of Gros Michel bananas to a port on the meandering Changuinola River at Farm 6. The river did not rush straight to the sea as it does today. During the late 1890s and the first decade of the 20th century, the cold mountain river meandered across the center of the Changuinola flood plain, where the Changuinola Airport is today. From the Farm 6 banana port, bunches of fruit were loaded onto barges tied up at the riverside.

A tugboat powered by a Wolverine gas engine pulled barges with bananas through the canal to Knapps Hole, now called Starfish Beach, a calm cove in Almirante Bay, where the bananas were loaded onto ships. Communications were expedited via a telephone line on the bank of the canal and then went underwater at Boca del Drago, passed by Knapps Hole and on to Bocas Town. A few old rail telephone posts can still be seen along the canal.

END OF THE CANAL

After shipping millions of bananas through the canal, plus transporting tons of supplies, numerous employees and even sick patients going to Hospital Point, operations were terminated when the Changuinola River bridge was completed in 1908. A metal plaque in the middle span of the bridge reads: BUILT BY BALTIMORE BRIDGE COMPANY, 1908. The same plaque was mounted on the bridges over the San San and Sixaola Rivers. Thus, by 1908, bananas, passengers, cargo and all rail traffic from Changuinola and both sides of the Sixaola River went directly to the new banana port of Almirante. Vegetation, *Raphia taedigera* palm fronds and fallen trees quickly covered the canal, ending a strategic and historic lifeline.

OPENING THE CANAL AGAIN

Fishing and excursions to the beaches, bays and islands was nice if you lived in Almirante, but the employees who lived out "on the rail line" in Changuinola were landlocked. Sometimes the Almirante Social Club and the Changuinola Golf Club organized trips on the old *Talamanca* excursion boat, and we would go to Zapatilla islands or Boca del Drago for wonderful outings. When I was director of the Changuinola Research Station, it occurred to me that I had the technology to open the canal once again so that we Changuinola people could go to the bays and islands like the Almirante employees did. Previously, Aaron King, the tough old general manager, ordered the chief engineer and his department to clean and open the canal, but the effort was a failure. That is when I decided to do the job on my own.

I called a meeting of my research laborers and monthly employees and proposed to open the canal. They were told that the project was not a Company activity and no one would get paid. However, I promised that if we ever reached the end of the canal at Boca del Drago, I would treat them to a party they would never forget.

The Company whistle blew at 11:00 AM on Saturdays and we stopped work for the weekend. About a dozen of us volunteers went in two large *cayucos* down the river and to the canal to start the arduous task of unblocking the old waterway that had been abandoned for 58 years. To nourish our bodies, Phyllis provided huge pots of rice and beans. I animated the men with Caballito, a strong, popular drink made from sugarcane. Spirits were high and no one complained although it was hard work pulling out fallen trees with a block and tackle pulley, treating tons of floating vegetation and removing fallen palm fronds and canes. The same group of men worked on Sundays.

After six months of hard labor, we reached Boca del Drago. It was a hallelujah moment, and we celebrated on the beach at Boca del Drago as I had promised. It was a party not to be forgotten. I might add that the manager shamed the chief engineer because that "research boy" and his volunteers got the job done. That was in 1966.

This opened up a new world of recreation for us landlocked residents in Changuinola. Some of the heads of departments had nice plywood *pangas* (skiffs) made in the Company shop, powered by 18-horsepower engines. Families enjoyed Boca del Drago and other aquatic places on weekends. As for me, I continued to use my *cayucos*, which were always reliable.

When the river flooded, the opening to the canal was blocked. Lucky for us boaters, the manager sent a backhoe on a barge to keep the canal open. After many floods, the canal stayed blocked some of the time, but the entrance at Boca del Drago was always passable.

Now there are paved roads between Changuinola and Almirante and almost no one uses the canal anymore, even if it is not blocked. That might be the saving grace for nature lovers and environmentalists. The old banana canal, sometimes called the Snyder Canal, is considered to be a prime birdwatching destination and is within a protected nature preserve. I have recorded five of the six species of kingfishers along this canal in one day. And I'm going back to find the sixth one.

Biarra River Canal, depicted on an old United Fruit Company map. Clyde Stephens collection.

Map of old Donato Point Canal and new canal. Clyde Stephens collection.

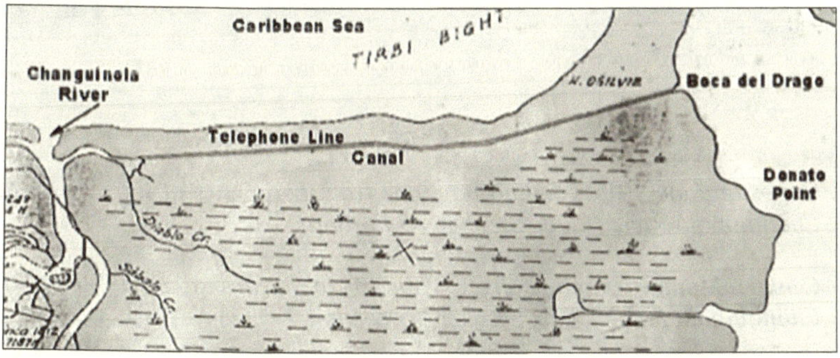

A Snyder canal. Linda Baker Lawrence collection.

Finished canal to Changuinola, 1903.

Opposite Page: Top: Moving bananas from train cars at Farm 6 on the Changuinola River. Clyde Stephens collection.

Center: Moving bananas from railcars to barges at Farm 6 on the Changuinola River. Clyde Stephens collection.

Bottom: Banana crew. Clyde Stephens collection.

If you like fruit have a banana. on me.
dont look for any white men in this
picture
D.E.S
From the Truck to the Lighters. Changuinola Plantation. Bocas del Toro R. P.

Tugboat pulling barges with fruit.
Clyde Stephens collection.

Michael Snyder (right rear, in vest); Seveline Matthews Snyder (front center, in white blouse and black skirt).
Larry Mahan collection.

MICHAEL THEODORE SNYDER, RIGHT REAR IN VEST, DIGGING CANAL IN BOCAS DEL TORO, COLOMBIA. POSSIBLY WIFE SEVELINE IN CENTER FRONT IN BLACK AND WHITE DRESS. 1890s. Louis Irving Snyder, Jr. collection.

Snyder Canal a century later.
Photograph: Alex Chester.

Purple Gallinule. Photograph: Alex Chester.

4. Taming the San San River

BEGINNING IN 2015, I was invited to give lectures at the University of Panama branch campus on Farm 13 in Changuinola; it was nice to come back home to an area where I had lived for 11 years. My host was Professor Rosa Villar, a dynamic member of the staff in charge of special presentations. I spoke in Spanish to students, professors, medical doctors, the Ministry of Education and other groups. On some days, I gave three or four presentations on different subjects, but the most appropriate PowerPoint was entitled: *La historia de la industria bananera en Bocas del Toro* (*The History of the Banana Industry in Bocas del Toro*).

In explaining why the university was surrounded by banana plantations, I told them that commercial bananas grow in silty loam soils that have been deposited by floods. This means that people who live and work on flood plains should expect floods. When I showed them several scary pictures of flooding in Changuinola, a student reassured me that there was a dike protecting them. However, as an experienced old engineer from Costa Rica always told us, "*El río va donde quiere.*" "The river goes where it wants to." Ricardo Gomez knew from experience; he had seen it all. When I projected some of my flood pictures on the screen, I focused on one in particular.

It was about a disaster that occurred in the early 1920s just a kilometer from where I was lecturing—and very close to the homes and offices of those attending.

Regressing a few years, I could not understand the function of a tall, massive concrete structure that was in a grassy cow pasture in Farm 15. Years later, I saw an old photograph that gave me the answer. A flood had completely washed out a railroad bridge crossing the San San River, a branch of the Changuinola River, that had flowed near where I was giving my program. No one attending the lecture could imagine floods of that magnitude in their community. I explained that before 1925, the San San River had been a fast-flowing, cold, mountain-water river that branched from the Changuinola River at El Silencio. When the river flooded, all banana people, plantations, houses, railroads and bridges were destroyed or badly damaged. Because of this devastation, management decided to block the San San River. In 1924 and 1925, a huge earth and rock dam was constructed across the waterway. That blocked Changuinola River water from flowing through southwest Changuinola via the San San River branch. As planned, the riverbed dried up, grew over with grass and left the deep riverbed that we see there today.

Professor Villar was fascinated by this historical development so near to her office, so I agreed to show her and the others the dam. We drove only a few minutes through Farm 15 on the road to Zegla and found the dam blocking the old San San River channel. The site of the bridge was at decimal coordinates: 9.400527°N, 82.532282°W.

Numerous vehicles pass within a few meters of that dam every day, but no one sees it because of dense vegetation. I was surprised that many houses were built below the dam in the old river bottom—a precarious place to live.

Next, we drove a short distance down a farm road and walked down a banana cableway, where I showed them two large concrete structures. Those massive bases once supported the bridge that was destroyed in the early 1920s. One bridge base was surrounded by cultivated bananas and the other one was in the cow pasture at the bottom of the San San River channel. One can see these bridge foundations near decimal coordinates 9.4054654°N, 82.5339215°W. I felt good that I could share some history with the professor and her students.

In 1925, environmental impacts from blocking a river were of no concern. However, if a project to block the river with a dam was proposed today, an environmental impact study would have been made, by law and by public demand. The results of such a study in the original San San River may have shown the following:

1. The source of fresh, cold mountain water from the Changuinola River would be suddenly blocked and would change the entire ecosystem down stream all the way to the Caribbean Sea.

2. Certain species of fish, eels and other fauna that migrate up the river to reproduce would be blocked by the dam.

3. Fauna and flora of the San San River would be impacted and possibly some species might become extinct.

4. Fresh mountain water would be replaced by salty, brackish water entering the river mouth from the sea. Former ecosystems would be replaced with new habitats with new species.

5. Because the high volume of rushing water from the mountains would be suddenly stopped by the dam, the mouth of the San San River would be sealed off by wave action, thus blocking fauna from entering or leaving.

Perhaps some ambitious graduate student might attempt

to make such a study, after the fact, and present it as a Ph.D. thesis. We can only imagine the conclusions.

San San River bridge near Empalme, early 1920s.

Mysterious concrete structure in a cow pasture.

5. Hospital Point: A Lost Story of Medical History

IT WAS ALL BECAUSE of Teddy Bendiburg Snyder, bananas and mosquitos that I learned something about Hospital Point. Teddy and I were friends in Changuinola in the plantations, where he was in charge of a banana farm. During one of our weekend excursions in 1966, he took me to a little beach on the western tip of Nancy's Cay, now Isla Solarte, in western Panama. I was told that his grandfather, Michael Theodor Snyder, became a millionaire in the banana business when he came to Bocas del Toro in the late 1880s, and that by the end of the 19th century, Snyder and another banana tycoon, Minor Keith, had a small hospital on the point of this island. I was very skeptical about Teddy's story because we had consumed several libations during the trip in a *cayuco*. Once we landed on the island, we could not walk past the shore to explore Hospital Point because the jungle was impenetrable.

Later, hundreds of old pictures fell into my hands because the office of a United Fruit Company manager was not interested in a large box of old, neglected photograph albums dating from the early 1900s. I grabbed them in an instant and found some of the story of Hospital Point. Old books, library records and interviews provided more details. These references gave credibility to Teddy's story, and my curiosity expanded. Prior to 1966, Hospital Point was a lost chapter of tropical medical history.

TROPICAL DISEASES

By 1899, the United Fruit Company had acquired major banana interests in Bocas del Toro plantations bordering the Chiriqui Lagoon and Almirante Bay. Rampant disease among the workers had to be conquered to make the plantations productive. Dr. Robert Earl Swigart, the company's first medical director, reported that one out of 20 white men died per year from yellow fever and malaria between 1899 and 1901. Dr. W. E. Deeks stated that 40% to 100% of lowland tropical populations were debilitated with hook-worms, which, combined with malaria, produced much acute morbidity. He also stressed the losses from dysentery and amebic dysentery. Pneumonia, kidney disease and tuberculosis were other very common causes of death.

With banana company officials also recognizing that about 25% of the French-Panama Canal labor force died, it was obvious that the United Fruit Company could not grow and export bananas unless it had a healthy labor force. To do this, the Company aggressively began organizing and controlling public health, reaching this objective by building Central America's first major medical center.[1] The site was on Nancy's Cay in the northwest corner of Panama, which was then part of Colombia.

The new hospital replaced a smaller facility built by Minor Keith and the Snyder brothers at the same location. In the beginning years, this site was known as Quarantine Island and accommodated only a few white administrative employees.

CUBA AND THE SPANISH AMERICAN WAR

The United Fruit Company had quickly learned the causes of yellow fever and applied controls discovered in Cuba before, during and after the Spanish American War of 1898. A Cuban physician, Dr. Carlos Juan Finlay, discovered that a female mosquito, *Stegomyia fasciata,* transmitted yellow fever, but he was completely ignored for 20 years before the war. After the Americans acquired control of Cuba from Spain, records revealed that fewer than 1,000 men died in combat as compared to over 5,000 who died from yellow fever. This quickly initiated action in Havana to study and control yellow fever. Dr. Walter Reed assigned Major William Crawford Gorgas to control mosquitos in Havana, and within a few months, yellow fever was wiped out.[2]

It was this historical action that inspired the United Fruit Company to do the same in Bocas del Toro and in other Company operations.[3] Under the direction of Dr. Swigart, a mosquito eradication program was implemented and yellow fever cases drastically declined. Patients were confined on Quarantine Island (now Hospital Point) and isolated under mosquito nets. Later, the same treatments and controls were applied to malaria cases. Thus, the Company was a pioneer in controlling yellow fever and malaria in Central America and elsewhere in the Tropics, before the start of the American Panama Canal in 1904. Ironically, it took several years to implement yellow fever and malaria control measures in the Panama Canal Zone. The Isthmian Canal Commission in Washington, D.C. was arrogantly opposed to Dr. Gorgas and his mosquito control mandates.[4]

SOME STATISTICS AND FACTS:

- In 1901, soon after the new hospital was built, Colombian Conservative forces invaded and occupied Hospital Point. They used it as an assault base against Liberal forces across the channel in Old Bank on Bastimentos Island during the War of a Thousand Days.
- After Panamanian independence from Colombia on November 3, 1903, a large delegation of the new government visited Hospital Point from March 24 to 29 in 1904. A story in *La Estrella de Panama* described the large annex behind the hospital where 63 laborers were treated.
- A large hospital for the labor class was constructed after 1904 and was known as the labor or "colored hospital." Prior to its construction, there were no morbidity statistics for people of color, most of them Jamaicans brought from the West Indies.
- Eventually, Hospital Point became a complex consisting of 16 buildings. After work, doctors entered a game room where they could play pool, smoke cigars and have a few drinks; In the laborer's house, also called the Bone House, games that were played with bone dice caused wild excitement on payday. The losers, totally broke, would row their dugouts back home to Bastimentos or Bocas. A Company launch left daily, making trips from Nancy's Cay to Bocas del Toro and Almirante; rainwater from the roofs was stored in elevated tanks and used for drinking, cooking and bathing.
- Since the early 1900s, General Manager John Morgan Kyes had lived at Hospital Point because he loathed living in pestilent Bocas del Toro. He commuted daily to his office in town or to review banana farms on the mainland.
- Carlos Reid was Kyes's cook in 1912 and wrote about life

at Hospital Point.[5] In 1913, Kyes moved to his new luxury home in the port of Almirante. The Hospital Point manager's house was then occupied by subsequent medical directors.

• Reid recalls that on its two daily trips, the Company launch would bring up to 10 persons from the mainland farms with snakebites, malaria and venereal disease.

• In 1914, a new surgery facility was built by Norman Sparks, the maintenance man. Underneath the main floor, there was a pharmacy, a medical laboratory and the administrative office to admit patients getting off the boat at the passenger dock.

RESIDENT AND HISTORIAN

Bruce Brown was born in Bocas del Toro in 1906 and spent 11 years of his youth living with his mother, Rebecca Swartz, on Hospital Point from 1908 to 1919. She was the housekeeper for the general manager, and later for medical doctors who occupied the luxury home perched on the west cliff of Hospital Point. Bruce and his mother lived in one of the four rooms for servants located underneath the main floor.

As a boy, Bruce had free rein of the entire medical center and knew everybody and everything that was going on. I interviewed him and his wife, Lilian Campbell de Brown, five times on Hospital Point in the 1990s. His memory was exceptional with numerous details and names.

Bruce Brown's step-father, Norman Sparks, was a carpenter and jack-of-all-trades maintenance foreman for Hospital Point, who helped to raise and motivate him. Bruce Brown later became a laboratory technician identifying stomach parasites in the medical laboratory when a new hospital was built in Almirante.

In 1918, a 110-volt generator powered by V-belts from a Morse Fairbanks gas engine was installed by W. E. McLaren. It was not until 1926 when Carl Friese, a local German merchant, was contracted by the Panamanian government to install a power plant for the city of Bocas del Toro. Thus, Hospital Point had electricity eight years before the capital of the province. When Hospital Point was abandoned in 1920, this plant was moved and installed by Norman Sparks at the United Fruit Company office building in the city of Bocas del Toro.

There was a working relationship between the United Fruit Company and the appointed Panamanian Port Doctor of Bocas Town, Dr. Paul Osterhout, an American. The United Fruit Company manager, S. G. Shermerhorn, wrote a letter dated March 7, 1906, offering kerosene to Dr. Osterhout to kill mosquitos in Bocas. It is not known when Kyes replaced Shermerhorn as manager.

MEDICAL REPORTS

- In 1903, 122 cases of smallpox were isolated and 7,000 people were vaccinated by United Fruit Company doctors, which ended an epidemic.
- The last case of yellow fever was recorded in 1906.
- The Annual Medical Report of 1908-09 showed that 582 white patients were admitted to the hospitals and nine died. This compared with 3,155 colored patients admitted and 91 died. A third of the colored patients admitted had malaria. In that same report, 18,508 treatments were recorded at the dispensaries, which included an outpatient dispensary on Hospital Point.
- Of all the admissions listed in the report, patients came from 55 countries, plus other miscellaneous nationalities

listed for 260 patients. Of the colored patients, 2,017 were from Jamaica, 99 were from the French speaking islands of Martinique, Guadeloupe and Haiti, and 56 came from the United States of America. The remainder of non-white patients came from India, Africa, Central and South America, and many islands in the Caribbean. Of the 582 white patients, 295 were from the USA and the remaining 287 came from 36 other countries. In the operating room, 47 types of surgeries were performed on 310 patients.

• In 1912, Hospital Point had the following medical staff:

• Dr. I. W McLean, Superintendent, Baltimore Medical College.

• Dr. R. H. Wilson, Assistant Surgeon, Starling Medical College.[6]

• Dr. J. E. Smith, Pathologist, Tulane University.

• C. B. Lawrence, Matron.

• In addition to the foregoing, there were 70 employees at Hospital Point with 24 job descriptions to back up the medical staff.

• Dr. I. W. McLean reported in 1912 that the medical department in Bocas del Toro was responsible for providing health care for 275 white employees and their 200 family members. There were 6,000 colored employees with 5,000 dependents who received medical care. In summary, a total of 11,475 people were dependent on the United Fruit Company for medical services. Non-Company patients from anywhere were always accepted and accommodated.

• In 1914, the principal causes of 86 deaths were as follows: tuberculosis, 7; chronic nephritis, 12; acute nephritis, 4; pneumonia, 23; pernicious estivo-autumnal malaria, 8; pernicious tertian malaria, 1; cerebral estivo-autumnal malaria, 2; typhoid fever, 1; pellagra, 2; heart disease, 9; beriberi, 2; external violence, 4; and miscellaneous diseases, 11.

• In his 1915 Annual Report, Dr. McLean listed the six principal diseases among laborers in order of importance: Malaria, venereal diseases, pneumonia, nephritis, uncinariasis (hook worms), and tuberculosis. Dr. Swigart submitted his last annual report from New York on October 1, 1914. Dr. Deeks reported in 1915 that Dr. Swigart suddenly died (no date was given).

• In 1918, Dr. Deeks confirmed that 10 United Fruit Company doctors had been commissioned to serve in the military in World War I. Three were medical superintendents and included Dr. McLean of Hospital Point who left in 1917.

DEATHS AND BURIALS

Each year, the medical director submitted an annual report with numerous statistics about patients at Hospital Point and the clinics in the banana plantations. Such data included race, nationality, age, sex, disease, treatments, cures and deaths. These records are available in the Countway Library of Medicine at Harvard University.[7]

I calculated from these records that over 1,200 patients died and were buried at or near Hospital Point until the medical center closed in 1920. The first marked grave was 1908, and the last burial was marked August 22, 1920, a week before the hospital closed.

Burials were not accounted for before 1908, except for two marked tombstones in the German Cemetery at Macca Hill, where two medical doctors lie. This was a period when yellow fever and malaria were rampant. One doctor was only

30 years old. It is not known if the two had been employed with the United Fruit Company. Their tombstones read:

- Clarence G. Probert, M.D., Born August 30, 1871, Died June 28, 1901.
- John Eccles, M.D., Age 51 years, 5 months, 24 days. Born in Madras, India, July 24, 1850, Died in Bocas del Toro, Colombia, January 18, 1902.

Before 1908, bodies from Hospital Point were buried in the public cemetery of Bocas Town and at the German Cemetery for white foreigners at Macca Hill. The United Fruit Company cleaned and maintained the Macca Hill cemetery. Some deceased white people at Hospital Point were buried between Hospital Point and Pine Creek. Only 17 stones are legible. Depressions in the ground indicate that many more burials were made but never got grave stones. If the deceased was an employee of the Company, the surviving members of the family typically moved away. That may explain why some graves were never marked. Also, most single men did not have any family in Panama to mark their grave.

Tombstones vary greatly. One elaborate monument with a Gallic/Celtic cross reads:

> Sacred to the memory of Charles Sheridan Moseley, born at Port Antonio, August 21, 1883, died at Bocas Del Toro, April 2, 1912, R.I.P.

Port Antonio was a busy banana port on the north coast of Jamaica, and it seems likely that the booming banana industry attracted Moseley to Bocas del Toro. He was 28 when he died. Nearby is a tall obelisk stone marking the burial of J. E. Campbell, from Cuba, Alabama.

Some Causes of Death Annual Reports noted the deaths of some Americans buried in the Nancy's Cay Cemetery as follows:

- R. N. Coates. Engineer, 2 years in residence, age 38, hemoglobiuric fever, 1913.
- J. E. Campbell, overseer, 28 months of residence, age 26, killed by gunshot, 1913.
- D. F. Willis, overseer, 8 months of residence, age 38, thrown from a mule, 1914.
- T. A. Timlin was a railroad man from Pennsylvania and chief storekeeper of materials and supplies when he died in 1916 at age 31.
- Twenty meters southwest of the main burial ground are three isolated graves.
- E. D. Nash, railroad superintendent, died of heart disease on November 11, 1911.

In January 1917, two wives of key employees died in a railcar accident near the Sixaola River railroad bridge that crossed from Panama to Costa Rica. Both women were buried side by side. One tombstone reads: Mrs. Effie Louise Roberts, died January 15, 1917. She, her husband Neill S. Roberts, and their son lived in a huge, isolated plantation house high on the hill in Margarita Farm on the Costa Rican side of the Sixaola River. He was a banana superintendent.[8]

The other wife, Audrey Coul White, was married to R. E. White, who was the banana superintendent of the Guabito farms on the Panamanian side of the Sixaola River. They lived in a spacious house in Guabito. Bruce Brown remembers watching Mr.

White come on weekends and lie down on his wife's grave and tearfully talk to her. Bruce observed the sadness by hiding in the bushes outside of the cemetery fence. When the bereaved husband left Panama, her body was exhumed and taken elsewhere.

BURIALS AND THE COLORED CEMETERY[9]

Coffins were built at the Hospital Point workshop by Norman Sparks, "Father" Lewin and Sam Joseph. Contracts for digging the grave and marking the burial were given to Sam Galero, Peter Hudson, Alfonso Ponier and Dawson, who earned five dollars per burial. Originally, a coffin was put on a two-wheeled cart and pulled by two men down the trail through Nancy's Cay Village to the colored cemetery just beyond Diego Point. However, local residents petitioned the Company to change the route because of the smells and unpleasantness of dead bodies passing by their homes so frequently. The Company agreed and the cart was replaced by a flat, open, copper-bottomed barge, built on a wooden frame. The boat was known as the Dead Skiff, and it held two coffins which were loaded from the cargo dock by the morgue. A pair of oarsmen rowed the skiff around Hospital Point to Diego Point.

Burials in the Company-owned colored cemetery were made in the clay hills above the mangrove shore beyond Diego Point. Many of the dead were bachelors and did not have families in Panama. Each grave was marked with a wooden stake with an identification number. The number referred to a file kept in the hospital office. Rot and termites destroyed the markers in less than a year.

Today, there is no sign of the colored cemetery. Hundreds of indigenous people now inhabit the entire burial ground and are not concerned that their dwellings are on top of 1,200 graves. Originally, some burials were marked with crock bottles and conch shells but those are long gone, too. The community where the graveyard was is now called Aldea de Solarte.

END OF AN ERA

The 1919 Annual Medical Report showed architectural plans for a new hospital in Almirante. Hospital Point had become obsolete and remote because the banana industry had shifted to Changuinola and both sides of the Sixaola River. It took sick patients from the far-off Talamanca Valley in Costa Rica a day by train and boat to get to Hospital Point.

On September 1, 1920, the Hospital Point Medical Center was closed by Dr. F. C. Watson of Vanderbilt University, who then opened the new hospital in Almirante. Later, he was replaced by Dr. Otto Brosius.

Bruce Brown reported that the dozen or so buildings on Hospital Point were dismantled and removed over a period of two years. That gave local residents from the nearby islands the opportunity to gather on Saturday nights in the abandoned big, white hospital where they could socialize, dance and listen to jazz music. There were several good musicians who played New Orleans-style jazz. Bruce Brown said there were always big turnouts for these events.

After the buildings were removed, the jungle quickly grew back over the point of the island. Phyllis and I bought the property in 1970, cleared it, built a home and retired there in 1991. A fitting end to that excursion in 1966, when my friend Teddy told me of a mighty medical facility that once stood at Hospital Point.

A virtual lifetime of research and discovery resulted in a bilingual booklet I wrote in Spanish and English, *La historia* *de Punta Hospital * History of Hospital Point*. Now both versions are availabe as individual titles.

Hospital Point, circa 1914.
Colorized and improved by Ken Huff.
Photograph: Public domain.

Hospital U. F. Co's

21. U. Fruit Co. Hospital – Doctors Residence & Wards.
Panama Studio Dockrell & v. Wedel, Bocas del Toro, R. P.

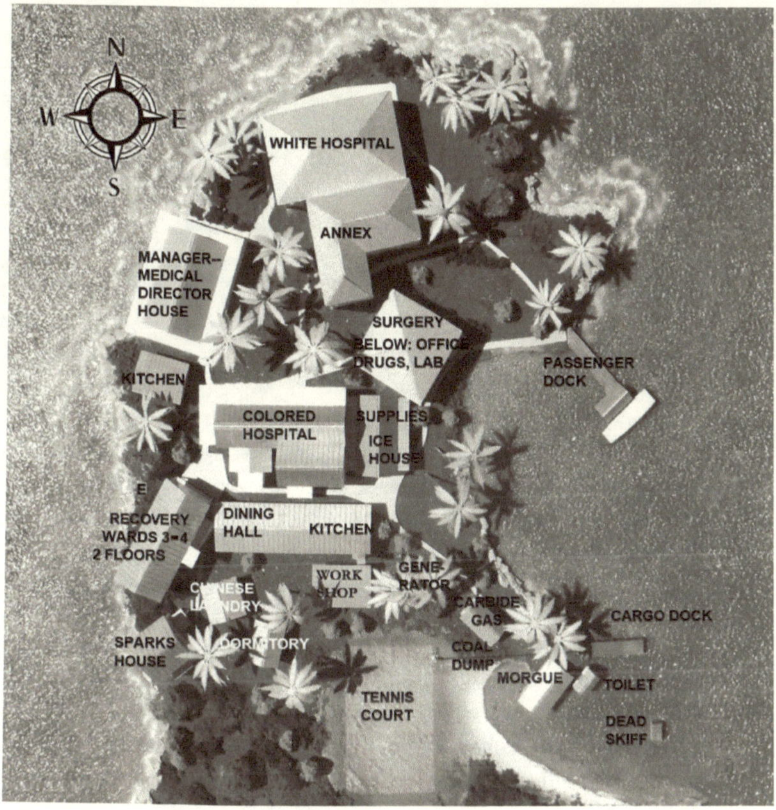

Model of Hospital Point as it would have looked in 1920.
Model design: Clyde Stephens.

Opposite Page: Top: In 1920, this hospital on Quarantine Island replaced a smaller hospital previously built by Minor C. Keith and the Snyder brothers in the late 1890s. Photograph: F. E. Read.
Linda Baker Lawrence collection.

Center: Laborers Hospital (left rear). Manager's house and later medical director's house (right). Photograph taken from the White Hospital porch on the point between 1911 and 1916. Clyde Stephens collection.

Bottom: New Surgery Building in 1914. Underneath was an office, a pharmacy and a medical lab. Clyde Stephens collection.

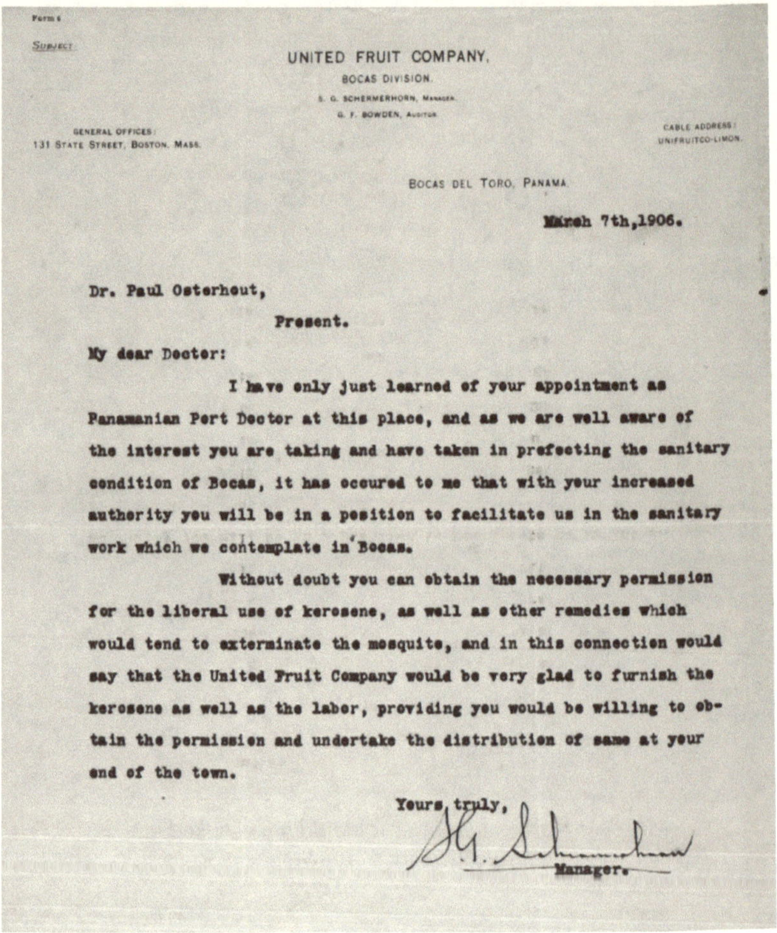

Form 6

Subject

UNITED FRUIT COMPANY,
BOCAS DIVISION.
S. G. SCHERMERHORN, Manager.
G. F. BOWDEN, Auditor.

GENERAL OFFICES:
131 State Street, Boston, Mass.

CABLE ADDRESS:
UNIFRUITCO-LIMON.

Bocas del Toro, Panama.

March 7th, 1906.

Dr. Paul Osterhout,

Present.

My dear Doctor:

I have only just learned of your appointment as Panamanian Port Doctor at this place, and as we are well aware of the interest you are taking and have taken in prefecting the sanitary condition of Bocas, it has occured to me that with your increased authority you will be in a position to facilitate us in the sanitary work which we contemplate in Bocas.

Without doubt you can obtain the necessary permission for the liberal use of kerosene, as well as other remedies which would tend to exterminate the mosquito, and in this connection would say that the United Fruit Company would be very glad to furnish the kerosene as well as the labor, providing you would be willing to obtain the permission and undertake the distribution of same at your end of the town.

Yours, truly,

S. G. Schermerhorn

Manager.

In this letter from 1906, Manager Schermerhorn emphasizes the importance of limiting yellow fever and malaria via mosquito control. Clyde Stephens collection.

Opposite Page: Top: The main living room in the manager's house where Rebecca Swartz was a housekeeper. Bruce and his mother lived in a room below. Clyde Stephens collection.

Center: Nurse in staff quarters. Clyde Stephens collection.

Bottom: L-R: Believed to be Sheppard Candy Schermerhorn (manager), John Morgan Kyes (manager), his son and his wife. Clyde Stephens collection.

Company launch at Hospital Point dock, circa 1910. Dr. I. W. McLean is at the far left. There were three scheduled trips daily to Bocas Town. Clyde Stephens collection.

Morgue, behind the public toilet over the water, handled over 1,200 bodies up to 1920. Cargo dock (far right). Clyde Stephens collection.

Opposite Page: Top: E. D. Nash, railroad superintendent, at the controls of a railcar in Almirante, circa 1908. Clyde Stephens collection.

Center: E. D. Nash's home in Empalme. Clyde Stephens collection.

Bottom: E. D. Nash's home (far right). Clyde Stephens collection.

50

24. United Fruit Co. – Residence Genrl. Supt. of Railway.
Panama Studio Dockrell & v. Wedel, Bocas del Toro, R. P.

Bruce Brown (left) and his half-brother Martin Sparks at Hospital Point, 1995. Clyde Stephens collection.

Leovigildo Julian Lorenzo restored Moseley's tombstone that tree roots had badly damaged. Clyde Stephens collection.

6. With C. H. Ellis through Central America and Panama, 1906, by Ed Loyd

OVER THE COURSE OF nearly a decade at Chiquita, I was fortunate to have a role that brought me into contact with numerous Company stakeholders, including the large, multigenerational group of *bananeros*. As the head of Corporate Communications and Investor Relations, there were always interesting challenges to be met. Yet, some of the most interesting had nothing to do with the business of the day. From time to time, the opportunity to learn more about Chiquita's early lore and history would arrive in the mail. When Chiquita moved its corporate headquarters to Charlotte, North Carolina, the widow of one Company veteran in the area donated her husband's collection of more than 120 glass banana stands. Other opportunities were missed—I recall the offer to acquire a significant amount of memorabilia and material from a descendant of Captain Lorenzo Dow Baker or to bid on a special auction in New Orleans of Great White Fleet items. Neither of these came to pass.

That is what makes the generosity of Ed Fendley, Jr., of Omaha, Nebraska, so special. Like so many others, he had a 47-year career as a fruit-and-vegetable broker, serving 42 of those years with C. H. Robinson & Co. He presented to Chiquita a specially printed photo album in honor of his father titled: *With C. H. Ellis Through Central America and Panama, 1906*

Quarantine Tour of Central America and Panama by Health Authorities as Guests of The United Fruit Company.

The album, which still contains the business card of Crawford H. Ellis as manager of the New Orleans Division of the United Fruit Company, had been a prized possession of Edward R. Fendley, Sr., who "lived his life for the banana industry," his son recalled. During Fendley's career in bananas he had been the manager of the Chicago office of the Weinberger Banana Company, led by Charles Weinberger, who was a strawberry-and-banana pioneer and the uncle of Sarah Weinberger Zemurray, the wife of Sam Zemurray. Afterwards, Ed Fendley relocated to Omaha, where he was for decades the branch sales manager for United Fruit Sales Corporation, a subsidiary of the United Fruit Company. Whether through Weinberger or another *bananero* connection, it is unknown how this rare album came into his possession.

The mission of the trip was succinctly recorded as:

> During the yellow fever epidemic at New Orleans in the summer of 1905, the citizens were very much enraged with the Fruit Companies, and denounced them as being the cause of the introduction of fever, believing that the fruit ports were all in a more or less unhealthy condition, and that the yellow fever prevailed there practically the entire year.

It was to disabuse the minds of the public of this impression that the United Fruit Company invited the health authorities of Texas, Louisiana, Mississippi and Alabama, and the Marine Hospital Service, to visit the different fruit ports and make a report on each place and see the progress which had been made at the different points to guard against fever epidemics. To their great surprise, they found the health conditions at the majority of the various ports to be most satisfactory:

> The object of the United Fruit Company in extending this facility was, in the first place, to preserve its business, and in the second place, to save New Orleans the immense amount of commerce between New Orleans and Central America, so as to avoid any sacrifice of this business in the future and cause it to be diverted to Eastern Seaboards.

Along with these medical leaders, several prominent board of trade and commerce officials were in the party led by Crawford H. Ellis—who would go on to serve for decades on United Fruit Company's board of directors and was the founder and president of the Pan-American Life Insurance Company. Always pioneers in public relations, the sailing party included members of the press corps and John N. Teunisson, as the official photographer and compiler of the book that was specially bound and presented as a memento to the 24 members.

The party sailed from the dock at the foot of St. Joseph Street in New Orleans at noon on January 20, 1906, on the SS *Anselm* with Captain Louis Rivara. They arrived in Limon, Costa Rica, the morning of January 26. This first part of the voyage encountered four days of bad weather—all the way to Swan Island by which time most had been suffering from seasickness, yet were still in good spirits. Dense fog had even forced them to anchor on their first day when reaching the lower Mississippi. Over the following weeks, they toured through Bocas del Toro and Colon, Panama; Bluefields, Nicaragua; La Ceiba and Puerto Cortes, Honduras; Belize; and

Puerto Barrios and Livingston, Guatemala, before returning to New Orleans on February 15. They arrived in Bocas del Toro at first light on Tuesday, January 30, 1906, and toured Bocas and the Quarantine Hospital and departed that same evening for Colon.

The following description was published in the album:

> Bocas del Toro is a town of about 3,000 inhabitants, and is situated on an island in Almirante Bay. The United Fruit Company and the Government of Panama are now engaged in raising the level of the town to about three feet above the tide water, and are putting in a system of tide drainage, which flushes the sewers twice a day. All cisterns in the town are well screened, and none are allowed to be erected without such provision. When the present sanitary improvements are completed, Bocas will be one of the healthiest places in Central America. The United Fruit Company has a very fine hospital on an island in the Bay, which has natural drainage. The ships do not go to the wharf at this point, but load in the harbor, about six miles from town; therefore, the danger of the ships bringing the fever from this port, even though it should become infected, is extremely remote.

The *Times-Democrat* of New Orleans on February 10, 1906, reported the following on the conditions in Bocas:

> Heretofore Bocas del Toro has had no quarantine regulations, and has had no local system on inspection or sanitation. The Republic of Panama had looked with jealousy on any steps taken by the Panama Canal Commission to extend authority over this port, and had even refused to permit the United Fruit Company to do what it would have liked toward making the town clean and habitable.

The latter Company has already taken whatever steps it possibly could toward protecting the health of its own employees, and maintains a splendidly located and equipped hospital on a cay situated a mile and a half from the main island on which Bocas is located. There [the patients] are cared for in the most approved manner; all employees of the fruit company, either from the town of Bocas or from any of the sugar plantations in the back country that are controlled or operated. In charge of that hospital is R. E. Swigart, formerly from Ohio, who has been in the Tropics in his present post for several years. On shore, the United States Marine Hospital Service has an acting assistant surgeon in the person of Dr. Austerhaus [sic], who is a veteran of the Tropics (Osterhout). His work is purely advisory, and is limited to informing the American authorities of the health developments on the island or near the mainland.

The following account of the day was part of the coverage of the *Times-Democrat*:

It was barely daylight Monday morning when the *Anselm* arrived in the anchorage off Bocas. By 7 o'clock S. G. Schermerhorn, local superintendent for the United Fruit Company, arrived at the ship's side in a natty navy launch. There were no formalities. There wasn't any quarantine officer in sight to examine the ship, and there was not evident any customs official to see that nothing was smuggled in.

By this time, the party on the *Anselm* was all on deck. The ship was lying in between the main island of Bocas, with Careening Cay on the outer side. There was barely a toss to the water, and the most squeamish individuals were able to look on the waves without an unpleasant sensation.

Bocas could be seen, half a mile away—stretching in either direction a long line of one- and two-story houses on stilts above the water—their entrances looking into the front street of the town skirting the water front. Cisterns could be seen in every premises, and the half-completed and then neglected seawall broke the spaces between the houses. Everything was wooden, everything looked temporary, everything looked barely of yesterday, raw, crude and apparently filthy. This much could be seen even from the vessel.

Then the party went ashore in launches; Mr. Schermerhorn, another specimen of young American manhood in charge of great interests, guided the party through the Company's storehouses, salesrooms and offices. These were all wooden, all on stilts over the water, all temporary, and destined to be replaced by new, staunch, handsome edifices when the town should be filled in and made anew.

After a brief survey of the town, the entire party took launches for Nancy's Cay, where the United Fruit Company's hospital is maintained. Here was a different world. The hospital is located on a bluff headland of the island looking toward the outer exit from the main bay and catching on the rocks beneath it a rolling surf that palms fringe the rocks, and behind the hospital ,up stand hills covered with tropical verdure but cleaned away beneath to prevent the hiding of insects of noxious sort. The hospital has wards for males and for females, for whites and for blacks, for infectious or contagious diseases, and for fever patients. The establishment has three or four separate buildings, for its different uses, is scrupulously clean, and is said by the physicians of the party to be conducted as science would dictate in every apparent way.

One objection was raised. The wire screening was only fourteen14-to-the-inch mark instead of 18 as required in New Orleans. Dr. Swigart said that this had been the only wire obtainable when the hospital was built, because the Federal authorities in Colon seized every foot of wire available in all the section. He said, however, that the 18-inch wire would be obtained and used about the hospital. At the hospital Dr. Swigart entertained the party at luncheon, Mrs. Swigart, her mother and Mrs. Herbert, the wife of one of the plantation managers from the Company, received the visitors. During the evening these ladies, with Mrs. Flynn, Mr. Flynn, Mr. Schermerhorn and several other attachés of the Company were guests at a dinner on board the *Anselm*. The opinions of the physicians following their day in Bocas could be summed up by the commentary of Dr. Goldthwaite of the Mobile Board of Health who stated:

> Bocas is at present unsanitary in the extreme. Every condition there is favorable to an outbreak of fever if the disease is once introduced there. However, the situation may be materially changed when the present plans of improvement, carrying an expenditure of $223,000, have been brought to completion.

To this, Dr. Anders of Mississippi added:

> There is a lack of cooperation between the citizens of the town and those officers of the United Fruit Company who have sought to bring about a better condition of affairs. However, the plans said to have been agreed on are most excellent, and when these are carried out, and the proper quarantine and sanitary code is established according to the agreement entered into by President Amador in Panama—then I believe Bocas will be just as excellent a little port as Port Limon in Costa Rica.

Dr. John N. Thomas of Louisiana noted:

> The hospital arrangement of the United Fruit Company to care for all the sick on their properties or in their employ are most excellent, and with the care the Company exercises on all their own properties, I do not fear any contagion from that source.

When their tour was complete, the entire commission submitted very positive reviews to each of their states and published an editorial that summarized their findings. They determined some ports, like Limon and Bluefields were "practically clean territory with which trade might be carried on virtually without restriction." For other areas, such as Bocas, they gained assurance of the improvements being made:

> President Amador, of Panama, agreed to appoint any officer detailed by the United States Hospital Service, as the sanitary officer of the Bocas del Toro district, giving him full power as quarantine officer and local health officer. This, with the improvements planned for immediate execution, should make Bocas safe and clean by the summer of 1907.

The goodwill that the tour achieved, accomplished "worthwhile definite results, and further, to have begun a general movement that may result in the sanitation of all Central America."

As for the album itself, it remained in the Chiquita corporate archives when the Company was acquired in 2015 by the Cutrale family, so it may have been moved to their new headquarters in Ft. Lauderdale, Florida, or to Switzerland, along with some ledgers of the Snyder Brothers Banana Company that I regret not copying.

The SS Anselm was one of the early stately ships of the Great White Fleet, being part of the fleet since the 1899 merger that created the United Fruit Company. It was on the Anselm that Bessie Preston, the daughter of Andrew Preston, created the red and blue burgee with a white diamond in the middle that would become the house flag and emblem on all of the Company's ships. Photograph: John N. Teunisson.

The entire medical party and civic officials on the deck of the SS Anselm on the day of departure down the Mississippi, January 20, 1906. Crawford H. Ellis is seated in the front row, center. Photograph: John N. Teunisson.

Aboard the SS Anselm at Bocas, January 30, 1906. Crawford Ellis, who led the tour for the United Fruit Company, speaks with Bocas manager S. G. Schermerhorn. Photograph: John N. Teunisson.

Admiring the view from Quarantine Island, as they called Hospital Point during the early years. The entire party prepares to return to the Anselm following a tour that impressed them with the caliber of the hospital facilities to care for the patients. Photograph: John N. Teunisson.

The entire party along the sidewalk of Hospital Point, January 30, 1906. Photograph: John N. Teunission.

7. Unforgettable Memories at Hospital Point

OUR FIRST CAMPING TRIP

In April 1971, six months after we acquired Hospital Point, we loaded our *cayuco* with a 25-horsepower Evinrude outboard engine and motored from Changuinola to Hospital Point in about two hours. Torrential rain pelted us through high waves across the Almirante Bay from Boca del Drago. At Hospital Point, there was no harbor and no dock or facility to tie the boat, except an iron pipe that I had driven into the marine rock with a sledgehammer. We waded ashore through the waves, taking items to set up camp.

The campsite was an old concrete slab where a hospital building had once stood in the early 1900s. Thick jungle had been removed so I could set up a canvas cover on top of several wooden poles. We tied guy-lines to anchors in the soil. After a couple of hours, we got in the tent and listened to the rain and wind outside. Linda was six and Carl was five, and this was a big adventure for them. It was hard work for Phyllis and me.

Phyllis cooked on a one-burner, hand-pump kerosene stove made by BAT in Sweden. I laid pieces of concrete on the bottom of the canvas so water would not flow onto our floor. It was *Semana Santa* (Holy Week), and we had a three-day weekend to enjoy. It was rainy and windy for the entire three days. When the rain subsided for a few minutes, Linda and Carl would dash out of the tent and run and yell like wild animals until the next downpour of rain came, usually within a few minutes.

When bedtime came, we changed to dry clothes that were stored in a waterproof bag. That made sleeping on air mattresses reasonably comfortable.

During those three days, we did no hiking, no swimming, no snorkeling and no fishing. I forgot about the work tasks that I had planned to do.

On the third day, we packed and returned home to Farm 8 in Changuinola. Rain, winds and waves never let up until we entered the old banana canal at Boca del Drago. It was a relief to ride on the calm water of that historic waterway—some 11 kilometers long.

The canal was calm, but crossing the flooded Changuinola River was precarious. The chocolate-colored river was churning with logs and debris, but our boat crossed the torrent without a problem.

Just the year before, in 1970, there were five major floods, including a flood in April that suddenly cascaded underneath our house. The water level rose a full meter high for several days but we remained safe upstairs in the stilt house. Hundreds of Company houses were flooded up to the roofs, people drowned, bridges and railroads were damaged and millions of banana plants were severely affected. Lucky for us, Hospital Point is safe from floods because the center of the former tropical medical center is eight meters above sea level.

RISING FROM THE DEAD

Mr. Enoch Flemming was born in Jamaica, but his father

brought him to Bocas del Toro when he was a boy in 1910. Like thousands of others from the Caribbean islands, his father worked for the United Fruit Company. Mr. Flemming lived in a shack beyond Hospital Point. When he visited us to do chores, he told me about deaths and burials of patients at the hospitals. One of his close friends was Robert, who died at Hospital Point.

On a sunny day, Mr. Flemming and I were cleaning the little beach at Hospital Point. I could tell that he wanted to tell me a story, so I stopped working and listened. In the most dramatic Jamaican accent, I learned about the return of his deceased friend, Robert, who slowly emerged from the water on that little beach we were standing on. He told me how he became immobilized for a moment, then started trembling uncontrollably. Robert came out of the sea and gently asked his old buddy how he had been, but Enoch was too paralyzed to talk. The visit was not very long, and Robert slowly and calmly walked back into the sea and disappeared. Enoch Flemming never saw Robert again. He summarized the experience with the following: "*Laud mi Gad, Mista Stephens, mi a tell yuh, mi see all a dat yahso pan dis beach.*"

Just a few steps away was the site of the Hospital Point Morgue that closed in 1920. This is where about 1,200 bodies had been held over a period of two decades. When a grave was dug and ready in the cemetery, the cadaver was loaded onto the Dead Skiff and rowed around the tip of Hospital Point. The burial ground was owned and managed by the Company and was farther along the south shore. Mr. Flemming and Bruce Brown showed me where the hundreds of burials had been made. The location was a shock. The entire graveyard was underneath the indigenous Ngäbe community known as

La Aldea de Solarte. This happened because newcomers had built their dwellings on top of the graves that were either unmarked or marked with bottles or conch shells. Mr. Flemming never told me if his friend Robert was resting there or if he got a watery grave instead.

OTHER GHOSTS AT HOSPITAL POINT

Virgilio Pineda left his home in Puerto Armuelles after he graduated from high school and flew to Changuinola in an ATOPAN spray plane. In 1990, he was 18 and joined the construction crew of Ariel Flores, who was building our retirement house on Hospital Point. After the house was finished the next year, Virgilio remained as our caretaker. At night, he played cards and parlor games with Phyllis and me.

Hector lived in a humble room at Solarte Village where he was a teacher in the local school that had grades one through six. The young teacher had no running water, no bathroom, no toilet, no electricity or lights. To relieve himself, he walked the board plank through the mangrove swamp along the edge of Almirante Bay, but those conditions were the same for hundreds of other Panamanian school teachers, who were assigned to teach in outback communities in the country.

Sitting alone in his one-room abode at night inspired him to visit us at Hospital Point, especially about supper time. After supper, he played cards and other games with us. When it was time to return home, he walked about 10 minutes in the dark, sometimes in the rain on a muddy jungle trail. He had a flashlight, but the walk was scary and precarious when the batteries failed. At times, he heard sounds, like bats swooping erratically along the narrow trail. And there were more spooky noises that rustled in the dense jungle as he nervously walked to the quiet village. Hector knew about Hospital Point cemetery not far from the trail and this was not at all a comfort to him. His nerves and his state of mind intensified each night as he walked alone on that ghostly trail back to the village. His fear reached a peak, and he asked Virgilio to walk with him, at least some of the way home. Virgilio heard those same sounds and did not feel comfortable either. After some time, Hector told Virgilio that he was convinced that spirits from the dead in the cemetery were haunting them.

Virgilio revealed this spooky situation to Silvia Chong, a devout Catholic. She talked with the Catholic priest about the haunting spirits coming from the dead in the cemetery, and he agreed to deal with this malady through some sort of divine intervention.

Virgilio brought the priest, in full religious regalia, to Isla Solarte, and they landed at our dock. He followed Virgilio on the spooky trail, and they continued to the cemetery about 100 meters on the other side of the island. The graves were on top of a hill surrounded by dense tropical forest. No one knows what the priest did to exorcise the evil spirits coming from the dead in that "white folks" cemetery, but Hector and Virgilio swore that the trail was no longer spooked. The priest had pacified the dead, banished the bad spirits and made the trail safe to walk on again, even at night. After that purification, card games at our house were so much more pleasant.

As a postscript, Hector's father visited him in 1994, but he had no place to sleep or eat, so I gave him permission to sleep in my hammock on the front porch. He paid for his keep by giving me a book of his poetry and short stories and telling me accounts about representing the Panamanian Government in North Korea, Cuba and Mexico. General Manuel Noriega was the dictator at the time, and the country was consorting with socialist countries around the world. I heard incredible accounts of how he traveled through China just to enter North Korea, traveling widely for the Panamanian dictatorship, and I learned much about international politics and intrigues.

THE PATTI PICNIC AT HOSPITAL POINT

In January 1973, Phyllis and I invited the Vito Patti family to go with us from Changuinola to Hospital Point for a picnic. Anselmo Chuito, an indigenous neighbor, had built a small open hut like he and his large Ngäbe family lived in. (Back then, they were known as Guaymi Indians.) Our native hut could sleep eight people on the upstairs chonta palm wood floor, which was protected by a tight guagara palm-thatched roof. On the lower floor, we enjoyed lounging in the cool sea breeze. The entrance was the concrete steps of the former medical superintendent's house that had been dismantled fifty years earlier. It was a beautiful day, and I was explaining the history of the old medical center to the Patti family. Suddenly, we heard someone calling from a boat. I walked down the hill and found Jose Thomas riding the waves in his boat beyond the reef. Jose was the owner of the former United Fruit Company office building in Bocas Town and had converted the old structure into the Gran Hotel Bahia.

Most of the hotel rooms stayed empty because almost no one traveled to Bocas del Toro at the time. But one room was always occupied day and night by what remained of the Tropical Radio and Telegraph (TRT). Bocas del Toro is the capital of the province and government officials made it mandatory that the United Fruit Company maintain Tropical Radio and Telegraph service in the Hotel Bahia where it had always been. This mandate by the

Panamanian government was made because there was no other way to communicate to or from the island.

Jose Thomas suspected that I was with Vito Patti and he delivered an urgent message to Hospital Point. I waded in the waves to the boat. He told me the message was for Vito Patti, who should go immediately to Bocas and communicate with the manager's office in Changuinola via Tropical Radio and Telegraph. Jose was not sure but he quietly told me, "mi tink di ole man dead yuh kno." He was referring to Vito's father, Vicente Patti, a respected Italian who had retired from the banana company and lived in Progreso, Honduras. I told Vito to go with don Jose to Bocas.

Vito's wife was from Mississippi and she was waiting apprehensively on the shore with me. When Jose returned, she saw the pained face of her husband and instinctively knew that her father-in-law had died. Instantly, she dashed into the water, fully clothed in a dress, and ran toward the boat. Vito jumped out of the boat and ran through the waves toward his wife. They met in the middle and passionately embraced in knee-deep water. Both were crying uncontrollably. He repeated over and over, "Daddy is dead, Daddy is dead."

I was on the shore nearby and choked up with emotion. As the host, I quickly terminated our island excursion and we returned to Changuinola with the Patti family in our boat. It was a sorrow that I will never forget—and a scene that Hollywood could never duplicate. RIP *don* Vicente.

Enoch Flemming (left rear) and Phyllis Stephens (far right). Hospital Point Beach, 1968. Cayuco: El Robalo. Clyde Stephens collection.

Stephens family at Hospital Point native hut, 1971. L-R: Front: Clyde's parents, Carl and Era Mae; Carl, Linda and Clyde. Rear: Clyde's brother Truitt, Joann and Phyllis. Photograph: Anselmo Chuito.

8. Natural Entertainment at Hospital Point

WHEN PHYLLIS AND I retired from Chiquita Brands International in 1991, we started a new life in the first solar-powered house near Bocas del Toro. Entertainment was provided by nature that surrounded us. At that time, there was no television beamed to the island although later some dim signals came from the high volcanic peaks of Panama and Costa Rica. But TV did not interest us anyway.

Over the years, there was never a dull moment, day or night. We enjoyed the marvels of nature around our house and in the adjacent rainforest. If it did not come to us, we only had to walk a few minutes to find plants and animals of special interest. Hospital Point was originally surrounded by pristine coral reefs only a few steps away. In those years, lobsters, crabs, fish, conchs, and octopus—but not turtles—were abundant, and even made it to our table.

After more than three decades, there are many memorable examples to write about, and the following are some of my favorites.

THE DIVE BOMBER EATS LUNCH

By pure chance, I was standing on the upper deck of our house looking at the sky. Suddenly, out of the blue, I saw a Bat Falcon with its wings folded in a swept-back position, and it was pointed toward the earth at a ballistic speed. In a split second, it made contact with its target—a barn swallow. Feathers flew and the little migrant was carried away for lunch except for one wing. That wing was severed and spiraled toward the ground like a gyrating helicopter blade. All of this drama gave me an adrenalin rush. When I found the wing, I realized it was mostly feathers and not a great loss for the small raptor. Bat Falcon is an appropriate name because it also eats bats at dawn and dusk.

WAKEUP CALL AT DAYBREAK

Hospital Point is the wrong place for late sleepers. At the crack of dawn, Red-lored Parrots fly to Hospital Point and direct their raucous noise and screeches right to your pillow. Usually it's time to get up anyway. These birds are colorful and very chatty. I have counted over 50 in some flocks. They are attracted to Royal Palm seeds and other fruit in our yard, and they come almost daily.

Since 2016, the Yellow-headed Caracara, a new vagrant raptor in Bocas del Toro that is completely out of its range, has added to the early morning wake-up call. A pair that nests on Hospital Point produces two offspring each year, and they contribute to the morning cacophony.

An indigenous lady enticed Phyllis to buy her tame parrot. We named the bird "Lorita" and she or he lived in the giant mango tree by the house for nearly ten years. When Lorita walked down a mango limb to the feeding tray, there was always food waiting. During one of our daughter Linda's visits, she looked up in horror when she saw the parrot walking on a mango limb right toward a giant boa constrictor that was stretched out close to 2.5 meters on the same limb. The big snake was waiting for breakfast and only had to open its mouth to swallow the parrot. Linda yelled and grabbed a long pole and urged the bird to go back.

After some quick thinking, we found two long, forked poles and manipulated the heavy boa off of the mango limb and released in the rainforest where it belonged. Lorita got the green light to continue on her morning walk to the feeding tray and never knew how close it came to being breakfast.

MONTEZUMA OROPENDOLA ALIAS EL CHACARERO

If the parrots don't wake you up, oropendolas will. They visit Hospital Point almost daily. Flocks of over 40 are common. These birds are omnivorous and eat lizards, frogs, insects, mangos and thousands of royal palm seeds. They forage in our big mango trees and find an abundance of food a few steps from the house. The species is a large, colorful and very comical bird. Their whimsical sounds are loud, bubbling, melodic notes that vacillate up and down the musical scale, followed by a funny gurgle. They are amusing when they pivot their colorful heads toward the ground and point their brown and yellow tails toward the sky.

These birds make many long, woven nests in the same tree. In Spanish, a Montezuma Oropendola is known as a *chacarero* because their long, hanging nests look like the male scrotum. Nests are constructed at the fragile tips of limbs so that heavier predators cannot get to the eggs and chicks. Watching these birds forage and put on their shows is some of the best entertainment that nature can provide.

THREE-WATTLED BELLBIRD THAT DOES NOT RING

Dedicated birdwatchers pay huge amounts of money just for a chance to see and hear the Three-wattled Bellbird. Hotels and bird guides cater to this group of nature lovers. But we see and hear them for free in the tall trees in our yard. The sound is unmistakable, but there is no hint of a bell.

When I bought a recording of the sounds of *Birds of Panama*, by George Angehr, I told a visiting Costa Rican ornithologist that our bellbirds at Hospital Point did not sound like the one on the recording. After considerable discussion, we found out that George Angehr used the sound of a Costa Rican bellbird from Monteverde. That population sounds different. The mystery was solved.

The Costa Rican bellbirds do have the reverberating resonance of a bell if you have an extreme imagination. But the bellbirds in Panama have a different and indescribable call. George Angehr told me that sounds vary from one population to another.

The female is nondescript but the male is a spectacle. He has three long wattles on his head that dangle like plastic earthworms. Sometimes, he shakes these wattles from side to side to make a display that gets attention. At other times he sticks his head between his legs, pauses, then comes up with a silent, wide-open mouth that is black inside. A few seconds later, a loud scratchy reverberation is emitted, occasionally followed by a harsh "irkh" in two frequencies. Even with the wildest imagination, none of these sounds resemble any kind of bell. Nevertheless, I hope they keep on chiming right near our house.

A TALE OF ONE HUMMINGBIRD

During more than 30 years, we have recorded seven species of hummingbirds near our house. The most common is the Rufous-tailed Hummingbird, *Amazilia tzacatl*. Sometimes five or six of these iridescent green and brown birds fight in vicious competition to hold a place at the feeder.

I marvel at how newly hatched chicks in the nest survive drenching rains when the adult is away looking for food. They build their nests under a few leaves, but heavy rains drench the two naked chicks waiting for their next meal, which consists of nectar in a juicy soup of small spiders and soft insects. This nutritious blend is regurgitated into the throats of the tiny birds. It is easy to watch this activity from selected spots near the house.

As a disciplined banana man, I checked my weather station every day at seven in the morning just as I had done at times during my career. During a walk to the rain gauge ten steps from the house, I noticed a flurry of excited hummingbirds buzzing around a small boa constrictor in a bush along the path. In a micro second, the snake snatched a hummingbird out of the air right before my eyes. What a rush I had!

I got the camera and recorded the next maneuvers. First, the constrictor coiled around the tiny bird to kill it. Then, the boa slowly maneuvered the body of the bird so that it went headfirst down the throat. This took several minutes before a bulge in the belly of the boa was obvious. Phyllis and I had our breakfast on the front porch and watched the reptile begin to digest the hummingbird. We knew that we were fortunate to be near the top of the food chain. It was consoling to know that we were not part of the law of the jungle that was unfolding just a few steps from our breakfast table.

DANCE OF THE MANAKINS

Isla Solarte has two species of manakins. The Red-capped Manakin is an incredible showman that is active near the house, but the most common species is the Golden-collared Manakin. Each male competes in a wild dancing ritual that must convince the female that he is the one to pass his genes to her. To do this, he must build what I call a dance floor, better known as a lek, a chosen spot on the ground in the shaded forest near several other leks. Each male has his own private lek, and each dance hall is groomed immaculately by cleaning all the leaves and debris from the floor of the forest site so that the ground is bare. He selects a place where there are small bushes with thin twigs.

When a female flies to a community of perhaps five leks, she perches on the sidelines and waits for the show to begin. This is the moment the males have been waiting for and they start performing. Each jumps to the ground of his private lek, then excitedly perches on the side of a tiny bush, followed by rapid noisy wing snapping and bill clacking. He displays a beautiful golden beard under his chin and breast and follows this with other antics. The competition is wild, but the female finally chooses her next lover. The transfer of sperm only takes a second or so and it is all over. I am glad I am not a manakin.

A GREAT MYSTERY OF LIFE

One of the great mysteries in the American Tropics has been the unexplained migration of *Urania fulgens*, aka the Urania Swallowtail Moth. It looks like a swallowtail butterfly, but it is a fast-flying moth that travels during the day, sometimes by the millions. Their wings are black with brilliant green metallic bands. Thomas Belt, an Englishman who was a gold mining engineer in Chontales, Nicaragua, observed this moth migrating during his time there from 1868 to 1872. His book, *The Naturalist in Nicaragua*, detailed the moth's erratic and mysterious flights. He was mystified.

An explanation was not found until one of the world's greatest natural scientists, Dr. Neal Smith, decided to solve the mysteries about this moth. His complicated studies started in 1964. In one of his early publications, he reported hundreds of millions of the moths migrating south except in Panama, where flying east is necessary in order to get to South America.

Smith started from the beginning and analyzed the chemical contents of the egg, larva, pupa, adult and the host plant of the larva, which is a common vine, *Omphalea diandra*. They all had one thing in common, and that was a toxic compound that is produced by the vine. This was an "Aha!" moment for Dr. Smith, and he theorized, then later proved, the following biological mysteries after many years of research:

> When *Urania fulgens* larvae weaken the vine by feeding on it too long, the plant defends itself by producing a toxin that makes the insects sick and later kills many of them. Other moths develop their own defense by migrating elsewhere to Omphalea vines that have lost their toxicity. It is an incredible "cat and mouse" game that works out for both the insect and the host plant.

When Phyllis and I married in Honduras in 1963, the United Fruit Company transferred us to Changuinola. Fishing in the quiet lagoons of the Changuinola River was wonderful. In 1964, we were awed by the tens of thousands of *Urania fulgens* adults that were flying east across the waterways where we caught snook and tarpon.

My attention turned from serious fishing to moth migration. I asked Phyllis to use a butterfly net and collect moths from the front of our wobbly *cayuco*. Some moths landed on the water as if to rest. These individuals were easier to catch. It turns out that they were all female moths heavily laden with eggs. When Phyllis attempted to catch a specimen in the water, the moth would suddenly take off like a duck.

After we were transferred to Golfito, Costa Rica, in 1973, Lacsa Airlines cancelled our flight from Golfito to San Jose because of awful weather. Consequently, Fabio Campos, the new general manager, approved a special vehicle to take several of us travelers by road to San Jose via the Pan-American Highway. I will never forget the thousands of *Urania fulgens* struggling to fly across the cold, windy mountains as we headed cautiously across *Cerro de la Muerte* (Death Mountain), an alpine ridge 3,341 meters high. This behavior was not normal for most insects, but the masses of *Urania fulgens* were determined to cross the highest stretch of the Pan-American Highway in Central America and continue down the isthmus towards Panama and maybe South America.

In the mid-1980s, I met Neal Smith in his office at the Smithsonian Tropical Research Institute in Panama City, and we discussed his studies of the Urania Swallowtail Moth. I was awed at what this scientific giant had accomplished. He asked me to report Urania migrations when I lived in Puerto Armuelles. One of my most memorable reports described a huge flock of moths that landed near our house on teak wood trees to roost for the night. The next morning, tens of thousands of these moths flew east toward Panama City, a spectacle that I reported to Dr. Smith.

After 32 years of retirement on Hospital Point, I still get excited to see local populations of these moths flitting across my yard. I hope to see at least one more massive migration heading east toward South America.

BATS IN OUR BELFRIES

It was Hollywood and Frankenstein who terrified the world with vampire bats sucking blood from humans, who have a weakness for believing such fantasies. It took a serious biology professor, Dr. Maurice Thomas, from West Palm Beach, Florida, to show my family that bats are harmless, beneficial and certainly interesting. When we helped him stretch the mist nets to catch bats at Hospital Point, everyone participated in learning the truth about these flying mammals. My beautiful daughter-in-law, Myra, has always been scared of critters, even butterflies, but after she stroked a few bats with her fingers, she gained confidence. The family found out that bats are fascinating and harmless if handled correctly.

Dr. Thomas took many measurements for his studies, then released the bats unharmed. During his several years of seasonal visits that began around 2002, he identified approximately 16 species of bats on Isla Solarte, mostly near our house on the Point. They are all interesting, but I will only comment on a few.

Uroderma bilobatum is a common bat that hunts and eats fruit at night. When we hang a bunch of bananas in a tree to ripen, these bats accept the invitation to feast on the fruits of our labors. During the day, they rest under the giant fronds of the talipot palm, *Corypha umbraculifera*. This plant is originally from Sri Lanka and is one of the world's largest palms. I planted two of these palms in 1991 at Hospital Point, and they provide a home for numerous bats. *Uroderma bilobatum* is known as the "tent-making" bat because it cuts fibers of the leaf lamina to make it collapse at an angle and form a cozy tent.

Dr. Thomas refers to these tents as "birthing chambers" because he always finds a baby bat clinging to its mother. Male bats of this species were never recorded on Hospital Point.

A nectar-feeding bat, *Glossophaga soricina*, is like a hummingbird extracting food from flowers with its long tongue. When it visits a banana flower to feed, claws on itsfeet scratch young bananas and cause bat scratch scars. I studied these bats in banana plantations for two years in Changuinola. These little bats are common on Isla Solarte and are sometimes preyed on by another bat. I refer to *Phyllostomas hastatus* that has a wingspan of over half a meter. Being omnivorous, the latter bat feeds on flowers, fruits, insects, mice, birds and other bats. I actually saw one fly into the net while chasing a nectar-feeding bat. Both were caught in the net at the same time. That was high drama. *Desmodus rotundus* is the dreaded blood feeder, but humans do not need to worry for the most part. To get a warm blood meal, the shy bat walks stealthily on all fours to find a vein on a person's arm or leg dangling from a hammock or on someone sleeping on the ground. Razor-sharp teeth cut a vein without the host even knowing it. The bat has an anticoagulant to keep the blood flowing so it can lap blood like a cat drinking milk.

Cattleman Luis Flores asked me to help with a vampire bat problem at his Flat Rock cattle ranch on Isla Colon. He wanted to kill them all, but that was not my recommendation. During those visits to Flat Rock, I learned a lot about bat colonies that roosted in the caves of the Mimitimbi River. We have the vampire bat at Hospital Point, but nobody worries about it.

POISON DART FROGS

I think *Dendrobates pumilio* is one of the most studied animals in the Tropics. A common name for this colorful group is poison dart frog but there are other names, too. The scientific name was recently changed to *Oophaga pumilio*, which describes the tadpoles as "egg eaters." This is because the mother feeds her tadpoles with unfertilized eggs. You do not want to eat these beautiful frogs because you will die. The poison in their skin derives from ants that they eat. Indigenous people in the forest learned long ago that they could poison their darts and arrows with this toxin and kill even large animals in the rainforest, including monkeys and deer.

Each island in the archipelago of Bocas del Toro has different colored frogs that vary drastically. On Isla Solarte, our frogs are a bright orange or reddish-orange. Some have small black spots. While we eat breakfast on the front porch, it is common to see one or two come out of the vegetation and jump across the floor as if waiting for a handout.

Here in the islands, we are lucky so far that a horrible fungus disease has not decimated our frogs like in other parts of Central America. So many people from all around the world re constantly studying the disease and the many complicated aspects about this little animal. I promote their studies in our forest and learn from the biologists, especially doctoral candidates working on a thesis about this amphibian. That is how we learn about nature and how to conserve these treasures that hop on Hospital Point. These frogs have also generated a big economic benefit through frog tourism—tourists pay lots of money to see them.

Much of the community was outraged that giant developers destroyed huge habitats of *Oophagus pumilio* on Bastimentos Island. As a reaction, a t-shirt was designed and sold as a protest. The name of the developed area is called Red Frog Beach, but the t-shirt called it Dead Frog Beach and showed bulldozers crushing a frog and made the eyes and tongue bulge out of its head. The shirt made a point.

THE URGE TO KILL

Brown Pelicans amuse us as they fly back and forth along the Point. To add to this event, a not-so-amusing phenomenon takes place on occasion. A couple of dogs living by the dock go berserk, chasing the big birds that fly along the rocky shore. The dogs leap high into the air, hurting themselves when they fall off the cliff and hit rocks below. When the pelicans land to rest on tall coconut trees along the cliff, the dogs leap as high as they can under the palm and bark in a frenzy. The sophisticated pelicans sit on their high perch and look down scornfully to witness one of the most mindless forms of animal behavior that I know of.

BATTLE OF THE BUMBLE BEES

There is an orchid species that grows on trees near the house. *Catesetum viridiflavum* has beautiful green yellow flowers that are either male or female. Male flowers produce an alluring perfume that attracts only one species of bee, *Eulaema cingulata*, a bumble bee. The insect does not come for nectar because there is no nectar in the flower. Instead, the male bee gathers odiferous volatile chemicals in the orchid flower and mixes the perfumed ingredients with other potions gathered from other species of flowers and fungi. These volatile mixes are compacted in a spongy, oily tissue in the hind leg pouch. This complicated mixture of perfumes identifies this male as a member of the *Eulaema cingulata* species. Once he has the mixture, he goes to a mating perch and waits. Somewhere out there, a female detects his volatile perfume and approaches the nuptial site. After she determines that he has the precise mixture of potions that identify him as the same species, she finds him and they mate. The orchid has cleverly become a matchmaker that enables this sexual union between the bees.

One might think that is the end of the story until you realize that the orchid has evolved this irresistible lure to gain something in return. That swap, of course, is pollination. Once again, the chronicle gets complicated. Instead of the bee taking pollen from the male flower to the female flower for pollination in an ordinary way, the orchid has evolved an unimaginable scheme.

As a prelude, it is imperative that the male bee visit the male orchid flower first, simply because the male flower has the pollen. When male bees visit orchid flowers, you can witness one of the most vicious battles in nature. Each bee tries to enter a flower but must defend itself against competitors that try to enter the same flower.

The intruder darts back and forth and intimidates the occupant inside the flower. It appears that the male on the outside is trying to sting the defending male inside, but this species does not have a stinger. It is an awesome battle that holds your attention.

When the male bee enters the male flower, he touches a spring trigger mechanism that catapults a floral botanical missile to the top of his thorax. One end of this missile has glue that sticks to the thorax. Attached at the other end of this missile are two sacs of pollen. And that is what the male flower is designed to do—spread his genes in the pollen. He has done his duty using some clever botanical engineering.

After the visit to the male flower, the bee flies away and finds a female flower that looks different. When the insect enters the female flower, she is waiting with a device that releases the two pollen sacs that were glued to the bee by the male flower. Thus, the pollen has been delivered to the female flower and pollination is achieved. The bees have mated and the orchids have pollinated. Both the animal and the plant have accomplished their missions employing one of the most complicated evolutionary developments ever.[2]

LEAPING LIZARDS

For many years, it has been a ritual to gather on our front porch at Hospital Point about 6:30 to see the sunset and talk about pleasant activities of the day. Then, about 6:45, at dusk, a nightly spectacle begins. Numerous geckos emerge from the house upstairs, go down a pipe and pause. They flex their tails and posture. After hesitating for some time, each one finally gets the courage to make the great leap from the pipe to an old frangipani tree. Some lizards hit the tree trunk as planned; others miss and have to climb back up. Eventually, these geckos go up the frangipani tree and assemble for some mysterious purpose unknown to us. We may never know the reason until some smart graduate student writes a thesis and reveals this enigma about the leaping lizards at Hospital Point. *Bienvenidos*, graduate students. What a way to finish the day.

Bat Falcon
Photograph: Birds of the World - Cornell Lab of Ornithology https://birdsoftheworld.org/bow/home.

Red-lored Parrot
Photograph: Clyde Stephens.

Montezuma Oropendola
Photograph: Alex Chester.

Three-wattled Bellbird
Painting: Painted on my Hospital Point house wall by Melissa Lintern, an artist from England traveling through Bocas del Toro.

Rufous-tailed Hummingbird
Photograph: Alex Chester.

Golden-collared Manakin
Photograph: Taken by an unknown visitor.

Urania fulgens
Photograph: Courtesy of C. R. Hernandez, Smithsonian Tropical Research Institute.

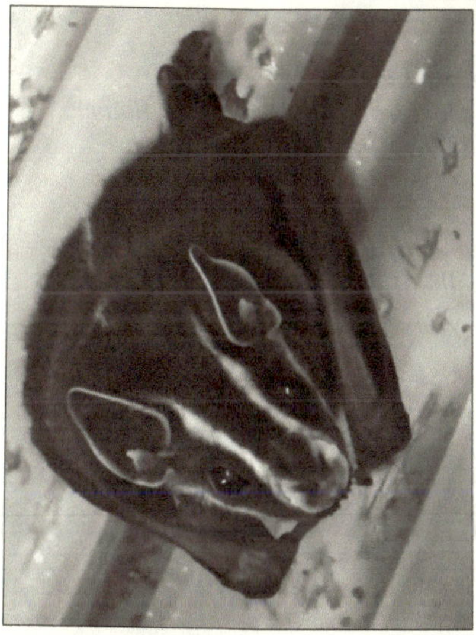

Uroderma bilobatum
Photograph: Maurice Thomas.

Bumble bee
Bee with pollen sacs glued to thorax by the male orchid flower.
Photograph: David Rubic.

Eulaema cingulata
Bumble bees pollinating orchids.
Photograph: Clyde Stephens.

Poison Dart Frog
Hospital Point, Isla Solarte
Photograph: Linda Stephens Radcliffe.

Female Red Frog variation
carrying tadpole (on its head) to water.
Hospital Point.

A mutant
never seen before.
Near Hospital Point.

Corypha umbraculifera
The Talipot palm provides birthing chambers for bats on Hospital Point: Leo under a Talipot palm. Photograph: Leovigildo Julian Lorenzo.

Human entertainment.
Indigenous Ngäbe natives, our neighbors.

9. Ticks and Spider Bites

IT TOOK 60 YEARS before I got a spider bite. My hand instinctively removed a dry Heliconia leaf on a plant at Hospital Point, but that was a big mistake. A hidden mother spider protecting her egg mass pierced both fangs into my hand near the thumb. It hurt. Swelling began quickly and advanced up to my elbow. I pondered about what to do. My decision was to wait and see.

My logic was that no doctor would have an anti-venom for an unknown species of spider. After four days and several strong pain medicines, the swelling and pain gradually subsided and went away after a week. I have been more cautious since then about disturbing spider egg habitats.

But I am like a magnet for ticks. Even though I take precautions to protect myself from these dangerous vectors of diseases, a few small ticks manage to grab me as I walk along trails. One day, while I was sitting in the office of a lawyer friend in Bocas Town, I felt an itch and reached down and pulled a small tick off of my ankle. Without anyone looking, I smashed the fat little arachnid under my shoe. That should have been the end of this story—but it wasn't. Day by day, swelling and redness increased up to my knee. I began to worry and wonder about amputation, and that is when I took my large, outdated (1984) Lange Medical Manual and went to Changuinola to see the former United Fruit Company physician, Dr. Raul Davila. This was in the 1990s, and Dr. Davila was in charge of the entire Social Security hospital in Base Line. After work hours, he had a small clinic along the railroad tracks to see patients like me. I normally paid the usual $10 consulting fee, but he told the receptionist there was no charge. I insisted on paying, but she would not take my money.

When I told Dr. Davila why I was there, he said that in all of his 40 years as a doctor, he had never seen such a problem. I told him that the infection might be similar to Lyme disease that is caused by a spirochete that is transmitted by a tick that lives on deer in the United States. He knew nothing about the disease, but I read descriptions from the medical manual in English and translated to him in Spanish. This was a new case for him. He wrote down the recommended prescription, which was a heavy dose of tetracycline. Lucky for me, there was a local pharmacy in Changuinola that sold the drug inexpensively, and I started the treatments immediately.

Apparently, the spirochetes were controlled. Infection and swelling subsided day by day, and my leg recovered normally. That was a close call. In his honor, the Changuinola medical center is named Hospital Dr. Raul Davila Menaso.

10. Black Physicians in Early 20th Century Bocas del Toro

by Ariel Rene Perez Price[1]

RACE RELATIONS IN EARLY 20th-century Bocas del Toro is an orphan area of study. Only a few researchers have genuinely delved into this topic from a sociological and historical perspective. Perhaps the most notable exception being the monumental work of Philippe I. Bourgois.[2]

At the turn of the 20th century, Bocas del Toro was a multicultural place plagued by racial tensions, exacerbated by the imposition of a labor apartheid set in the upper-class context of Boston's White Anglo-Saxon Protestant society, where the United Fruit Company was formed.[3] The archipelago witnessed the interaction between the local Black elite, known as Creoles,[4] who were displaced by a new group of Afro-Caribbean workers destined to the rigors of the plantations.[5]

This chapter seeks to explore the contributions of a different and overlooked group of Afro-Caribbean migrants. They were a professional class, who were mostly educated in the United States during the time of depredating segregation. Bocas del Toro was their home, where they decisively contributed to the development of the medical profession in the recently proclaimed Republic of Panama.

LIFE AND MEDICINE IN AN INTERNATIONAL PORT

The dawn of the 20th century saw Bocas del Toro become an

international port,[6] immersed in the economic boom sparked by agricultural exports and the foundation of the United Fruit Company in 1899.

The productivity of the banana plantations prompted a diverse immigration that increased the population to 13,000 by 1911.[7] The commercial relevance of Bocas del Toro was obvious. By 1917, there were seven consular offices in the capital of Bocas del Toro on Isla Colon: the United States, Germany, United Kingdom, Portugal, Norway, Costa Rica and France.[8]

In 1900, the United Fruit Company established a modern medical facility on the western tip of Nancy's Cay, known as Hospital Point.[9] The sanatorium was initially conceived for the control of tropical diseases that could place the port at risk of quarantine and for the care of its white employees for whom surgery and hospitalization services were prioritized.[10] Hospital Point's first superintendent was Dr. Robert Earl Swigart, who applied sanitary procedures used by Dr. Walter Reed in Cuba. This resulted in the final eradication of yellow fever in Bocas del Toro by 1906.[11] In 1912, Dr. I. W. Mclean, a graduate from Baltimore College, replaced Dr. Swigart.[12]

In 1904, Hospital Point inaugurated a segregated area for Black employees.[13] However, the demand for medical services exceeded the capacity of the Company, which by then had 6,000 Black employees between Afro-Caribbean and Afro-Americans, who were also accompanied by 5,000 dependents.[14] The medical services in the archipelago placed the Black population and anyone without ties to the Company in a vulnerable situation. There were no public hospitals under Panamanian jurisdiction until 1919.[15] At the time, the most prevalent diseases in Bocas del Toro were malaria, pneumonia, tuberculosis, nephritis, uncinariasis, and venereal diseases.[16]

THE BLACK MEDICAL PROFESSION IN BOCAS DEL TORO

Segregation and the lack of sufficient medical attention, created the conditions for the arrival of first-generation physicians of African descent to Bocas del Toro. These professionals were pioneers of their race who faced difficult and adversarial circumstances while trying to get their medical training. This was at a time when educational options were scarce for non-whites in the Americas. They were also the product of an upcoming wealthy Creole class from the British Caribbean.

In the southern states of the United States, after the failure of Reconstruction, Black physicians became indispensable in the communities where they practiced their trade. In Bocas del Toro they brought medical care to a majority, who had previously depended on quack doctors, folk medicine and the magical powers of conjurers to treat disease. Carlos Reid, in his memoirs, compiled by Dr. Stanley Heckadon-Moreno, confirmed this analysis:

> However, in Bocas Town there were some quack doctors and real doctors, such as Vaz, Rawlins, Price and Dr. Osterhout, a German, and there was also Dr. Thomas. Old Bank had no doctors, so people had to make the trip in a canoe to Bocas and seek assistance from these men.[17]

Little is known about these pioneers of the medical profession on the Isthmus, but Carlos Reid contributed valuable information about them. Searches in the archives of the Catholic Church of Bocas del Toro of the death records between 1912 and 1913 also helped to shed light on more information.[18] The documentation contains the names of the physicians who certified deaths, confirming in many of the cases, those mentioned by Reid and unveiling others such as Dr. Veitch. Further research based on the transnational Black press of the time and other primary sources provided enough information to piece together this forgotten chapter of Panama's medical history.

DR. CHARLES ALFRED VAZ

Charles Alfred Vaz arrived in Bocas del Toro during the last decade of the 19th century before the arrival of the United Fruit Company. However, at that time, banana exports were already the main source of the economy driven by fierce competitors such as the Snyder Banana Company and the Camors, McConnell & Co.

Dr. Vaz was born in Spanish Town, Jamaica, in 1863. In addition to his Afro-Caribbean lineage, some of his descendants reported that he had a Jewish heritage.[19] He traveled to Canada for his medical studies at the progressive Bishop's College which had been operating as a medical school in Montreal since 1871.[20] Several sources confirmed that he lived for thirty years in Bocas del Toro holding multiple positions in government, such as official doctor of Bocas del Toro, municipal doctor, quarantine doctor and head of the Board of Health and Sanitation.[21] He was a prominent member of The Order of Knights of Pythias and The Independent United Order of Mechanics.[22] He died in 1924.

DR. JOSEPH EVANS THOMAS

Another physician on Reid's list, about whom we discovered detailed information, is Dr. Joseph Evans Thomas. The son of Jamaican parents, Thomas was born in Pearl Lagoon, Nicaragua, in 1875. He studied in Kingston at the renowned York Castle High School[23] before relocating to the United States, where he obtained his Medical Doctor degree at Howard University, Washington, D.C., in 1899.[24]

Howard is an Historically Black University (HBCU) founded in 1867 by the First Congregational Society of Washington shortly after the American Civil War. Thomas arrived in Bocas Town at the turn of the 20th century, establishing a private practice located at No. 47 First Street. He married Rosilda Lopez and had three children: Guillermo Celestino (1903), Ruby Ethel (1907) and Jose Gordon (1908).[25] He passed away unexpectedly at the age of 40 in 1915. He was buried in Haul Over Beach Cemetery and, according to a newspaper account, his coffin was draped with the British flag while the ceremony was presided over by both Reverend M. C. Surgeon from the Methodist Church and Reverend Wait-Smith, a member of the Anglican congregation.[26]

DR. A. O. RAWLINS

Dr. A. O. Rawlins, a physician from Jamaica, practiced medicine and surgery in Bocas del Toro from the beginning of the 20th century until at least 1930.[27] According to an article published in *The Gleaner* in 1915, Rawlins obtained part of his education at the Tuskegee Institute in Alabama.[28] Further investigation should be conducted on the studies of Dr. Rawlins since Tuskegee was merely a vocational institute until well into the 20th century. However, since 1892, a Training School of Nurses was installed by the Institute under the auspices of the John A. Andrew Memorial Hospital to provide "instruction to young men and women of color who desired to learn the art of caring for the sick."[29] Perhaps Rawlins attended this course before progressing on to one of the many HBCUs established in the South of the United States.

DR. F. GORDON VEICHT

Dr. F. Gordon Veicht, a native of St. Ann, Jamaica, was another physician who immigrated to Bocas del Toro. He studied at Meharry Medical College located in Nashville.[30] The institution, founded in 1876 by the Freedman's Aid Society of the Methodist Episcopal Church, aimed to "train aspiring caregivers to serve not only newly freed African-Americans but also all who were deprived of and needed medical attention."[31] There is evidence of Dr. Veitch's involvement in war-related efforts on the archipelago as one of the physicians of the Friendly Society War Contingent Committee, which helped recruit volunteers in Bocas del Toro to join the British West Indian Regiment in World War I.[32]

DR. PAUL OSTERHOUT

The only non-Afro-Caribbean physician mentioned by Carlos Reid is Paul Osterhout, whom he wrongly described as a German. Our research shows that Osterhout was a white American born in Texas, who arrived in Bocas Town in 1895. He was a graduate from Jefferson Medical College in Pennsylvania and was appointed as quarantine physician (1901 to 1910) and consular agent by North American authorities in 1909. Osterhout was also appointed official doctor by the Panamanian government in 1905. There is evidence that he lived in Bocas Town until 1917.[33]

DR. JOSE ANTONIO PRICE

One of our own so far, our research has shown a clear pattern of natives from the Caribbean who accessed medical education by attending progressive and HBCUs in North America. Those individuals arrived in Bocas del Toro amidst a massive migration of plantation workers following the collapse of the sugarcane industry in the Caribbean and the debacle of the French attempt to build a sea level canal in Panama. A peculiar protagonist of this story is Dr. Jose Antonio Price, the only Panamanian on Reid's list and the youngest of the group.

Jose Antonio was born in Panama City in 1890.[34] He arrived as a child in Bocas del Toro as early as 1896.[35] Hilda E. Price, his mother, was an immigrant from British Honduras who relocated to Bocas Town where she owned a small hotel and other small businesses. Jose Antonio benefited from the opportunities offered by the cosmopolitan international port studying under the guidance of Jamaican teachers, but also learning Spanish, French and English before heading to Kingston where he finished his high school education.[36] From there he followed the same pattern as the rest of the Afro-Caribbean physicians researched in this chapter, arriving at Shaw University in Raleigh, North Carolina, around 1907[37] and then moving on to conclude his studies in medicine and surgery at the University of West Tennessee (UWT), College of Medicine and Surgery in 1913.[38] Shaw, founded in 1865, is the oldest HBCU to begin offering courses in the southern United States, while UWT was a proprietary institution founded in 1900 by Dr. Miles Vandahurst Link. Later, Shaw succumbed to the infamous Flexner Report, closing its doors in 1923.[39]

Dr. Price, upon his return to Bocas del Toro in 1913, started a prolific career in the archipelago that lasted for almost forty years until his death in 1951. Price installed a private practice in his own medical facility located at Fourth Street and Central Avenue called Hospital Santa Fe[40] and where he also established a drugstore named Farmacia Central.[41] Dr. Price held various government appointments in Bocas del

Toro including forensic doctor starting in 1914,[42] prophylaxis doctor between 1915 and 1920,[43] as well as director and surgeon of the Charity Hospital,[44] the first government sanatorium in the province created in 1919.

An important contribution by Dr. Price was his role in the process of healthcare nationalization, a campaign initiated by Belisario Porras during his second term as president. Until then, public healthcare was still largely in the hands of the Americans, who had taken over the sanitation of the Isthmus in 1904. This was done to avoid epidemiological disasters, like the ones faced by the French during their canal construction era. Dr. Price was then appointed as official physician of the eastern section of Chiriqui[45] and official physician of Darien[46] during Porras's last term as president. The documented evidence shown in this investigation aims to corroborate that Jose Antonio Price was the first Afro-Panamanian to obtain a degree as physician and surgeon at the beginning of the Panama republican era that began in 1903.[47]

ALEXANDER MORTIMER SURGEON, AN ELUSIVE MEDICAL PIONEER OF BOCAS DEL TORO

At Haul Over Beach Cemetery in Bocas Town, an old Masonic artifact marks the grave of Alexander Mortimer Surgeon. The oral history preserved by the descendants of the patriarch of one of the most historical families of Bocas del Toro claims that Surgeon was the first trained physician to set foot on the island. A document without formal sources was delivered to me recently by one of Surgeon's descendants shedding some light on this pioneer of medicine in the archipelago.[48]

According to the document, Surgeon was born in Jamaica in 1843, and arrived in Bocas del Toro in 1893, establishing a drugstore and his own private medical practice on Columbus Island, where he taught the inhabitants of the village to boil water for drinking in order to prevent illness. He married Martha Elias Surgeon with whom he had 13 children. Surgeon became involved in politics fighting on the side of the Liberal Army during the War of a Thousand Days. He died in Bocas del Toro in 1903. No reference to his studies has been provided and so far, we haven't been able to find any other written sources corroborating Surgeon's activities in Bocas del Toro.

A MISSING CHAPTER IN PANAMA'S MEDICAL HISTORY

The existence of a Black medical class in Bocas del Toro, mainly educated in HBCUs of the United States, substantiates once again that the archipelago was a vital place with stronger links to the Circum-Caribbean than to the continental mainland of Hispanic heritage. It is a forgotten history of diversity on the Isthmus and an important piece of Panama's medical and republican history that needs to be further researched.

The evidence summarized in this research portrays a diverse group of pioneers of medicine on the Isthmus, a group that also contributed to the foundation of Panama as an independent nation. These doctors provided medical care for a vast population that was the engine of one of the more productive locations of the newly founded Republic of Panama, which was Bocas del Toro.

Dr. Jose Antonio Price.

Ariel Rene Perez Price.

Dr. Alexander Mortimer Surgeon.

Dr. Joseph Evans Thomas.

11. A Boy from Bocas
by Luis "Ibu" Alvarado[1]

WHEN I WAS A boy, it seemed for many years that *la Compañía* was an overpowering, eclectic society that resided somewhere up there in the heavens, because *la Compañía* was mentioned in just about every conversation and was ever-present in our home. There were many all-encompassing aspects about *la Compañía*, such as: *La Compañía* is going to replace the steam locomotives for modern diesel-electric ones; *la Compañía* is going to hire hundreds of laborers to plant bananas in previously abandoned farms; *la Compañía* is going to reduce its labor force by 50%; *la Compañía* is planning to use airplanes to spray the bananas to combat the Sigatoka fungus; *la Compañía* this, *la Compañía* that. When I started 1st grade at the Company's Farm 8 School, a two-minute walk from our house, and I learned to read, I finally discovered that *la Compañía* did not reside in Heaven, but in Boston. This revelation came to me from reading one of my favorite magazines: *Unifruitco*, the Company's monthly magazine.

ANCESTRAL INVOLVEMENT WITH BANANAS

Bananas are in my genes. My maternal grandfather, William Charles Kinkead, was a British subject of Irish-Scottish ancestry. His family had emigrated to Jamaica, and he first came to Panama to work on the Panama Canal during the French construction era when Panama was still part of Gran Colombia. He later settled in Boquete, Chiriqui, where he married my grandmother, Marcelina Sanchez, a *boqueteña* and daughter of a Spanish immigrant, Ñopo Sanchez. Soon after, they set out on the Caldera foot trail across the mountain range to Bocas del Toro, where he was hired to work in the province's banana plantations, most likely the Snyder Banana Company in Chiriqui Grande. They remained there for two years. Then in early 1905, the family returned to Chiriqui with an infant daughter, my mother, stowed in a straw basket lashed on the back of a pack mule. They settled in Boquete, where they acquired a farm in Jaramillo and cultivated coffee. On my paternal side, my father, Teofilo, was the first Alvarado to be associated with the banana industry.[2]

BEGINNING OF THE SAGA

In 1928, as a young 22-year-old, my father was employed by the Chiriqui Land Company (CLC), the UFCo subsidiary in Panama, in Progreso, Chiriqui, in the Accounting Department. He got this job primarily because he could speak English that he had learned in a Methodist school in David and had then enrolled in a correspondence accounting course. Shortly afterwards, he was transferred to the Agriculture Department, where he began his banana internship. A year later, he married my mother and was assigned a house in Progreso, the Company headquarters at the time. Most, if not all, farm overseers were either British or American, and when their two-month vacations were due, he was assigned as a replacement during their absence, providing him with experience running different farms.

It was in the early 1940s, as World War II raged and the Allies had lost the Philippines to the Japanese Empire, that Manila hemp ropes were needed for the Allied navies' ships. The U.S. government asked the United Fruit Company to plant and process abaca fibers to produce Manila hemp rope for the war effort. The Bocas Division was the first designated region for this endeavor because experimental abaca had already been planted in Changuinola in the 1920s. The Company had undertaken the challenge and began converting abandoned banana farms to abaca and erected an abaca stems processing plant in Base Line, the center of the farming activity in Changuinola.

From one day to another, my father was informed that he would be transferred to Changuinola. Mr. Davey, the abaca superintendent in Changuinola had suddenly died, and my father was selected to replace him immediately. The harvest phase was just beginning, thereby the urgency of the move. Even though the transfer was a promotion, uncertainty of what to expect loomed over my parents. They were unsure what that move would mean for the education of their daughter and three sons, in an isolated, remote community such as Changuinola. An impromptu call from Mr. Myrick, the general manager for both Chiriqui Land Company in Puerto Armuelles and the Bocas Division, assured them that while a school was being built in Farm 8, my brothers would be taken to the Company American School in Almirante, an hour ride in motorcar; and my sister Ileana, who had finished 8th grade, would be sent to a boarding school in Panama City under the UFCo auspices; and lastly, that my mother would be employed as the Spanish teacher in the American School in Farm 8, as soon as the school was operational. With those assurances, they decided to accept the job and quickly packed and left within two days.

FLYING OVER THE MOUNTAIN TO THE CARIBBEAN SIDE

When on July 12, 1943, my father and mother, together with my sister and brothers, boarded the single-engine Company plane and departed from Puerto Armuelles, Mr. Myrick was there to see them off. Before taking off, the plane's captain, Marcos Gelabert[3] warned that they would fly over a very turbulent passage across the Continental Divide and instructed the children to sit still and not move.

After the turbulence ceased beyond the Boquete Pass, my mother recalls the sudden panoramic view of the blue Caribbean and the green valley of Changuinola in the distance. But the children were still and speechless. The tension was broken when Carlos asked in a loud whisper: "Mami, can I scratch my head now?" That made everyone relax as they descended over Changuinola. But the airplane kept on flying to the West, leaving Changuinola behind, heading to a grass strip in a pasture in Finca Costa Rica, a cacao farm on the Costa Rican side of the Sixaola River. This was the closest place to land near Farm 8. Esteban Lopez, the Chiriqui Land Company public relations specialist, and Noley Brown, the manager's personal chauffeur, met the family at the landing strip in Motorcar No. 189. An hour-long trip delivered them-to Farm 8, at the house Mr. Davey's widow had vacated just the day before, prior to being repatriated to British Honduras, her native home. My family remained living there for the next 19 years, during which time my father was involved in the abaca, cacao, and lastly, the banana era.

LIVING IN A COMPANY TOWN

Company residential areas had housing assignments based on an employee's job description, or rank, not on nationality or race. The highest-tier employees consisted of the managerial, administrative, and technical positions (division manager to farm timekeeper) and were considered first-class employees and designated to live in *la Zona*. Assigned houses were fully furnished, had solar-heated water, 24-hour electricity, and a full-time *yardero* (yardman). In addition, before refrigerators became available, the Company supplied 4-foot-square galvanized, insulated, wooden ice boxes, with blocks of ice delivered every Monday via cargo train from the Almirante Ice Plant. And, the most cherished benefit that I remember, was train transportation every Saturday evening to attend the movie at the Almirante Social Club on a special railroad car: *El Telire*.

The type, size and specific location of the house assigned depended on rank. Living in *la Zona* came with certain exclusive benefits such as a clubhouse, golf course and swimming pool, and an American School for an employee's school-age dependents. The mid-level employees, foremen and leaders, had their own slightly more congested residential area, *el Cuadrante*, consisting of two-story houses with indoor plumbing. The third-ranking employees, the laborers, were assigned to a residential compound known as *los Campos*. This area, at the time, had smaller houses built on stilts on a well laid-out grid on drained grounds with sidewalks, but no indoor plumbing. Toilets and showers were built separately between the houses. This area included soccer and baseball playing fields and a large, open-air movie screen for projecting the weekly walk-in movie, where each patron would bring a wooden stool or chair to sit on. If it was raining, you watched the film under an umbrella. The Company also provided a clubhouse for the area aptly called Club Bananero. Fond memories prevail to those who had the opportunity to listen and dance to popular local calypsonian groups of the day as were The Beachers, Los HB, and Los Ases del Ritmo.

FORMATIVE YEARS IN A UFCO DIVISION

If ever anyone comes up with a checklist to determine an authentic *bananero*, I think I can check most, if not all, the boxes: I was born in a Company hospital in the Company banana port town of Almirante; attended a Company primary grade school; witnessed hearing famed locomotive engineer "Papa" Lewis make the whistle on his steam locomotive No. 46 wail when carrying mourners to a funeral, or in joy when carrying guests to a wedding; caddied nine holes for 25 cents on the Company golf course; was employed as a part-time lifeguard at the Company pool at 15 cents an hour, which was better than delivering papers on horseback three times a week earning 2½ cents a paper, when subscribers paid; attended, and danced, in grown-up dances at the clubhouse celebrating three different Independence Days: *3 de noviembre, 15 de mayo*, Fourth of July, plus New Year's and Carnival; and sailed on the Great White Fleet ships while attending high school and college in the US. All in all, my life was a total immersion into that unique banana culture.

THREE ERAS IN BOCAS

Growing up in Changuinola consisted of three eras: Abaca, Cacao and Banana. Each of which, I'm sure, impacted my character. Distinct as each one was from the other two, they were just as exciting.

ABACA ERA

From about the age of four until I was seven, everything revolved around the abaca production and processing. The farms were fittingly named after places in the Philippines: Manila, Filipinas, Mindanao, Luzon, Davao, and Cebu. My father, as the abaca superintendent, ran the largest, single development of US extensive abaca projects on some 6,200 fertile acres in Bocas del Toro.

In the midst of these farm clusters, business communities flourished, resembling a modern-day mall. El Chino, located in Base Line, was the biggest. It was on the main railroad line from Almirante to Guabito and included a general store, restaurant, cantina, post office, tailor, watch repairer, basic dentistry services, a bowling alley with hardwood bowling balls and pins, and the crown jewel, Teatro Abaca, the only movie theater in the region. The three entrepreneurs that monopolized Changuinola's businesses were Jim Chong in Base Line, Sam Bailey in Lincoln Creek and Mayer Malca in Guabito.

During this time, *vaqueros* (cowboys) and *muleros* (muleteers) dominated my formative years and were my first role models. With the abaca, the Company brought in hundreds of mules from Alabama to be the mode of transportation for bringing the abaca stalks for processing to the plant. Under the tutelage of the *muleros* I learned to ride mules, specifically La Baya, a big cream-colored Alabama mule sidelined from work because of a work accident and assigned, unofficially of course, to me to ride on Saturdays. Eventually I graduated to horses, so then the *vaqueros* who tended the livestock the Company kept to provide the commissaries with meat and dairy products for the employees, became my role models. I tagged along with them when they rode past our house in Farm 8 every Saturday morning at dawn from Farm 6 to the Luzon pastures to tend to the herds. By then, I had my own horse, a gift from my parents on my 9th birthday, my gateway to becoming a dedicated *vaquero* understudy, riding, herding, lassoing, and assisting in branding and castrating steers. I had found my calling.

CACAO ERA

A few years after World War II ended, the demand for abaca's Manila hemp declined, and in 1954, the Company shut down planting and production. So my father was appointed as cacao superintendent. The cacao farms were along the Sixaola River on both the Panamanian as well as the Costa Rican side.

Some of the names that are still engraved in my memory are Dos Caños, Deborah, Margarita, Elena, Nievecita, Virginia, Catalina, Barranco, Fields, Isla Grande and Daytonia. I used to accompany my father on his periodic visits to these different farms on No. 176, his assigned 3-seater 2-stroke engine motorcar, or in No. 189, the seven-passenger motorcar, depending on the destination. His driver was the unforgettable George Chora Roberts, who took care of us when my parents attended social events at the club. Lunching on canned sardines, soda crackers and a Polar Cola, a Costa Rican soft drink at Babu's, who had the only store on the Costa Rican side of the Sixaola bridge, is part of my cherished memories. All the cacao farms were connected by the UFCo railroad network; however, once across the bridge into Costa Rica, access to some farms had to be conducted on either a rail pump-car or a horse-drawn *mesilla* (flatcar) because of the narrower rail gauge, which made those trips even more memorable at that early age.

On arrival, saddled mules were provided to accompany the farm overseer and foreman deep into the plantations. Besides overseeing the agricultural operations, these visits included commissary, medical dispensaries, playground fields, sanitation infrastructure, housing, aqueducts and the Cummings diesel generators that provided electricity from 6:00 P.M. to 10:00 P.M. I looked forward to these visits eagerly. Now, I realize they were also my first exposure to what is today known as Corporate Social Responsibility.

BANANA ERA

In the early 50s, when the cacao farms on the Panamanian side of the Sixaola River were targeted for conversion back to bananas, as were the whole Changuinola River Basin abandoned farms, my father was transferred to the banana operations as district assistant superintendent. Those were exciting times with the influx of Caterpillar heavy equipment to build roads and prepare terrain for housing communities, this time with indoor toilets, running water and 24-hour electricity. The huge Bucyrus-Erie draglines, that operated day and night digging drainage canals and forming lakes, captured my interest, but not so much as to divert me from horses and cowboying. Numerous lakes were constructed in the previously abandoned banana farms to rid the soil of the dreaded Panama disease fungus by flood-fallowing. After a year, the soils were planted again with the famed Gros Michel variety preivously used. As it turned out, it was a costly and ineffective venture. The banana plants still died. Flooding had not rid the soil of the Panama disease, but soil scientists noted that flood-fallowing had eliminated the devastating nematode species which had been in the soil. Unheeding the soil scientist discovery, the Company's senior tropical decision-makers decided to plant the treated farms with a different variety of bananas, Lacatan, from Jamaica, which was Panama disease-resistant. But unbeknown to them, these corms were infected with nematodes, and reinfected the soils. After that failure, this time a more weighted decision was taken to flood the soils again to rid it of the nematodes. After flooding, and once confirmed that the farms were nematode-free, they were then planted with another Panama disease-resistant variety, Valery. A year later, Changuinola was to become a showcase banana-producing division once again.

With the success of the Valery, a massive banana expansion in Changuinola took over, creating an economic bonanza in the region. It brought laborers, technicians, engineers, accountants, medical staff, and the loud and powerful radial engine-powered Stearman biplanes to spray for Sigatoka disease, a fungus disease that infects banana leaves. I was fascinated with that exciting and challenging world. I found flying unconventional and daring. My imagination now took a 90° turn, and now I was set on becoming—a pilot. I became an airport ramp-rat after school and on weekends. That went on until I graduated from 8th grade and was sent away to boarding school.

In the early 1960s, a major labor strike paralyzed the entire province for over a month, and my father was then transferred to head the Labor Relations Office in Almirante as a result of one of the Company-Labor Union negotiations to end the strike. Our family then moved from Farm 8 to Almirante, where we were assigned a house on the edge of an inlet on the Almirante Bay. Peaceful as the scenery was, with a view of the bay, in the center of everything Almirante had to offer, clubhouse, tennis courts, commissary, hospital, a movie theater and much more, it wasn't as glamorous as living in Las Fincas, as the banana farm settlements were called. Had I been an avid fisherman or boater, maybe I would have enjoyed living in Almirante. But I was more into hanging out at the airport around spray airplanes, mechanics and pilots. All this took place right when a major change in the Company's direction impacted our family life.

Just about that time, the Company was being forced to, or had decided to, or both, downsize its banana-producing holdings in the Tropics. A 1954 civil anti-trust suit filed against the UFCo based on the Anti-Trust legislation, which prohibits anti-competitive conduct that deprives benefits of competition, was exerting pressure on multi-national companies, and the UFCo was one of them. On the other hand, heavy rains had caused continuous flooding, strong winds had caused blowdowns, and the labor disputes had all taken its toll on operating costs and must have influenced the Company directors to rid the Company of that unwanted burden. Thirteen employees, my father among them, were selected from the Bocas Division and offered financing to purchase abandoned or in-production farms. The concept, which was also put in motion in the Puerto Armuelles Division, partly relieved the Company from being a land-owning banana grower to buyer. The contract with the independent producers specified that the purchase price per box would be determined once the Company's inspectors graded the fruit's quality. I can honestly say that of all the jobs I was given during my internship, being present at the docks during those all-night inspections was my least favorite.

The program spurned an economic boom in the region. Of the 13 farms, only 5 were in production, the other 9 had been abandoned and required a complete start-up preparation consisting of: soil preparation and construction of packing plants, installing a cable network for aerial transportation of the harvested fruit to the plant, all of this prior to planting, cultivating, harvesting and final boxing. An average of 200 employees per farm were required once the operation was in production. The Changuinola Airport became the gateway for persons from the rest of Panama seeking employment, making this airport the second busiest, after Tocumen, Panama City's international airport.

The family then moved from Almirante to Finca Terronal, named after my great-grandfather's property in Chiriqui. By this time, I was away at college, and being the youngest of four brothers, who had ongoing careers in engineering, architecture, and banking, I felt obliged to pursue a field related to managing the family's farm. The closest university curriculum that met most of what I needed was Agricultural Engineering. It wouldn't be much of a hardship, because of my familiarity and attraction to that environment. So, flying would have to take a back seat.

MANAGING FINCA TERRONAL

During 1970, as if natural disasters and labor unrest weren't enough to cause economic stagnation, OPEC's (Organization of Petroleum Exporting Countries) oil production boycott contributed in affecting operations. After four back-to-back floods, an infestation of nematodes spread throughout lower lands of the Changuinola River basin from farms that had not been rid of it, drastically affecting banana production. The West Indian bagworm plague appeared; a well-organized, government-backed labor movement defied and challenged the independent banana producers, promoting continuous labor stoppages. It seemed that the Seven Plagues of Egypt had descended on the Changuinola Valley.

That was the bleak scenario that prevailed when I returned to Changuinola. After a year, with no prompt remedy in sight, and knowing that the new military junta was preparing a new labor code to appease the unions, which had initially opposed the coup, we decided to close operations. The plantation was cut down and all vegetation eliminated, and the land lay fallow to prevent it from being a host to Sigatoka and other fungi. The land lay fallow. However, this time it was dry fallow. Soon an offer to purchase the property was made, and that ended our relationship with the banana world.

CONCLUSION

All in all, after many years, I've come to realize that growing up in that Company culture was truly a blessing. Just like my brothers and me, and the hundreds of other UFCo-ers, we interacted among different races, nationalities, languages, cultures, and religions. A certain dose of understanding and tolerance towards others must have become ingrained in those of us who were lucky to have been raised in that environment of truly valuable traits, as is the appreciation for the simple things in life. That was life in a Company division.

The four Alvarado children.
Top to bottom: Ileana, Teófilo, Eduardo and Carlos.
Luis Alvarado collection.

La Baya was instrumental in making me want to become a mulero first, and then a vaquero.
Luis Alvarado collection.

Puerto Armuelles Accounting Department, 1929. Teófilo Alvarado is 4th from the right, middle row. Luis Alvarado collection.

Alvarado house, Farm 8, Changuinola.
R-L: Pao, my older brother, Hilda Hubbard and an unidentified girl on the steps. Luis Alvarado collection.

Farm 8 School, Reading Class.
L-R: Barbara Miller, Ricardo Ceballos, Adolfo Ceballos, Patsy McClendon, Pepe Miller, Gary McClendon, Chase Hubbard, Terry Wood, Linda Wood, Ibu Alvarado, Frank Morris, and Fulvia Ceballos. Luis Alvarado collection.

Teófilo Alvarado (2nd from left) visiting the American School. L-R: Teachers Florence McLaughlin, Carolyn Hubbard and Emma Kinkead Alvarado. Luis Alvarado collection.

Social Life in Almirante, 1960.
Standing, L-R: Roy Wells, Blanca Wells, Rachael and David Zebede, Yisvy Nessim, Joseph Nessim, Mrs. Nessim, Marybelle Cantrell. Seated, L-R: Zulay Flores, Emma Kinkead Alvarado, Teofilo Alvarado, Guy Cantrell and Luis Flores. The Zebedes and Nessims were local merchants, the rest were banana people. Luis Alvarado collection.

Ford Motor Company car converted to run on rails.
No. 189 was assigned to the Abaca Superintendent, Teófilo Alvarado.
Luis Alvarado collection.

Common mode of transportation throughout the cacao farms in both Panama and Costa Rica. Cacao trees in the background. Photograph: Fr. John Kennedy.

The Caboose, el Telire. Photograph: Fr. John Kennedy.

Locomotive No. 46, one of five used in the abaca processing operation. Base Line water tower (center) and the abaca processing plant (right). Photograph: Fr. John Kennedy.

12. From Italy to the Western Hemisphere A Brief History of the Messina Family

by Paul Messina[1]

THE NICOLA MESSINA FAMILY lived in Almirante, Panama, from 1952 to 1970. My father worked there for the United Fruit Company as a physician. This chapter describes how my family got to Panama by way of Guatemala, our lives in Almirante, the challenges and difficulties we faced, and what happened after we moved to the United States.

OUR PREDECESSORS

My family on both sides was from Italy. My father, Nicola Messina, was born in Perugia in 1909. His father worked for a branch of the Italian government and later moved his family to Salerno. After high school, he graduated from the medical school at the University of Naples as a physician and surgeon in 1936.

My mother was Margherita Padovani Messina. Her father, Lelio Padovani, was Jewish. He converted to Catholicism in order to marry Rosalinda Marinelli, the daughter of Vincenzo Marinelli, a Neapolitan artist.

Lelio Padovani was an architect of major buildings in Naples, such as the medical school, the science and mathematics departments of the University of Naples, and the main office of the telegraph company. During World War II, my grandfather's company and most of his possessions were appropriated by the Fascist government.

At the height of the war in 1943, a few miles inland from Salerno in a village called San Cipriano Picentino, where a friend of the family had a house, I was born. My parents temporarily moved to that village to get away from the heavy Allied bombing at Salerno. A few months after I was born, we returned to our Salerno home.

My parents had a very difficult time during the war, even though my father was exempted from the army because three of his brothers had died in military service. My father took care of wounded soldiers, and my mother talked often about having to wash sheets soaked in blood. The money they had in savings or investments lost almost all its value.

My older sister, Maria, remembers being hungry all the time during and after the war. Maria said our mother told her that food was very scarce and hard to get. A few farmers had food, but they sold it at exorbitant prices.

Almost all of my mother's family escaped being sent to concentration camps, but at least one of my cousins was interned by the Germans. He was liberated in 1945 and managed to get back to Naples by walking all the way from Germany to Italy, nearly starving to death.

In the late 1940s, the Cold War began and my parents, especially my mother, feared there would be another world war and Italy would be in it. Those fears, coupled with adverse economic conditions, led them to emigrate to the Americas. They chose Guatemala because they were told that my father's medical degree would be honored.

My mother's family had a history of emigrating to other countries and that might have inspired her to press for the move. Soon after World War II, about 1947, one of her sisters emigrated to Brazil with her family. Two cousins moved to Israel. Her grandfather, Vincenzo Marinelli, had an adventurous life: he left Italy in 1849 to escape jail, or worse, because he was in favor of overturning the Spanish monarchy. He went to Greece where he worked in the court of the king, then moved to Egypt in 1854, where he worked in the court of the Khedive, the ruler of Egypt at the time.

OFF TO THE NEW WORLD

To obtain funds for the trip to the New World, my parents sold all of their belongings, including the grand piano that my mother loved to play. They were renting an apartment and thus had no real estate to sell; even physicians were not well off in those days in Italy.

We traveled to Guatemala in third class aboard the Italian passenger and cargo ship *Leme*, leaving Genova on March 9, 1950. Women and children slept in bunks in large dormitories and men were in separate facilities. Everyone ate in the same area.

The trip to Guatemala took 34 days, with stops in 14 ports. We crossed the Panama Canal to the Pacific side and stopped at harbors in all of the Central American countries before finally arriving at Puerto San Jose, Guatemala. We stayed at a *pensión* in Guatemala City for a couple of months while my father tried to get his medical degree recognized. When he learned that the degree would not be accepted, he applied for any appropriate job, including one as a physician at the United Fruit Company banana plantations. With no income, my parent's savings were dwindling fast, and they considered returning to Italy.

Luck intervened when my father was offered a job at the Guatemalan Social Security Institute to set up the country's first industrial safety regulations; he had some training in accident prevention in Italy, qualifying him for this role. My father also made a little extra money providing medical advice and care to fellow Italian émigrés, while my mother earned some money as a seamstress, using a foot-operated sewing machine. I attended a school run by Jesuits called Los Infantes.

Life in Guatemala was good overall, but my father missed working as a physician. Then, for the second time, serendipity intervened. In 1952, the United Fruit Company contacted my father and offered him a job as a physician in the banana port of Almirante, in Bocas del Toro, Panama.

Dr. Gustav Engler, the head of the Almirante Hospital, had seen my father's application when it was first submitted in 1950, but there was no position open at that time. Dr. Engler knew that the medical school of the University of Naples was of high quality and was impressed by my father's excellent grades. When an opening for a physician arose, Dr. Engler contacted him.

ARRIVAL IN PANAMA

On November 7, 1952, we flew from Guatemala City to San Jose, Costa Rica; it was the first time any of us in the family had flown. The plane stopped at every capital city in Central America, so the trip took most of the day. We spent the night in San Jose and the next day flew to Changuinola, Panama, on the Costa Rican airline, LACSA. The airplane landed at the Base Line airport, a small, grass strip in the middle of the United Fruit Company banana and abaca plantations.

Dr. Engler was there to meet us. We boarded a "motorcar," the term used for a pickup truck with railroad wheels instead of rubber tires. The vehicle had benches and was fairly comfortable. The motorcars used the same tracks as the trains that took bananas from the plantations to Almirante, where the bunches of fruit were loaded onto ships. At that time, there were no roads from Changuinola to Almirante.

I still remember the ride through the jungle to our new home, with Dr. Engler identifying some of the plants in the dense, tropical growth on both sides of the tracks. When we reached Almirante, the motorcar stopped right in front of our fully-furnished United Fruit Company house. It was much larger than our little apartment in Guatemala City. Maria and I ran with excitement and found our own bedrooms.

MY CHILDHOOD YEARS

Growing up in Almirante was delightful. I soon became friends with John Munch, the son of George Munch, the general manager of the Almirante Division of UFC's Chiriqui Land Company. Janet Campbell, Raul and Ricardo Matamoros and Ramon Aizpurua were my neighbors and classmates. Many of our activities were outdoors, including baseball, fishing, boating, and archery.

John and I often used his boat to navigate nearby rivers and mangrove swamps, getting bitten mercilessly by sandflies and mosquitoes. John loved to fish for sharks from the dock by his house. I was there when he hooked a very large one. I helped bring it in and, with the help of others, we strung it up from a tree branch. When I was 10 years old, Dr. Engler asked me if I would like to learn to sail and be his crew. He had a 25-foot sailboat of the Vertue class and was knowledgeable about all aspects of sailing. His son Tommy, only six years old, was too young. Dr. Engler was an excellent and patient teacher. I learned all the useful knots, how to splice ropes and basic maintenance of the one-cylinder, 8-horsepower inboard engine. When I was 12, he took a long vacation in Europe and told me I could use the boat in his absence. I was thrilled by his trust in me.

Spending time with Dr. Engler while sailing was always an educational experience. He was knowledgeable about many topics, and when he talked with other guests about them, for example classical music, I would listen and learn. I recall that on some days when there was a breeze, he would say to me "*Buena brisa*?" (Good breeze?) with a broad smile as an invitation to go sailing, and we always would if he was free. In memory of him, I named one of my sailboats *Buena Brisa*. I bought her in Los Angeles and kept her in Marina del Rey when I was working at Caltech. Boats soon became my passion and sailing has been my favorite pastime all my life. When I was almost 14, I saw the drawing of a boat in *Popular Mechanics* that was simple to build. It was a 16-foot punt. My parents gave me the money for the materials, and I spent six weeks of my summer vacation building it with help from my mother. A mast and leeboard were added so that I could sail as well as row. I spent many hours enjoying this boat in Almirante Bay right in front of our house.

Looking at photos from my childhood, I was reminded of how active social life was in Almirante. There were frequent parties at people's houses and events at the social club on the waterfront: a weekly movie, dances with live music and costume parties. The Club had a bowling alley on the ground floor, and my friends and I formed a team and competed in an informal league. A bar and restaurant were part of the infrastructure of any Company club. If you lost in a game of *chingona* dice while playing for drinks, you could sign *un vale* (chit) that would be deducted from your monthly paycheck. The same applied to meals that you ate in the restaurant. Like almost everything in United Fruit Company towns, the Club was financed by the Company.

The Company operated three launches: a small one, the *Limon*, about 30 feet long, and two larger ones close to 70 feet in length, the *Changuinola* and the *Talamanca*. Several times a year, groups of families could use one of the launches to go to the Boca del Drago beach, to Bocas del Toro and occasionally to the more distant Zapatilla Cays. In October 1958, the Messina family unexpectedly grew. At age 47, my mother gave birth to Elena. This addition was not planned, but we were all delighted.

PRACTICING MEDICINE IN ALMIRANTE

My father was pleased to be practicing medicine again. He liked treating people with so many different conditions. Seven physicians treated all of the patients who came to the hospital—both employees of the United Fruit Company and non-employees. The Company doctors had to be general practitioners to be able to attend to everyone, although some had expertise in specific fields. Dr. Engler did more surgeries than the others, and my father had expertise in chest diseases.

When we arrived in 1952, there was a hospital ward with 16 beds for tuberculosis patients because TB was a major disease in this area. Ten years later, that ward was repurposed because the disease was almost eliminated as a result of my father's guidance.

In addition to my father and Gustav Engler, other physicians at the Almirante hospital in the 1950s and 1960s were Drs. B. M. Cino, Carlos Morales (dentist), Ludwig Jaffe, Raul Davila and Velarde.

Dr. Engler was an excellent director and manager of people, as well as an exceptional physician. He was a Fellow of the Royal Society for Tropical Medicine and Hygiene and of the American Tropical Medicine and Hygiene Society. My father also held those titles, probably because Dr. Engler nominated him for them.

Dr. Engler died suddenly of a heart attack in January 1965 on a Sunday morning while walking to see his patients at the hospital. My father replaced him as director of the hospital and the medical department. In his resume, my father states that he was in charge of the hospital's 210 beds, six field dispensaries in the banana farms and a medical center. A total of 120 people reported to him, including eight physicians, a dentist, a pharmacist, and nurses, X-ray and lab technicians.[2]

EDUCATION

The American School that the Company provided was taught mainly in English, with only one period taught in Spanish. Italian was always spoken in our home, so when we enrolled, Maria and I had to learn English quickly; we were given 1st grade books of the "Run, Spot, Run" type and got through the first three grades in only a few weeks. After that, we were able to study in classes of our age group.

The school had two classrooms: one for grades one through four and one for grades five through eight. Being in the same room with students in four grades taught by one teacher was beneficial. One could hear what was being taught to the other grades, enabling us to learn what we would be taught the next year.

Teachers during the 1950s included Miss Florence McLaughlin, Edith Smith, and Mildred Swanson. Thanks to them, we received an excellent education. After graduation from the 8th grade, almost all of us enrolled in boarding schools in the United States. Choosing a boarding school for me posed a problem. My parents wanted to send me to the United States, but they had no contacts there. Luckily, advice came from a physician from Virginia who had spent a few months in Almirante doing research on tropical diseases with my father. He suggested the Thomas Jefferson School in St. Louis, Missouri. I applied and took some tests that they sent to my 8th grade teacher. In lieu of visiting the school for an interview, I wrote an essay and was accepted. It turned out to be an excellent choice. My son David also attended Thomas Jefferson.

When I left the boarding school in St. Louis, I went to Ohio where in 1965, I obtained a Bachelor of Arts in Mathematics from the College of Wooster. The next year, I married Mary Thayer, and we eventually had two sons: Mike and Dave. We divorced in 1987.

At the University of Cincinnati, I entered graduate school and earned a Master's Degree in Mathematics, followed by a Ph.D. in Mathematics in 1972.

Back in Almirante, my parents decided not to send my sister, Maria, to boarding school because of unwarranted fears that she would not be safe, even though several of our friends had sent their daughters to the U.S. Instead, Maria got her high school education from a U.S. correspondence school. After she earned her diploma, she worked as a secretary for the United Fruit Company Merchandise Department for nine years. Maria enjoyed her job, but she wanted to explore other options and live in a bigger city. In 1970, she went to a fashion school in Boston to study fashion merchandising.

Elena attended the Company school through the 7th grade, but she wasn't able to start the 8th grade because the school had closed. She would have had to travel daily by train from Almirante to Farm 8 in Changuinola like she had done in the 5th grade. This caused my parents to decide that my mother and Elena would move to the U.S. and live near me so that Elena could attend high school.

The next one to join us in Cincinnati was my father, who took early retirement and arrived in 1972 because in 1970, the Panamanian government had nationalized the United Fruit Company hospital and medical department, and my father was no longer allowed to work there.

LIFE IN THE UNITED STATES

In early 1971, Maria met Jim Troutman through a newspaper ad. He lived in Brewer, Maine, but came to Boston to meet her. They fell in love immediately and married in 1971. She moved to Brewer, and they had three children. Maria passed away in 2020.

My father decided that the process for obtaining a license to practice in the United States was too onerous at his age, so he worked at a clinic doing medical tasks that did not require a medical license. When he turned 72, he fully retired. My parents lived in Cincinnati for nearly two decades.

I lived in Cincinnati until 1973 then moved to the Chicago area to work for the Argonne National Laboratory (ANL). I worked for the ANL and The California Institute of Technology for most of my career, conducting research in computer science and mathematics. Retirement from ANL came in 2019 and I moved to Annapolis, Maryland.

Elena graduated from the University of Cincinnati in 1980 with a degree in engineering science. She wrote software for industrial robots at Cincinnati Milacron, was a developer and manager for geometric modeling tools at a firm which is now part of Siemens, and worked in robotics at the National Institute for Standards and Technology, retiring in 2021. She married Richard Beagle in 1982, and they have a son, Quinn. Elena retained her Messina surname.

My father died suddenly in Cincinnati on January 29, 1989, at age 79 of a ruptured aorta. As requested, he was cremated, and I scattered his ashes in the Gulf of Naples in Italy. Six months later, on August 8, 1989, my mother died of a heart attack. She wished for her body to be used for medical research or education, and it was donated to the University of Cincinnati Medical School. After two years, her ashes were given to Elena.

LOOKING BACK

Although I was only six years old when I left Italy, I retain a strong emotional attachment and go there several times a year. There are relatives to visit on my mother's side as well as many friends and colleagues whom I met through work.

I have always been grateful to my parents for the courage they had to move to other countries. Because of their courageous and adventurous spirit, I was able to live in countries with different cultures and make friends with varied backgrounds, especially in Almirante.

British troops landing at Salerno, 1943.
Photograph: British troops and vehicles from 128 Brigade, 46th Division. © IWM NA 6630.

A work by Vincenzo Marinelli.[2]
Photograph: Wikipedia.

Dr. Nicola Messina, Margherita, Maria and Paul.
Italy, circa 1948. Photograph: Courtesy of Paul Messina.

Costa Rican airliner, LACSA, on grass field at Base Line
in Changuinola after a downpour, 1950s.
Note the soccer field in the foreground.

Jungle along the railroad.
Photograph: Fr. John Kennedy.

Motorcar in Changuinola.

Margherita and Maria Messina at Company house.
Photograph: Courtesy of Paul Messina.

Left: Dr. Engler's boat, the Argo.
Right rear: Almirante Hospital, opened in 1920.
Photograph: Courtesy of the Engler family.

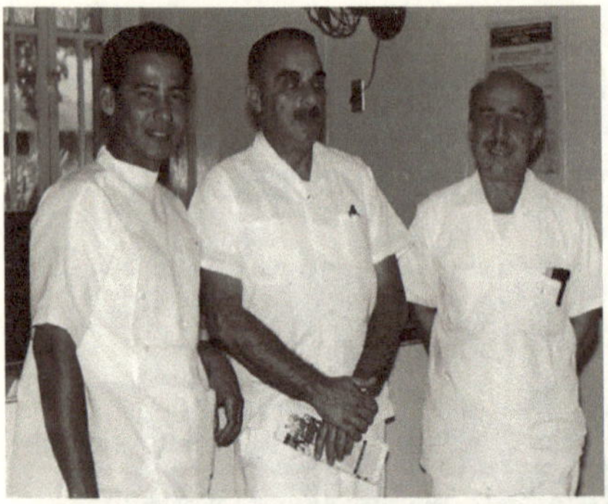

L-R: Drs. Carlos Morales (dentist), Gustav Engler and Nicola Messina. Photograph: Courtesy of Paul Messina.

Almirante Club entrance. Photograph: Fr. John Kennedy.

Party. Dr. Messina (striped shirt and hat). Photograph: Courtesy of Paul Messina.

Excursion on the Talamanca, circa 1957.
Left bench: Dr. Jaffe, Dr. Messina and Paul.
Center bench: Zita Engler,
Right bench: Maria, Mrs. Messina and Tommy Engler.
Photograph: Courtesy of the Engler family.

The Talamanca at Boca del Drago, circa 1957.
Photograph: Fr. John Kennedy.

Almirante American School, 1955. L-R: Paul Messina, Ivonne de Castillo, Virginia Bogantes, unidentified boy, Mildred Swanson (teacher), Maria Messina (with flowers, graduating) and Ramon Aizpurua. Photograph: Courtesy of Paul Messina.

Almirante, 1969, built over pestilent swamps by the United Fruit Company. The hospital is on the right. Clyde Stephens collection.

13. Sharks by John Munch, Dr. Erin Dillon and Clyde Stephens

THE FOLLOWING ARE THREE vignettes about sharks in Almirante Bay—catching them, studying them, and marketing them.

CATCHING SHARKS IN ALMIRANTE

by John Munch[1]

As an unruly boy growing up within feet of the water in the 1950s, Almirante Bay was a fascinating source for daily adventure, and sharks were an irresistible part of the bay's attraction. I first began to meet and know individual sharks when I started skin diving with a mask and snorkel. I convinced my father that scuba tanks were essential for really getting to know the sharks. He approved but insisted that I needed a dive buddy. My buddy was Tom Dunn, the superintendent of materials and supplies for the Company.

Tom had already been a mentor when it came to fishing. He and his wife, Alice Louise, would often invite me to join them on Sundays to cruise in their small boat on fishing trips around different parts of the bay. The craft was powered by an 18-horsepower outboard engine. Somehow I convinced Tom that he too needed an aqualung, as scuba tanks were then called. That also meant buying a 3-stage compressor to fill the air tanks, a process that took about 90 minutes per tank.

In any event, scuba tanks made it possible to dive deeper along the edges of reefs. A favorite location for diving was a reef between Almirante and Bocas Island (Isla Colon).

A partially sunken metal barge marked the reef and provided a grand habitat for lobsters and fish of all kinds. Tom and I dove there quite a few times, and we soon became friendly with several sharks of varying sizes who seemed to be gently territorial, claiming specific areas along the reef.

When we initially encountered each other, both diver and shark would speed off in opposite directions. Over time we got to know each other quite well, and neither of us would take flight. They were nurse sharks. Once, when several of us went free diving at the reef, we were actually scared out of the water by a group of a dozen or so sharks that appeared to be makos. I don't know if they were aggressive or just curious, but they did circle us.

Despite the friendship I developed with several sharks, one of my biggest regrets from my youth stems from the large number of sharks that I wantonly caught and killed to no useful purpose beyond youthful blood lust. I would set out a line to fish for a shark at night using a thick manila rope, the end of which would be tied to the cleat on the dock in front of our house. Large hooks could not be found, so a number of "shark hooks" were fabricated for me at the Company machine shop in Almirante. The leader between the hook and the rope was generally twisted, heavy wire from wooden boxes of White Label Scotch whiskey, what banana people drank back then. But there was a problem with this design and strategy. I never knew if a shark had taken the bait until the next morning when I would either find a broken rope, a straightened-out hook or a drowned shark.

That arrangement wasn't very exciting so I soon devised a system where I would tie a smaller line from the main rope and run it to my bedroom upstairs. The smaller line would itself be tied about five feet from the cleat end of the heavy rope, with the other end running up to the frame of my bed. When a shark would bite and run with the bait, my bed would be yanked across the room, waking me—and the rest of the family—so that I could run down to the dock and deal with my catch.

Over the years, I caught at least one hammerhead shark. On the nights that I fished for a shark, I almost always caught one unless it escaped from the hook, meaning there were plenty of sharks in Almirante Bay in the 1950s.

After a shark was pulled ashore, we boys would put a large piece of wood in its mouth, and the animal easily broke it in two with its mighty jaws. Although we had no use for the shark, local people asked for permission to cut out the liver.

Even though it was reported that nurse sharks are more common today, I know now that things have drastically changed in the bay. Along with my remorse and regrets about killing sharks, it is clear that we humans are a terribly invasive species as evidenced by views from Google Earth showing massive losses of habitats caused by the explosive human population in the Almirante Bay area and elsewhere in Bocas del Toro. Virtually none of the little farms existed when my family left Almirante in 1959. There was no massive, human habitation in the mangrove swamp across from the Company hospital. I saw a few small farms when I went fishing or hunting up Western River or on Shepard's Island. Now there are slums and developments where there was just wild bush. It is probable that there has been a similar loss of underwater flora and fauna. All of this is habitat loss that affects the entire food chain.

Education starts with the young. Considering all of the wanton damage I caused as a youth, I would have been a prime target-student for learning about conservation and protecting ecosystems. My youth is gone, but still I mourn the loss of my shark friends.

STUDYING SHARKS

by Erin Dillon[2]

As part of my dissertation, I reconstructed historical shark baselines in Bocas del Toro using the fossil record and exploring how their communities have changed over the last several thousand years. I've also been diving into the human history of Bocas to gather evidence regarding how humans have interacted with and portrayed sharks in the region, both recently and historically.

Currently, I've been combing through the literature and gathering archaeological data from middens (usually shark teeth or vertebrae), historical accounts of sharks, survey data, and modern fisheries catch data. The goal of this work is to supplement the palaeoecological data we have in two ways: 1) Assess how fishing may have driven declines in shark populations and 2) Evaluate how people's perceptions of shark presence and abundance have changed as an additional indicator of shark declines over time.

We found a tiger shark tooth at a 7,000-year-old fossil reef site on Isla Colon, but that species has yet to be sighted in modern surveys conducted in Almirante Bay. John Munch's photograph of a tiger shark in Almirante in 1954 is recent evidence of their presence and is particularly interesting since tiger sharks would be at the top of the food web.

We also collected shark fossils where mangroves and ancient reefs were dug up with machinery and developed by the builders at Little Bight and at the adjacent, defunct Sunset Paradise.

Our data reconstructing shark communities in Bocas del Toro echo John Munch's experiences catching sharks as a child in the 1950s. We used shark scales preserved in modern and fossil reef sand as well as historical documents to set a shark baseline before human impact and quantify how much the shark community had changed over the last several thousand years. We found an overall decline in shark abundance (about three-fold) as well as a shift in community species composition. In particular, nurse sharks are more common today than they were in the past.

The photo of the sawfish in Almirante is incredible. Today, they're critically endangered and rarely spotted in general, so that photo provides a pretty powerful context.

MARKETING SHARK MEAT

by Clyde S. Stephens[3]

In the 1980s, an ambitious local man managed to get $200,000 from the Inter-American Development Bank and build a shark-processing plant on Isla Carenero on the point across from Bocas Town. The plant was modern, with stainless steel infrastructure. To get sharks, he contracted with five small boats that employed indigenous men and other locals at Punta Valiente, Chiriqui Lagoon and Almirante Bay.

The enthusiastic entrepreneur came to my room in the Hotel Bahia soon after production began and proudly gave me a sample of his dried shark product. It came in a neat little package with a label that identified the product as Bocas Bacalao or some similar brand name. Traditionally, *bacalao* is salted, dried codfish from the North Atlantic that is widely distributed in Latin America and the Caribbean. It comes in wooden barrels from Scandinavia, Portugal and Spain, and sales are high during Catholic holy days when fish is eaten. This was the market the owner had planned to exploit, plus sales throughout the year in Panama and other countries in Central America. The label on the package did not identify the contents as shark. I complimented him on his new product, and did not see him again until I retired to Hospital Point in 1991.

After a couple of years, he went broke, could not pay the bank, and the plant shut down. The problem? There were not enough sharks to supply the business. The bank picked up the dysfunctional pieces of the factory after he took what he could salvage for himself to start another venture—a lobster-processing plant on the east side of Carenero. As usual, the entrepreneur was optimistic, but the lobster plant failed for the same reason as the shark enterprise—not enough lobsters. A few years later, he got on a bus near Chiriqui Grande destined for David across the Continental Divide. When I asked him what he was doing, it was no surprise to learn that he was a fish merchant and that fish catches were in short supply in Bocas del Toro.

Also during the 1980s, other investors developed a large shark business near David. That area was selected because studies showed a high population of sharks in those waters. The factory produced and distributed shark meat in attractive packages that were marketed in local stores and supermarkets. To avoid the stigma of eating shark meat, the product was called *cazón* instead of *tiburón* or shark. To deal with the slight smell of urine, an enclosed recipe gave instructions to marinate the filet in milk, onion, lemon and other spicy ingredients.

Phyllis found it amusing when she bought a pack of *cazón* in a local grocery store in Puerto Armuelles, where we lived from 1982 to 1987. She asked the cashier if it was shark. The reaction was defensive, and the vendor emphatically denied it.

Phyllis followed the instructions on the package and served a most delicious dish of *cazón*. We both liked the shark recipe and bought more, but after a few years, the shark factory closed for a lack of sharks.

Manager's house, Almirante, Panama, 1913-2010.
Clyde Stephens collection.

Tiger shark at the Munch house, 1954. L-R: Tommy Engler, Mike Smith, John Munch, Paul Messina and Stevie Richards. Photograph: Courtesy of Paul Messina.

A small-tooth sawfish near the Munch house, 1955. Photograph: Courtesy of Paul Messina.

Erin Dillon looking for fossils on a coral bed near Bocas del Toro. Photograph: Courtesy of Sean Mattson.

Above: UC Santa Barbara Professor Aaron O'Dea searching for shark fossils at Sweet Bocas, circa 2013. Photograph: Courtesy of Felix Rodriguez.

Left: Clyde Stephens in his cayuco on the Snyder Canal in Panama. Clyde Stephens collection.

14. Peace Corps Volunteers Bocas del Toro, Panama by Joseph Looby[1]

JOHN F. KENNEDY, IN his inaugural speech, stated: "Don't ask what your country can do for you, ask what you can do for your country." Today, in Washington's political arena and beyond, this appeal to public service seems to have lost currency.

Responding to this challenge, I ended up in Panama in December 1964, and was stationed in Bocas del Toro, in the northwestern end of the country. My new home was in a banana port called Almirante that served as headquarters for the Chiriqui Land Company, a subsidiary of the United Fruit Company. That is the company that gave us the term Banana Republic to the English language. For those of us who can still remember, think of Carmen Miranda, the Chiquita Banana TV entertainer and spokesperson back in the 1950s.

The assignment was fraught with challenges from both ends of the political spectrum. I was charged with working to help develop a solution for the oldest agricultural cooperative that had encountered a huge problem. The world price of cacao, the local farmers' cash crop, had sunk so low that it no longer covered the 522 members' expenses to produce it. This particular group of cooperative members and their neighbors were viewed with suspicion by the management of the Chiriqui Land Company, aka United Fruit Company.

Some members lived in Patois Town on the wrong side of the fruit company's railroad tracks that conveyed Chiquita bananas from Changuinola to Almirante, where I lived. Labor unrest was an issue at the time, and the Patois Town inhabitants supplied most of the dock workers who loaded boxes of bananas from railcars to the awaiting steamships.

Soon after my arrival, the American manager of the banana company, Aaron King, who was also a Wharton School graduate, accused me of being a communist. Perhaps it was to test my reaction to the accusation; perhaps it was because he couldn't imagine why a Wharton School graduate would want to waste his time on such a thankless and arguably hopeless task.

Ironically, some of the townspeople harbored the thought that I MUST BE either a United Fruit Company spy or a CIA agent sent to protect US business and political interests there. I say MUST BE because some simply could not imagine why anyone with a university degree would come to their town to just live, observe, and work alongside them, with their benefit in mind. That was a rather novel concept. Gene McCullers, the Peace Corps Volunteer, whom I replaced at the cacao co-op, must have experienced some of the same suspicion. To add to the admixture of distrusts, an anti-Peace Corps notice circulated in the country.

FAST FORWARD NINETEEN MONTHS

In June 1966, after many projects and the usual frustrations later, it was time to say farewell to the town and its people I had come to appreciate and admire, and yes, love. As I was assigned to this post to replace the very first Peace Corps Volunteer, I requested a replacement for me, this time, someone with experience in construction who could work to complete a residential housing project that I had initiated.

FAST FORWARD ONCE MORE

By the summer of June 2008, the volunteers who had formed the Peace Corps Panama Friends Organization decided to return and have a reunion to celebrate the 25 years of service and the 45th year after the Peace Corps' arrival in Panama in 1963. I went back to a fantastic reunion and learned that when Panamanian dictator Omar Torrijos ordered the Peace Corps to leave the country in 1971, the hasty departure resulted in a huge loss of records that we had kept since 1963. I had planned to return to Almirante to spend five days looking into the various projects that had I worked on and to come to grips with the burning questions that plague many Volunteers. Did you do anything that would stand the test of time? Did you change anyone's mind? Was anything at all made better because of the Peace Corps' efforts? For those of us who went to explore a foreign culture, the gains for most of us were and continue to be immeasurable, but the nagging question was whether our hosts could say something comparable about our efforts. Quantitative benchmarks are not sufficient. Prior to this return trip in 2008, I could never say with certainty whether the work we did resulted in real and valuable benefits to the community. Most Volunteers understand that it takes a generation to really know for sure whether introduced changes endure.

What follows is an abbreviated report on several projects that I sent to the Peace Corps to help recuperate their Panama records and to answer certain questions that I have asked over all these years, without satisfactory answers.

CACAO COOPERATIVE—COCABO
(*Cooperativa Cacao Bocatoreña*)

The original assignment I had was to follow Peace Corps Volunteer Eugene McCullers as business advisor to a cacao cooperative facing the deeply depressed world price of cacao in 1964. It was the oldest cooperative in Panama. Fortunately, the co-op survived and is thriving today. Organic cacao now commands the highest price of the three classifications of product that the co-op sells. In 2015, through the Fair-Traded Equal Exchange Program, a worker-owned co-op, a 3.5-ounce bar of 80% dark chocolate that is processed in Switzerland using raw cacao beans grown in Bocas del Toro and sold through the Almirante-based Cooperativa (COCABO) is sold for $4.25 retail in our Weaver's Way food cooperative in Philadelphia, Pennsylvania.

IN SEARCH OF AN ALTERNATIVE

Back in 1966, when raw cacao bean prices on the New York Market were so low as to make growing cacao unprofitable, we looked for an alternative cash crop that the co-op members could produce and sell. Some 650 heat-treated, nematode-free corms (bulbs) of Giant Plantain were provided to me by Clyde Stephens, who headed up research for the Chiriqui Land Company in Changuinola. The plan was that each mother plantain would produce four to six sprouts at the base of the plant. After the fruit was harvested in nine months, co-op members would transplant the extra sprouts to their own cacao farms. In several years, their unprofitable cash crop of cacao would be replaced with Giant Plantains that could be marketed in Panama and international markets like Miami and New York. The government provided farm land that was overgrown and had to be chopped wit*h machetes* before we could plant the plantain corms.

My tour ended shortly after the Giant Plantain corms were planted in the jungle setting, up the Western River across the bay from Almirante. After I left, inquiries about the project went unanswered until I was informed by Alfredo Vaz that indigenous neighbors dug up the plantain corms and planted them elsewhere.

When cacao co-op members and their manager, Roberto Dailey, went up the river to check on their plantain project, they found only holes in the ground where we had planted the Giant Plantain corms. Thus, the plantain project came to an end. We can only hope that the people who misappropriated the plants got some benefits.

THE ALMIRANTE SELF-HELP HOUSING PROJECT

This was not part of my assignment established between the Peace Corps and the Government Ministry responsible for fomenting the establishment of agricultural cooperatives and credit unions throughout Panama. It was, however, an idea whose time had come. Some townspeople thought it would be a good idea to build a cemetery on a hill overlooking Almirante Bay. I suggested that it might be better to use self-help efforts of the living to build substantial homes on that beautiful site. Through their own efforts, they could build some valuable equity on their new property that they could enjoy and pass on to the next generation.

The project got underway when the government provided plans for housing units. Then came the need to clear the land and start making the blocks for the construction. As my time in Panama was short, I asked the Peace Corps to send a volunteer with constructions skills, of which I had none. The replacement was identified, and Dave Long arrived with a new group. He was the perfect fit for this project, and he demonstrated his skills by building his own residence near the planned site for construction.

Other Self-Help members built their houses through this plan of teamwork. The project was very successful. Since then, residents have enhanced and upscaled these nice homes on the hill that might have become a cemetery.

OTHER PROJECTS AND ACTIVITIES

There were other activities along the way. Dave Long and I helped to raise funds for the Almirante Boxing Club. Another activity was coaching local high school boys' basketball teams. One of the basketball players became a priest and was ultimately named as the Auxiliary Bishop of Panama. Another player became the Dean of the Economics Division at the University of Panama. One player moved to the United States and had a career in the United States Army. Other Peace Corps Volunteers in Bocas del Toro were Tony Masso and Neal Jacobs, who created the playbook for attracting tourists to Bocas del Toro in 1965. Neal worked on fund raising for creating the Bastimentos Marine National Park. He designed, manufactured and sold two lines of t-shirts with sales proceeds going to the park.

During my time in the Bocas del Toro area, I observed the old tradition of catching, keeping and selling turtles. There were efforts to protect sea turtles, but conservation was not effective during my time there. Turtles were kept in pens made with cane or bamboo poles stuck in the bottom of shallow water.

FINAL THOUGHTS

My 19 months in Panama was an experience that I could never have had otherwise. This assignment expanded my view of the world and of people I had previously believed were so different from me. One of the goals of the Peace Corps was to take what you had learned with you and use it. Because of that, I became a changed person.

PROFILE OF A PEACE CORPS DOCTOR

When I first arrived in Almirante, I was introduced to Dr. Gustav Engler, a great man who was an international authority on tropical diseases.[2] He was the medical director of the United Fruit Company hospital in Almirante. Dr. Engler was also the local Peace Corps doctor.

Soon after we met, Dr. Engler invited me to go for a sail in Almirante Bay. I had never been on a sailboat, but he put me at the tiller when the spinnaker was unfurled to maximize the speed of the boat. What an introduction to Bocas del Toro and my new home in Almirante. About a week later, he asked me to consider sailing from Almirante to Jamaica, some 700 miles to the north. I had heard he was an expert sailor, and it sounded like the trip of a lifetime, so why not. Neal Jacobs, another Peace Corps Volunteer living in Bahia Azul, about 40 miles east of Almirante by sea, was discussed as a third member of the crew.

At the time, who knew that only several weeks later, Dr. Engler would suffer a fatal heart attack while making rounds at the hospital. The what-if question has remained in my head ever since. What if Dr. Engler, Neal and I had gone on that trip to Jamaica?

On the Sunday Dr. Engler died, I had breakfast with him and his wife *doña* Zita following the Catholic Mass we attended at the Iglesia San Jose. We ate Viennese pancakes and listened to Strauss waltzes playing on his big high fidelity system. What a shock to learn that he had a heart attack after excusing himself from the table to attend someone at the hospital.[3]

The following information was posted by their granddaughter, Gina Engler, who graciously gave permission to use her findings. She is the daughter of Dr. Tomas Engler and Angelina Guerra de Engler.

ABOUT GUSTAV
by Andreas Engler Waldner
Gustav was born in 1909 in Vienna. He studied at the Public High School of Grinzing of Vienna and graduated in 1927. Gustav studied at the University of Vienna Medical School and graduated in 1933. He completed one year of Internship at a hospital in Grieskirchen in Upper Austria. Thereafter he was chosen to join the staff of Professor Winkelbauer, head of the Surgical Department of the Vienna University Policlinic and was the youngest surgical resident at age 25. He obtained appointments to the Court of the Prince of Lichtenstein and the Sascha Film Studios.

Gustav left Austria in 1938 because he faced persecution and extermination as a Jew. His father died shortly after from intestinal cancer, and his mother went to live with her sister in Vienna. Gustav's mother and aunt were deported to Auschwitz and were killed in 1944 in the gas chamber.

Gustav's refuge started in Italy for two months in 1938, where he stayed with friends. Thereafter, he went to Sweden and obtained an appointment as a ship's surgeon aboard the *Axel Johnson* of the Johnson line.

He worked for two years as ship surgeon, traveling from Gottenberg, Sweden, to the west coast of USA-Canada, crossing the Panama Canal. Due to the need for ships during the World War II effort, his ship was called up for war service and Gustav stayed in California.

In California, Gustav remained several months and did postgraduate work at Stanford University. In the meantime, he established contacts with the United Fruit Company and was appointed physician at its tropical divisions in Panama. He took a bus from California to New York, headquarters of the Company and from there sailed on a United Fruit Company ship to Almirante in March 1940. He worked at the Almirante and Puerto Armuelles hospitals from 1940 to 1945 and then became the medical superintendent of the Almirante Hospital from 1945 to 1965.

As soon as he secured a visa, he was able to bring his fiancée, Zita, to Panama. She arrived on October 10, 1940, and they married as soon as she got off the ship. They had tried to marry on one of his stops in London while working on the *Axel Johnson*, but there was no time.

Gustav passed away on January 10, 1965. According to his wife Zita, she stated: "...It is important that you will have a true idea what Gus was really like, and knowing him so well, I could mention two points that might shed some light on his spiritual state of mind...1) Gus was a man who kept studying all his life. Whenever he came home from his hospital calls, which was almost every night, he would study for at least two or three hours before he went back to sleep in order to keep up to date on the medical progress of the world. Yet one night in conversation with a friend I heard him say, 'I have been studying all my life and now that I am 55 years old, I realize how little I know'. 2) This was the way Gus gave his life to his patients on that Sunday he died... After breakfast Gus left for the hospital to make his morning rounds. On the way to the hospital he had the heart attack, but kept on walking, made all the visits to his patients, walked back to the house, sat down on his bed and tried to give himself an injection. I called two of the other doctors, he was given oxygen but within 20 minutes he died of thrombosis. I believe he was ready to go."

Anti-Peace Corps cartoon: "Keep an eye on them." Photograph: Courtesy of Joe Looby.

Joe Looby, age 22, Peace Corps Volunteer. Photograph: Courtesy of Joe Looby.

Joe Looby (cayuco captain) and co-op manager Roberto Dailey in Almirante. Photograph: Courtesy of Joe Looby.

Left: Cacao drying on the sun dryer. Right: Bagged to ship to New York. Photograph: Courtesy of Joe Looby.

Norman Wood (front) with cacao beans ready for export to New York. Photograph: Courtesy of Joe Looby.

Dave Long advises Self-Help Co-op house builders. Photograph: Courtesy of Joe Looby.

Self-Help Co-op members and their block-making machine. Photograph: Courtesy of Joe Looby.

Self-Help house, later upscaled. Photograph: Courtesy of Joe Looby.

Mr. Kepsy's pen, Almirante, 1965. Photograph: Courtesy of Joe Looby.

Mr. Pinnock, turtle hunter / fisherman, Boca del Drago, 1965. Photograph: Courtesy of Joe Looby.

Man with sea turtle and fish (front of the boat) between turtle pens. Note the toilets over the water. Cedar Creek, Almirante, 1965. Photograph: Courtesy of Joe Looby.

Almirante Boxing Club. Photograph: Courtesy of Joe Looby.

Dr. Gustav Engler with his wife Zita. Photograph: Courtesy of Gina Engler.

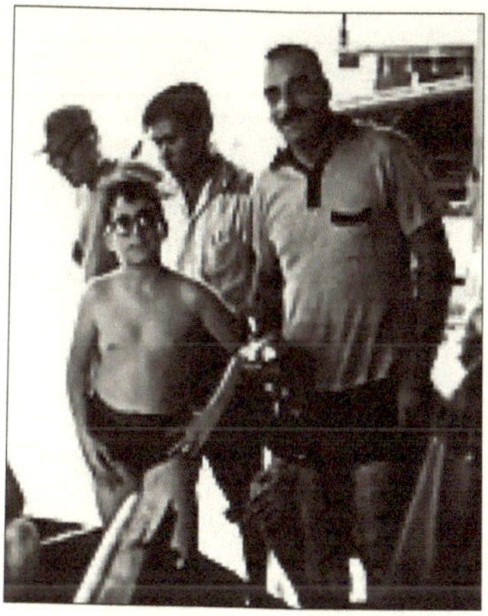

Dr. Engler with his son Tommy. Almirante, late 1950s. Photograph: Courtesy of Gina Engler.

United Fruit Company Hospital, Almirante, 1950s. Standing L-R: Oscar Valle, Dr. Gustav Engler, Dr. Ludwig Jaffe, unidentified man, Raymond Stanton, unidentified nurse. Photograph: Courtesy of Gina Engler.

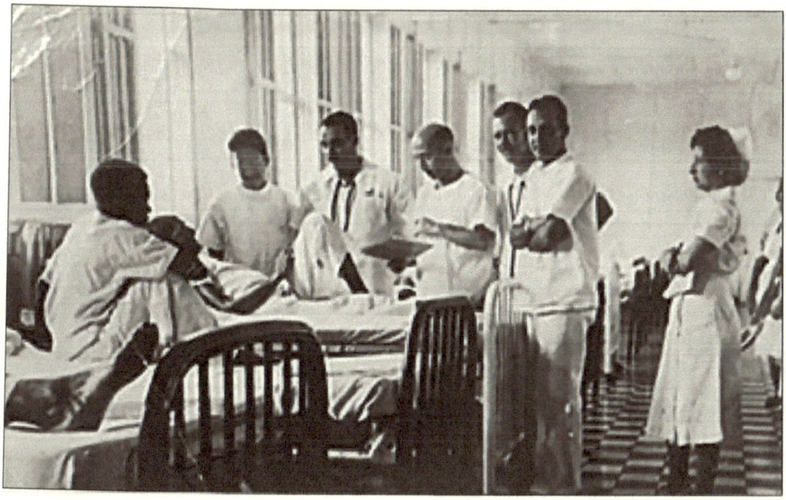

15. Family Planning and the Wart

IN THE LATE 1950S, great scientific thinkers warned that human populations were exploding at alarming rates around planet Earth. In 1968, Paul and Anne Ehrlich wrote *The Population Bomb* that shocked many of us into reality. I remember a young Harvard professor giving a lecture at an international meeting of entomologists. His subject was on insect population dynamics, but he also wanted to remind the audience that we humans better start thinking seriously about the consequences of too many *Homo sapiens* on this Earth. His attaché case was displayed on the podium so that the audience could clearly see a prominent message that read: "THE POPULATION BOMB IS EVERYBODY'S BABY." I was impressed and practiced what he preached long before I was married in Honduras in 1963.

Traveling to Haiti on a banana research trip, I soon observed what *The Population Bomb* warned about. The land there has been slashed and burned for centuries and topsoils have eroded into the sea. What remains today are rocky outcroppings and infertile substrata that can hardly produce any kind of crop. Instead of reforestation, remaining trees and other vegetation have been chopped in desperation. Traditional food plants have minimal nutrition that make malnutrition even worse. Diseases, stomach parasites, insect pests and most any conceivable malady cause misery and hopeless poverty in almost every level of the population. The land can only support a limited number of humans under these conditions. And that is what my colleague, Dr. Leonce Bonfils, tried to explain to his people so things would change for the better.

Dr. Bonfils is a Haitian entomologist whom I had met in Miami at meetings of the Entomological Society of America. He had been financed and sent to Iowa State University, where he received a doctorate in entomology. Upon his return home, he was obligated to serve as an educational advisor in outback, poor villages for two years or more. His attempt to improve the lot of his people was radical in their eyes but he pointed out educated realities nonetheless.

He delivered his messages in schools, at village meetings and by visiting families in their homes. He dealt with educators and religious leaders but did not seem to be moving forward. It became obvious to Dr. Bonfils that he must introduce family planning. He explained that fewer children per family would mean more money, better education, health, clothing, happiness and opportunities for the planned babies.

Birth control pills had recently been introduced to the world, but the Catholic church was adamantly opposed to any form of family planning. That was not God's plan. Also, it meant fewer members for the church. But this did not stop Dr. Bonfils from his mission. Then suddenly, all of that ended one afternoon. The elders and men in the village informed him that it was the mission of a Haitian man to sire as many babies with as many women as possible. They added that he was an educated lunatic who was destroying their traditions, and he must leave at once and never come back.

The agitated mob literally hoisted him on their shoulders, took him to the edge of the village and physically threw him into the brush and bramble. He could not even return for his personal belongings.

That was in 1961, and over sixty years later, things have only grown worse in Haiti. One merely has to look at a Google map taken from a satellite to see where the border is between Haiti and the Dominican Republic. The west side of the international border in Haiti is devastated right to the border line. The Dominican side east of the border is just a little different—but not much.

Untold billions of dollars from around the world have been donated and invested in Haiti trying to improve their situation, but there has been no progress. That tiny land mass can only sustain a few million people, but the population in 2022 was close to 12 million and growing, despite massive migration to other countries.

Back in Panama, I reflected on the 1960s when I encountered a Mr. Baker with his two wives and two sets of children living in large, palm-thatched *ranchos* on a sand spit at the mouth of the Snyder Canal at Boca del Drago. He proudly told me that he had another child "outside the house" in a banana farm in Changuinola where he worked. I pompously showed him my young son and daughter in the boat and told him that was my limit because I had gotten a vasectomy in the Almirante Hospital. He had never heard of such a procedure.

One Saturday afternoon after work, Victor, our yard boy, approached me in the garden in a very serious mood. "*Jefe, que voy a hacer? Tengo siete hijos y mi mujer va a tener otro. Además, tengo otra fuera de la casa.*" (Boss, what am I going to do? I have seven children and my woman is going to have another one. Besides, I have another one outside the house.) He was 28 years old.

I modestly explained all of the common forms of birth control for men and women, something he had no idea about. I concluded that none of those methods were very practical for an indigenous couple who came from the remote Cricamola River. After some pondering, I explained to Victor about the ease and benefits of a vasectomy. He liked the idea, so I made an appointment for him with Dr. Raul Davila in the Company hospital in Almirante, and felt proud that I could help a young man who was desperate to find a solution for his family planning problem.

Many months later, I saw Dr. Davila at a Company picnic at Boca del Drago, and he asked me what happened to my employee. I was puzzled and asked the doctor to explain what he meant. He said that when Victor came to the hospital for the vasectomy, he had asked him why he was there, a routine question before any surgery. Victor quickly cleared his throat and told Dr. Davila to remove a wart from his hand, which he did. I was stunned to learn this and a bit disappointed.

The next week, I asked Victor why he did not get the vasectomy. He raised his voice more than normal and told me: "*Jefe, este no es un asunto suyo.*" (Boss, this is none of your business.) I immediately backed away and said no more.

Later that year, Victor quit his job and said he was moving to a little farm in the hills above Almirante. I wished him well and said goodbye. But before we departed, he lingered on to tell me something else. "Boss, before I leave, I wanted you to know that I made another appointment and got the vasectomy. That was the best decision I ever made in my life, thanks to you." I was ecstatic, congratulated him for his courage and wished him well with his new life.

On parting, and with a twinkle in his eye, he added: "*Además, puedo estar con otras mujeres sin problema.*" (Besides, I can be with other women without problems.)

I hope the end justifies the means.

Much of Haiti has been devastated by overpopulation.
Map: © TerraMetrics.

Left: The Baker family, 1960s. Clyde Stephens collection.
Right: My children, Carl and Linda, with an indigenous person.
Clyde Stephens collection.

16. Reservations, Please

AS THE RESEARCH MAN in charge of banana technical services in Changuinola, one of my duties was to work with visiting scientists. After one visitor, Dr. Steve Johnson, and I finished our company business, he suggested that we visit the capital of Bocas del Toro. This was 1972, and I knew that Bocas Town was not very active. Nevertheless, I wanted to be sure that we had a hotel reservation. I asked Bob Bruce in the manager's office to reserve us a room in the Hotel Bahia, formerly the old United Fruit Company general office building constructed in 1905.

Steve hired a small Cessna airplane, and we flew from the Changuinola Airport in Base Line to Isla Colon, where government offices were located, a 15-minute flight. When we landed in Bocas Town, I looked for a baggage boy to help with Steve's heavy Samsonite suitcase filled with banana samples for a chemical analysis. There was no one in sight, nor was there a taxi or any vehicle, for that matter. We walked nine blocks to the Hotel Bahia to register, but there was no sign of human life in the hotel. I called loudly and got no answer. We walked upstairs and called for anyone. Astounded, we sat down not knowing what to do next. Then we heard someone singing and whistling in the far corner of the building. We rushed back and knocked on the door. A man opened it and asked what we wanted. I announced that we had a reservation and wanted to sign in and get into our room. "Ah, you come back tonight when *Mista* Thomas come. He take care of you then." Steve and I looked at each other in comical awe and wondered how to proceed. I raised my voice a bit and made it clear that we had a reservation and demanded that someone help us get in to any room. I suggested that he call *Mista* Thomas, but he said *Mista* Thomas did not have a phone at home or in the hotel, and he let us know that he was a cook, not a hotel clerk. Then he spotted a boy riding by on a bicycle and yelled at him to come back. He told the boy to go tell *Mista* Thomas at the far end of town that Clyde Stephens had arrived and wanted to get into his room. The message from the boy when he returned was that *Mista* Thomas would arrive at the hotel that night at seven o'clock and not to worry. Just go upstairs, open the door on the left and make ourselves at home. At least, we had a room with no key but we were making progress.

There were very few people in the streets and only an occasional vehicle that went around pedestrians. No one was in a hurry. As we looked at the bay, a local man made our acquaintance and was curious as to why we were in Bocas. He volunteered to say that he heard that a rich American was going to build a big hotel on Hospital Point on Nancy's Cay across the ship channel. I let him talk but did not reveal that I had been the owner of Hospital Point since 1970. I did mention that we were going to Hospital Point.

He told me that Mr. Grenald lived around the corner and had a small *cayuco* with a 5-horsepower engine. Mr. Grenald agreed to take us to Hospital Point so I could show Steve where the old United Fruit Company had a medical center that closed in 1920.

Six months earlier, I had paid Mr. Charles Samuels $45 to chop the bush along a trail so I could see some of Hospital Point. When we arrived at Hospital Point, we jumped ashore and found Mr. Samuels. He came rushing out from his shack to greet us with a big smile that showed very few teeth. With great animation, Mr. Samuels started sharpening his *machete* and proclaiming in an excited voice that it was such a coincidence that we came right when he was just starting to cut the bush that I had paid him to chop six months earlier. "*¡Qué casualidad!*" he exclaimed several times in both languages—"What a coincidence!" He went on and on with the same banter while chopping the trail. "*¡Qué casualidad!* What a coincidence!"

Mr. Samuels chopped ahead of us so we could thrash through the bush a little easier. Meanwhile, we had to keep an eye on his vicious dog, Patron, who kept sneaking up on us from behind. He scolded Patron for nipping at our heels but to no avail.

After making our way around some of the brushy point, Steve and I were hot and sweaty and went snorkeling. The entire reef was similar to a pristine coral garden like *National Geographic* magazine would feature. Lobsters and huge reef crabs were everywhere. It was a wonderful afternoon, but the sun was going down fast.

Mr. Grenald took us back across the ship channel but not directly to the hotel. Suddenly, a machine gun started firing right in front of us and caused a rush of adrenalin—and terror. It was getting dark, and we could not see what was happening. But Mr. Grenald did not slow down, he kept going toward the sea wall. This was the time when the Panama military dictatorship had control of everything, and we were already nervous about that. Then an excited man yelled "*Allí está*" followed with more machine gun fire. I thought he meant "there they are"—meaning us. We were very scared.

When we got closer, Mr. Grenald pointed out that we were in front of the military barracks where there was a large contingency of soldiers in residence. Some of the troops were fishing and one of them had caught a shark that was too big to pull over the sea wall. That is when they started shooting to kill the shark struggling in the water. Mr. Grenald put us ashore nearby. We walked to the hotel quite rattled but relieved to know we were not the target. As promised, *Mista* Jose Thomas came at seven o'clock and collected the $11 for the room with two single beds. He said there was a little restaurant in a private house way down the street if we wanted supper. At the restaurant, I noted that turtle was on the menu. A large lobster dinner was two dollars. After a couple of bottles of Balboa, we began to relax and laugh about the events of the day.

On our way back to the hotel, *Mista* Thomas rode by on his bicycle and wished us a good evening. We walked down the middle of Main Street (*Calle Tercera*), where it was perfectly safe to stroll. I pointed out to Steve that he and I were the only guests in the hotel. Bocas Town was so peaceful and quiet in those years—no reservation needed.

17. Bluefields – Panama: Going Back in Time

IN 1967, WE LIVED at Farm 8 in Changuinola. Our yard boy, Victor Salazar, was provided by the Company. Victor was the gardener and did odd jobs around the house. He even bonded with our two young children, Linda and Carl.

Victor told me that he was born and raised in Bisira, on the Cricamola River. I arranged for us to take a three-day trip around the Chiriqui Lagoon, a body of water with lots of rivers flowing into it. This was the region where commercial bananas were first grown in the 1880s, and I wanted to explore the area where the former plantations were located.

My 20-foot-long *cayuco* was powered by an 18-horsepower Evinrude outboard motor. We left early, went down the old Changuinola River Lagoon and motored through the banana barge canal to Boca del Drago. Our route continued east, and we passed Bocas Town and finally arrived at Split Hill on the shores of the formidable Chiriqui Lagoon. The waves and choppy waters prevented us from seeing the far shore and the high mountains behind. Victor guided me to the mouth of the Cricamola River. We were drenched in salt spray, and it felt good to wash in the fresh river water. We continued up the Cricamola to where the Snyder Brothers grew Gros Michel bananas in the late 19th century. They used small rail lines to get the green fruit to the river bank to load it on riverboats. I saw an old railroad bridge that had collapsed into the river because of floods. This was the kind of banana history that I had come for.

As we traveled upstream, there were more hazards because of river rocks and whitewater rapids. Victor pointed out where to stop, and we tied the boat to a tree at Bisira, Victor's village. We hiked about a kilometer away from the river, where Victor's mother lived alone in a palm-thatched hut in their pasture with a few cows. He had not seen her in two years. As we approached, his mother came to the doorway. Victor said, "*Hola, Mamá.*" She nodded and replied with a couple of words in their Guaymi language. That was all. There was no emotion or conversation. After staring for a minute in silence, Victor said to me, "*Vamos a ver las vacas.*" (Let's go see the cows.) He never went back to see his mother on that trip. We slept in a nearby hut close to the school. The next day, we returned downriver to the Chiriqui Lagoon and stopped at the island of Patterson Cay, where there was a store over the water, stocked with the basics. We bought gas and some fruit juice. I conversed with the woman behind the counter, who told me that she was the granddaughter of a German man, Lange, who had married a native Guaymi woman. Her other grandparents were Americans by the name of Smith, who came to the area in the 1800s. They too had crossed with the Guaymi Indians. I was fascinated by these backgrounds and the genetic admixture.

We arrived at the entrance of Bluefields, a large protected bay with incredible raw beauty. There were not many houses along the shores and coves. At the far end of Bahia Azul, as the region was called in Spanish, we met an elderly Guaymi man. He spoke to Victor in Guaymi and to me in a pigeon English dialect. He had an Anglo surname, like many other indigenous people there. Some of the local indigenous Anglo names that I learned about were: Trotman, Smith, Hooker, Baker, Taylor, Julian, Binns, Thomas, Jesse, Knapp (changed to Naap) and Surgeon. Very few had Spanish names. Anglo names were acquired in the early 1800s when English, Caribbean and American venturers and traders came to the Valiente Peninsula and Bluefields Bay.

The old man agreed to watch the boat so we could hike on a muddy trail to the Caribbean shore. After walking for 20 minutes, we arrived at one of the most beautiful beaches in the world with thousands of tall coconuts as far as the eye could see. The Baker Company had coconut plantations on those beaches in the early 1900s.

When we returned to the boat, I gave the man a couple of dollars and asked him where we might find a place to sleep that night. He suggested that we cross to the other shore of Bluefields, where there was a new school that the Americans had built under a USAID program called *Alianza para el Progreso* (Alliance for Progress). We found a local family in their palm-thatch rancho near the school, and they gave us permission to sleep there. The school had not been officially opened, and school teachers had not yet arrived either. I blew up my air mattress and slept there. Victor slept on the floor, the custom of these indigenous people. I noted that there was no toilet, but I suspected that an outhouse would serve the purpose later.

On the third day, we headed back to our home base in Changuinola, a trip of over four hours by water. I was awed by what Victor had quietly shown me of his indigenous heritage.

THE SECOND TRIP, WITH AIR CONDITIONING

In the 1980s, I was a technical advisor in the research department for United Fruit Company and traveled to Changuinola on occasions. The general manager was Karl Friedrich Koch, a man I had known since our days together at Manzanillo in the Dominican Republic from 1961 to 1962. Fred invited me to go with him on a boat trip and I accepted.

A large percentage of the banana workers in Changuinola were Guaymi Indians—today known as Ngäbe. By law, each worker got a month's paid vacation every 11 months. This meant they had money to pay transportation costs to their remote communities along the shores and up the rivers of the Chiriqui Lagoon and the Valiente Peninsula. That is, if shopping costs and too much drinking didn't wipe out their vacation pay.

The problem was getting back to work on time in the plantations in Changuinola after the month-long vacation was over. By then, most of the workers were broke. It was common for many workers and their families to get back to the work place six months later, because they had no money to return.

Fred found a way to solve this serious labor problem by bringing in a rather large, air-conditioned passenger boat that had previously been used to ferry off-shore oil workers from the Louisiana mainland to the oil rigs in the Gulf of Mexico. To cater to the indigenous workers, he named the boat the *Chobrá*, meaning angelfish in their language.

Fred was a hands-on manager and wanted to experience the trip for himself. So we left Almirante with about fifty indigenous employees, many with their spouses and children. I guessed that this was probably the first time any of them had ever known air-conditioning. Some lively Ping Ding folkloric music blasted on the speakers. There were comfortable seats, but some sat on the floor to stay together as a family. The toilet provided a special fascination for them. As usual, the women were dressed in long, colorful embroidered smock dresses. Sometimes, the whole family was bare-footed but occasionally the father wore shoes or rubber boots.

Outside of the port, the captain revved up the engine and we quickly gained speed. That really added excitement. The first stop was at Split Hill, where several passengers got off. The vessel headed diagonally across the Chiriqui Lagoon that was rough, as usual. After nearly an hour, we arrived at the mouth of the Cricamola River, where a dozen or so employees got off and boarded a large *cayuco* waiting to take them to their communities up river.

The next stop was Patterson Cay, then on to the protected Bluefields Bay, where there was a dock at Punta Alegre. We then returned to the Chiriqui Lagoon. The *Chobrá* cautiously steered around the treacherous Punta Valiente rock islands but did not stop at Cusapin because of the hazards. We cruised through the heavy seas of the rocky Caribbean shore to the isolated and protected settlement of Tobobe. This was the end of the line, and the last passengers got off and some came aboard. Then we headed back to Almirante, a trip that took several hours across rough seas.

The return trip was loaded with employees and their families, who came aboard at each stop along the way. They had finished their vacation and were ready to go back to work in Changuinola. The absentee problem had been solved.

RETURN TO BLUEFIELDS AND PUNTA VALIENTE IN 2021

One of my favorite months in Bocas del Toro is October when normally turbulent waters grow calm. That is the time to explore and enjoy areas not normally accessible for people like me. The locals, who were born and raised there, have no choice, and they come and go regardless of bad weather.

On October 15, 2021, Leovigildo Julian Lorenzo, my talented indigenous employee, and I left our home at Hospital Point on Nancy Cay at 7:00 A.M. and arrived at Punta Valiente in only one hour and ten minutes because the sea was nearly flat all the way. We had one passenger, Arnold Smith, who was returning to his home on Punta Valiente. As the boat approached a steep mountain, Leo stopped near several small rock islands to look at the rookery of Brown Boobies. Some of these large birds were feeding their big fuzzy chicks in nests on the rocky surface. The islands were covered with a green, salty vegetation that was partially shaded by coconuts growing in the rocks.

Leo made a beach landing at Punta Valiente helped by several villagers who grabbed the boat and pulled it ashore. A couple of men were cleaning fish in the surf where little children were swimming. This beach was about 25 meters long, located between steep cliffs on each side. Everyone cordially greeted Leo who had grown up there. A dozen or more boats were stored above the high tide mark, some with 40-horsepower motors but most with no engines. Several of these boats were typical native *cayucos* used only for paddling and sailing. The first building on the beach was the church where Leo's father-in-law, Roy Smith, a community leader, was the preacher. We walked up the steep, narrow and eroded trail that zig-zagged to Leo's house. Luckily, I did not slip or fall. The view from his house was breathtaking, looking out over the village, the surrounding rainforests and the Caribbean. He unlocked the house that he had personally built, opened the windows and gave me a tour. It was the best house in the village of about 30 dwellings. I was pleased to see that some of the houses had solar panels. They had leapt into the future. This was even more amazing considering that income here was minimal. Right next door was another structure where six related Smith families lived communally. One section had a palm-thatch roof and other parts had metal roofs. All rooms were connected to each other. I greeted Leo's mother-in-law, Filomena Hooker, and others in the dining area. Soon, a large bowl of fish, casava, plantain and other tropical vegetables was served to us. While we ate, a couple of naked little children played around the table. As a courtesy, I left a ten-dollar bill on the table for the cook, who was Leo's sister-in-law.

I listened to stories about two dozen American military men stationed at Punta Valiente during World War II. Old-timers remembered the legends of how the Americans landed a small bulldozer on their beach. The tractor made a narrow zig-zag road up the creek ravine in order to access the top of the steep ridge overlooking the Caribbean. They then built a concrete bunker and lookout on top of the forested ridge high above the village. Their duty was to spot and report German submarine activity. Men were always on duty on the ridge, but those not on duty stayed below in wooden houses that they had built near the little beach entrance. Leo told me that he had hiked to the top of that ridge and had seen the concrete structures now covered with jungle.

German submarines were a serious threat during the war. A United Fruit Company ship, the *Sixaola*, was torpedoed and sunk not too far from there on June 12, 1942. Twenty-nine crew members were lost. When the German submarine, the *U-159*, surfaced, the commander interrogated some of the survivors in lifeboats. As a humane gesture, he told them how to find Bocas del Toro, gave them two packs of German cigarettes and then disappeared.

In the 1990s, there was no school in the village of Punta Valiente, but some of the children walked on a rugged jungle trail for over an hour to attend the primary school in Puerto Alegre in Bluefields. They would trek up and down forested ravines near the rocky shore and in the rain, an incredible task for students so young. Getting wet and muddy was part of their life in getting an education. One of those students was Maria Smith, who graduated from the 6th grade when she was 12. Maria had hoped to continue her studies, but the walk to the high-school in Cusapin took two hours each way on steep, forested trails. Some of the male students made the 4-hour walk daily, but girl students were not encouraged to go. Parents, especially fathers, considered that six years of primary school was enough for their daughters. That was their tradition.

It was in this village that Maria, age 13, attracted the eye of Leovigildo Julian Lorenzo who was 17. Their first baby was delivered by her mother when Maria was 14. Leo's mother-in-law had some doubts about him, but his father-in-law gave Leo a favorable assessment and approved of the union of this young couple. Motherhood of young girls was common and usually accepted by the community. Leo had been raised there by his paternal grandparents, and he was no stranger in Punta Valiente. This helped the villagers to give their blessing to Maria and Leo.

When we left Punta Valiente, I wanted to see Cusapin, where the largest population of indigenous people lived along the beach of an extremely dangerous rocky coast. Most of the year, enormous waves crashed in from the open Caribbean

Sea. Sometimes boats flipped and people drowned. Cusapin was the seat of government and had schools through the high school level. There was a paradox of fundamentalist churches interspersed with cantinas, where men with enough money to buy booze, enjoyed lots of sin. Income was mostly from diving for lobsters, working in the banana plantations in faraway Changuinola or working for a governmental agency in Cusapin.

We dropped off Arnold Smith, age 19, who had gone with us to Cusapin to get his Covid vaccination. He wanted to go to the University of Panama in Changuinola, where students were required to be fully protected against the virus. While Leo filled up the gas tank, I walked the trail along the beach to the far end of Cusapin. There was mostly silence along the way, no noise, not much talking and people stayed inside their humble houses. The beach was filled with flotsam, and several buildings had been destroyed by the raging sea. But the day I was there, there was hardly a ripple. It was October.

About noon, we entered the calm and beautiful bay of Bluefields and tied up at the far end. We found a man to watch the boat, and he began to quiz me about who I was and asked my name. I was not sure what he was up to. Soon, he came out with one of my books: *Bosquejo Histórico de la Provincia of Bocas del Toro, Panamá* (*Outline of History of the Province of Bocas del Toro, Panama*). I was amazed and humbled that an indigenous person in this far-off place in Bocas had read my book.

Leo and I walked about 20 minutes on a concrete sidewalk all the way to the beach, pristine with old, gnarled sea grape trees, tropical almonds and hundreds of coconuts as far as the eye could see. The community's political representative had built the sidewalk for his constituents. I had walked the same trail with Victor Salazar in the 1960s when it was muddy and difficult to transit. Strangely, not a single tourist waas in sight during that trip. Valiente, far off the beaten path, had no tourist infrastructure, and most of the Ngäbe liked it that way. It was their comarca, an autonomous reservation where they governed themselves with the assistance of the Panamanian government that provided certain social and medical services and public schools.

We returned to the entrance of Bluefields Bay and tied up on the calm leeward side of Toro Cay Point. This is where a Frenchman, Louis Victor Georget, settled in the 1880s. He married Mina Smith and had French-speaking children who were born here. His plantation produced Gros Michel bananas sold to buyers who exported his fruit to Mobile or New Orleans. His wife died here in 1895, and he moved to Chiriqui Grande where the banana business was booming. Later, Georget went to Bocas Town, where he was an executive with the United Fruit Company.

Leo knew Toro Cay Peninsula well because he had built a large palm-thatch rancho there for the Smith family. This is where Maria Smith was born. They lived in a good wooden house and cultivated fruit trees, tropical vegetables, rice, beans and corn. The land was fertile, and life was good there.

The Smith brothers had bought this property from the Georgets long ago. But in recent years, some of the young Georgets regained the property in an expensive legal battle with the Smiths who claimed the land. There are still bitter feelings.

Leo and I went back to the Chiriqui Lagoon, turned left and found a littered shoreline community at Bucori, where Leo visited his maternal grandmother. It had been several years since he had seen her, and it was a happy reunion. She was attending a Mama Chi or Mama Tata religious service, as the founder of a movement started in 1961 was known. Mama Tata's divine decrees included the following:

- Disengagement from the Latino society, including schooling, merchant activities, and general contact.
- Abolition of fences and therefore the end of property disputes.
- Abolition of alcohol and the balsería[1], a harvest and sports festival that can become rowdy.
- Treatment of all fellow Ngäbe as "brothers and sisters."
- Abolition of fighting and killing over land, wife-beating, and maltreatment of children.
- Saturday, the day of the vision, is a day for rest and prayer.
- All Ngäbe should behave like good Christians.

The Mama Chi followers were aggressive, and even expelled some Christian missionaries, like the distinguished Reverend Ephraim Simeon Alphonse, who had lived among them for decades. Some of their practices were against common Panama laws, like no birth registration.

As I waited in the boat while Leo visited his grandmother, one of the family members stayed to talk with me, a talkative chap, twenty years old. I asked questions about his education. His answers were shocking but understandable. He was not allowed to go to school because of Mama Chi, and there was no school in their community because religious beliefs forbid education. The young man could not read, write or even count numbers; nor could he sign his name. He seemed intelligent and normal, and I felt regret for his being deprived of the basic needs and expectations in this life.

I asked if he was a believer of Mama Chi, and his response clearly showed his doubts. He confessed that when he chopped the bush with his *machete* and earned some money, he would walk to the next village and have a few drinks in a cantina there. His family opposed his occasional drinks, but he did it anyway as an escape and possibly as a protest. When Leo returned, we motored a short distance to Patterson Cay, a small island off-shore where we drank a carton of pear juice. The store looked the same as when Victor Salazar and I visited that identical building over the water in the 1960s. Nothing had changed very much except for the six small children playing on the floor of the store.

Leo pointed to the contents of a box on the floor near my feet. My eyes opened wide when I saw a single, large lobster among many undersized ones and a couple of big crabs, destined to far-off Chiriqui Grande, completely across the Chiriqui Lagoon. My mouth began to salivate, so I bought the one-kilo lobster for ten dollars. I didn't care about the price at that point. Leo cranked the engine about 4:30 PM, and we were home on Hospital Point by 5:40. There were no waves or wind. The ride on a flat sea in Bocas del Toro in October is one of the pleasures of life, especially when the trip is followed by a big tasty lobster for supper.

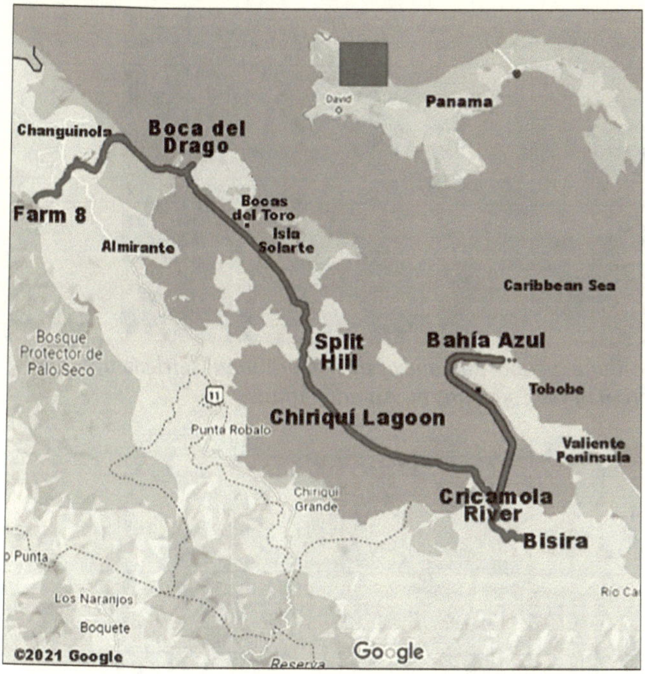

The trip I made with Victor in 1968, from Farm 8 (top left) to Bisira (bottom right). Google maps with permission.

Clyde in the cayuco he used on this trip. Clyde Stephens collection.

Ngäbe-Bugle people and one of their traditional settlements. Collection of photographs by Amado Santos.

Leo and me on the front steps of his house, October 15, 2021.

View from Leo's house. Clyde Stephens collection.

Victor at the Valiente Peninsula, 1967.
Photograph: Clyde Stephens collection.

Punta Valiente. Sources: Mindat.org,
an outreach project of the Hudson Institute of Mineralogy,
and OpenStreetMap contributors.

U-boat Captain Helmut Witte. Source: WW2gravestone.com/people/merten-karl-friedrich/

U-boat 159 arriving at Lorient, France, July 12, 1942. Source: U-159 arriving: File:Bundesarchiv Bild 101II-MW-4386-01, Frankreich, Lorient, U-159 und U-107.jpg - Wikimedia Commons.

The Sixaola at port in camouflage paint, circa 1918. Source: USS Sixaola in harbor, circa 1918, Sixaola (ID 2777) (navsource.org).

Outline of History of the Province of Bocas del Toro, Panama. Clyde Stephens collection.

Dancing on the 23rd anniversary of the founding of their comarca. Source: mupa.gob.pa/comarca-ngabe-bugle-cumple-23-anos/.

Louis Victor Georget.
Photograph: Courtesy of Virginie Georget.

Reverend Alphonse and his family.
Photograph: Courtesy of the Missionary Alphonse Guaymi Scholarship Fund.

18. Striking It Rich at An Oil Well

AFTER DECADES OF EXPLORING and searching, I finally struck it rich at the oil well that had eluded me for so many years. It had been hidden all this time in a dense jungle of tropical growth that had quickly enveloped the Sinclair Oil Company drilling site after its well was abandoned in 1948. The American manager in charge had declared publicly that if there was no oil there, he would eat his papers. We can only speculate if he was honest or not because oil was never found.

Maggie Guy, Silvia Chong and Jean Slolovan, secretaries in Almirantc in 1958, told me how they had been wined and dined aboard an oil drilling barge of The Offshore Company that had been contracted by Champlin Oil & Refining Company. Their big rig was anchored in Almirante Bay not far from the port. A luxury speedboat brought the ladies from the Almirante Club dock to the drilling site.

Although the giant barges were not meant to be beautiful, some of the employees lived in comfortable quarters aboard those vessels. Their dining room and bar were well-appointed and the food was first class. Jean Slolovan writes in *Bananeros in Central America*:

> A group of us were invited to have dinner with the crew on an offshore oil drilling rig. We arrived by Company launch and

> were hoisted up to an air-conditioned dining room for an excellent meal. This was followed by an orientation tour to watch the drill being changed.[1]

In 1968, I was searching for diseased banana plants in the wild on the upper Sixaola River on the Panamanian side of the international border with Costa Rica. There was no railroad because it had been removed after all the banana farms in that region died from Panama disease by the 1930s. Afterwards, those soils were planted with cacao, and this is where I was—in the Sibubi and Senosri cacao farms.

In this remote corner of Panama, I came across a man who wanted to tell me all about the company that had drilled for oil in this region. He described how lots of people signed on to good jobs that paid big money when the oil company was exploring there. The locals hoped that an oil company would return soon and employ them once again. Frankly, I was skeptical about his unlikely story until I saw a point on a United Fruit Company property map that was marked "oil well." The marked point was high on a ridge above the Sixaola River. That is when I became curious about other oil wells that were drilled in Bocas del Toro. Later, I found a reference that named Senosri Number 1 as an oil well site where the Caribbean Gulf Oil Company had drilled in 1962.[3] Two other wells had also been drilled in that region.

Over the years, I heard about other oil explorations in Bocas del Toro, but most stories were not well substantiated. However, several sources were credible and eventually took me to the oil well I was seeking.

Carlos Reid, a Creole from Bocas del Toro, commented on the oil exploration:

In 1924, the Sinclair Oil Company came, started operation at Big Bight. It was rumored that they found oil, yet, after a little time, they moved out. They came back again in 1945, made a big showing, employing lots of men and women. Their head office was in Bocas Town. Everybody felt that the time had come, but their time really came when they moved out again. Nothing more is said about them.[4]

Teofilo Alvarado related the following:

In the 1950s, a subsidiary of Sinclair Oil Company explored Isla Colon and the ridge above the Sixaola River for oil. They brought in lots of elaborate equipment, shops, supplies, offices, housing, boats and their own labor supply of trained people. Then came offshore drilling rigs. The boss had a super speed boat. They led a rich life. At a big farewell at the governor's house in Bocas, this *yanqui* boss departed by telling the public that Bocas would soon be converted into another Maracaibo. Everyone believed him.[5]

One of the most fascinating accounts came from a well-known family in Careening Cay, (*Isla Carenero*). Music teacher Felix Alex Russell wrote to his son, Luis Carl Russell, in New York on January 14, 1947, reporting the latest news from Bocas del Toro. The son had become one of the most recognized names in American jazz as a pianist, playing with Louis Armstrong and other jazz greats. He also had his own orchestra.[6] In this letter, his father writes about Sinclair Oil developments on Isla Colon. The following is an excerpt from that letter:

The Sinclair Oil Co. is operating here. They have an expectation of striking oil. If they do, the place will boom. They are making

> an airbase up Haul Over. They land their planes for the present on Careening Cay just below where we lived when you were a kid. They intend building a driving road to Big Creek.
>
> The U. F. Co. (United Fruit Co.) advised them not to pay more than 18 cts per hour for labor and laborers refuse to work at that rate. So far no real start is yet made—anyway the outlook is not discouraging.[7]

It is of interest that Sinclair Oil landed their airplanes on Carenero. Obviously, these were seaplanes that landed on the water. Which side of Carenero has not been confirmed but waters are calmer between Carenero and Bocas Town as compared with the rougher east side of the island that faces the ship channel and Isla Solarte. When this letter was written, the Russell house was on the beach of the East shore of Carenero and was still there in 2022.

In 2021, Simon Abadi Balid Piterson, a personal friend and merchant who owns La Casa Rosada in Almirante, sent me a picture of the Sinclair oil well when it was being inaugurated on Isla Colon. This amazing photograph was copied from his father's album and shows a large crowd of about 36 dignitaries in formal suits while a Catholic priest christened the oil well in the rainforest. Champagne, toasts and speeches followed. Only one woman was present. Simon's father, Abraham Abadi Balid, a prominent merchant and councilman in Bocas, is in the photograph. Dates on the pictures show 1947. Stewart Redwood lists the well as: Bocas del Toro No. 1, Sinclair Oil Company, 1948.

A similar photo from Joyce Robinson that was given to Stanley Heckadon-Moreno records the date as circa 1955 and notes that the oil well was at Big Creek, somewhere along the road between Bocas Town and Boca del Drago. That date and location have now been corrected.

The trail was getting hotter, and I was getting closer to the location of the Sinclair oil well on Isla Colon. But none of the photographs, books or interviews gave the location, that is, until I talked with Mr. Federico Pardo on September 9, 2022. He was also a good friend who was born in Bocas Town in 1947, the year the oil well was inaugurated. His father told him that the well was near La Gruta in the middle of the island. That was all I needed to hear. I thanked Mr. Pardo and immediately started making plans to find that oil well the next day. But before I left him, he told me that when he was about eight or ten years old, he watched oil company barges and boats drop explosives into the sea north of Careening Cay and Bastimentos. The deep explosions shot huge columns of water out of the sea in full view of the people looking from the shore on Avenida H Norte.

The next day, my caretaker, Leo, and his five-year-old son, Adonis, and I rode a bus from Bocas Town to La Gruta. I found Ernesto Diaz, who knew where the oil well was, and he guided us there. The walk was about 30 minutes to the north-east through dense jungle interspersed with small pastures. We took the original limestone rock road built by the Sinclair Oil Company in 1947. Much of the road was covered in bush, grassed-over and full of mud holes, with a few smooth river rocks that had probably come from the Changuinola River. A portion of the road went through a small pasture, literally a cow trail, but most of it was in the jungle.

The Sinclair road disappeared under layers of leaf litter, limbs and rotten tree trunks shaded by dense secondary forest. Suddenly, Ernesto pointed his long *machete* at a steel pipe sticking up through lush vegetation. "*Por allí está*," (It's around there) he said. Emotion came over me. After all these years, I had finally located this oil well that had been abandoned since 1948. I showed the others the photographs that Simon Balid had given me and matched the steep terrain behind the oil well site. It was hard to imagine the tall tower looming over 36 dignitaries dressed in hot, sweaty dress suits, at the inauguration. We checked out the concrete drainage structure built outside the oil well pipe now grown over with vines. It was hard to believe that I had just turned over another historical stone of exploration. It was the end of the road for Sinclair Oil, and for us.

As we hiked back to La Gruta, the sky opened up with a deluge that soaked us to the inner core. We took cover under some wide leaves of Heliconia and Calathea, but to no avail. As we waited in the downpour, Ernesto told me about himself and what he was doing there.

According to Ernesto, a politically-connected banana farmer in Changuinola, Luis Flores, originally a military pilot from Honduras, suggested to General Omar Torrijos that he should bring agriculturalists from Los Santos province to show the Bocas people how to cultivate their lands. The general liked the idea, so he flew 28 farmers, a few with families, in to Bocas airport, and promised them 100 hectares of land with a title. Ernesto arrived in 1971 and acquired about 50 hectares with no title but was told not to worry. He was a young man from Pedasí and started chopping down the forest to cultivate his new land.

I asked him how they got from Bocas Town to their newly donated lands in the middle of the forested island. He said the new landowners had to chop down the secondary jungle that had grown on top of the old Sinclair Oil Company gravel road that began at the northwest shore of Sandfly Bay and extended about six kilometers to their assigned lands. Thus, the Sinclair road became an active road again after lots of hard work.

Ernesto said that agriculture in Bocas del Toro was very different from their farming on the Pacific side of Panama, where there were long periods of no rain. It was the opposite on the Caribbean. Most of the original 28 members sold their free land and left because they could not adapt. Some returned to Los Santos, but a few went to remote Darien where there was a land rush of squatters invading the virgin rainforests there. Only three of the original 28 stayed on their gifted lands at La Colonia Santeña. Ernesto is the only one left in the colony. He still has his plot of land nearby and is also the caretaker of La Gruta cave that tourists come to visit.

In parting, he told me that the other half of the road that continued to Boca del Drago was not built until the 1980s during the regime of General Manuel Noriega.

A SUMMARY

The best in-depth research about oil explorations in Panama, including Bocas del Toro, was written by Dr. Stewart D. Redwood in 2021.[8] With his permission, the following are summaries of his findings in Bocas del Toro, plus some of my notes:

- Bocas del Toro: 8 wells.
- The pioneer oil exploration company was Sinclair Panama Oil Corporation that explored from 1917 to 1928 in Bocas del Toro and Darien.
- Sinclair drilled two wells in Bocas del Toro: Isla Colon No. 1 in 1919 and Isla Colon No. 2 in 1922.
- Sinclair drilled Bocas del Toro No. 1 in 1948 (The Balid pictures say 1947.) This was in the center of Isla Colon near

La Gruta and La Colonia Santeña and is the oil well that I re-discovered in 2022.

- Union Oil Company drilled one hole in Bocas del Toro in 1956 (Yorkin No. 1). Yorkin River (aka Zhorkin) is the international boundary that flows into the Sixaola River. Stewart Redwood marks this well near the confluence of the Yorkin and the Sixaola Rivers.
- Champlin Oil & Refining Co. drilled the first two offshore wells in Bocas del Toro in 1958 using an early jack-up drill rig (Almirante No. 1 and Anita Bay No. 1). Almirante No. 1 was near the banana port and is where the banana company secretaries were wined and dined. The Anita Bay location is not precisely known, but Dr. Redwood shows it in the eastern Chiriqui Lagoon.
- Caribbean Gulf drilled two wells in Bocas del Toro in 1962: Senosri No. 1 and Soledad No. 1. Redwood marks Soledad up the Yorkin River. Senosri cacao farm is easy to find and is where I met the man who told me about oil drilling in that area. A United Fruit Company property map marks an oil well on a mountain ridge above Senosri and the Sixaola River. Teofilo Alvarado also mentioned this well on the ridge.

It has been over a century since interests in oil exploration started in Bocas del Toro. Millions have been invested in this form of gambling. Personally, I think the Panama lottery is a much better bet. To prove the point, Phyllis and I won the first prize on May 30, 1964, by playing 0113. Lots of other people in Panama also win first prize because the wealth is more evenly distributed among the masses. We took our first prize, $1,000, and enjoyed a month-long vacation in Colombia, Ecuador and Peru. When we returned home to

Changuinola on Rapsa Airlines, I counted seven dollars left in my pocket. The other $993 paid for everything on our 30-day trip.[9] That was a much better investment than gambling on oil in Bocas del Toro.

Almirante No. 1 well being drilled by an early jack-up rig in 1958 by Champlin Oil.[2] Photograph: Joyce Robinson.

Guests arriving at the oil site for the inauguration, 1947.
Simon A. Balid P. collection.

Opposite Page, bottom: Dignitaries at the inauguration of the Sinclair oil well on Isla Colon near La Gruta, 1947. 1. Abraham Abadi Balid C.; 2. Hernando Santos Kearny; 3. Reverend Gomez, CMF (Catholic Church); 4. Eduardo Thomas (former provincial governor); 5. H. Icaza. Simon A. Balid P. collection.

Sinclair oil well road is now a cow trail.
Clyde Stephens collection.

Sinclair oil well road overgrown, 2022.
L-R: Ernesto Diaz, Adonis Julian Smith
and Leovigildo Julian Lorenzo. Clyde Stephens collection.

Clyde Stephens at the Sinclair oil well on Isla Colon, 2022. Photograph: Leovigildo Julian Lorenzo.

Giant trees had grown back near the abandoned well.
L-R: Ernesto Diaz, Adonis Julian Smith and Clyde Stephens.
Photograph: Leovigildo Julian Lorenzo.

19. The Zipper and La Solucion

I DISLIKE SHIRTS WITH no pockets, but Phyllis solved that problem for me many years ago. She designed and made wonderful Guayabera-type shirts with four deep pockets. When the old remaining shirts grew threadbare, I found an indigenous Panamanian tailor in the local Bocas del Toro market to make me three perfect replicas of Phyllis's shirts for $15 each.

Recently, a malfunctioning zipper on my short pants caused me considerable inconvenience and cried out for a similar solution. The pants were still in good condition, so I decided to have the zipper replaced. Unfortunately, the tailor who made the shirts lost his cubicle in the market during reconstruction and disappeared. My overseer thought he lived in La Solucion,[1] and I decided to find him to replace the broken zipper. That was a challenge.

I know where La Solucion is because I take visitors in my boat to see the squalor and poverty of the indigenous Ngäbe people who live there. They left their traditional homelands in the huge, protected Comarca Reservation and came to Bocas del Toro to earn money. As a result, this slum community sprung up in the only place they could find to squat on, a red mangrove wetland swamp located across the street on the south side of the provincial capital's international airport terminal.

The mangrove habitats are usually protected because they are very sensitive and important habitats for numerous species of marine plants and animals. But newly arrived Ngäbe find ways to nail boards together to make a floor and walls and put some kind of roof over the structure. Keeping the shack above high tide is a challenge. They have no fresh water, no sewage disposal or electricity. Toilets are improvised right over the water. Sandflies and mosquitos, both vectors of diseases, constantly torment them. The air is stagnant with the stench of mud flats at low tide.

I glided my boat along this same red mangrove shoreline of La Solucion. Within two minutes at idle speed, we passed by multimillion-dollar luxury yachts tied up in a beautiful marina developed from the old United Fruit Company dry dock. It would be hard to find a greater contrast in two life styles so close to each other—a complete paradox.

To find the tailor, I had been told to ask local residents in La Solucion. I walked past the main entrance of the international airport terminal and soon found a narrow trail covered with logs, plastic bottles, cans and broken concrete stepping stones. My shoes got muddy very fast. Appalling and scary shanty houses bordered both sides of the trail. I got lots of curious stares from the residents. It wasn't long before I turned the wrong way. Out of fear, I decided to make a fast retreat and start again.

Eventually, I came upon a one-room dwelling with a tin roof and noticed there were several sewing machines inside. A young indigenous man was there when I walked in. After the usual greeting, he looked a little puzzled by an 88-year-old *gringo* speaking fluent Spanish to him. He said his parents were gone and told me where I could find them. Well, there was no way that I would try to find them in that jumbled community of shacks.

When I hesitated to leave, he asked what I wanted, and I showed him the problem. I was afraid I might have to walk back to town and buy a zipper, but luckily he had one on hand. I asked when he could do the job. He told me to go to the corner of the room and take my pants off. I did and threw the garment to him across the room. Meanwhile, I sat on his parent's crude homemade bed in the corner in my jockey shorts and waited, sweating profusely because the heat under the low tin roof was sweltering.

The man went to work immediately and finished the job in about 15 minutes. He threw the pants back across the room and told me to try the new zipper. I was amazed at his skill. He asked for four U.S. dollars, the currency of Panama, but I gave him a five-dollar bill in gratitude.

Before leaving, I asked where he learned how to sew. His parents are both *sastres* (tailors) and taught him the profession. Also, his brother, an aunt and an uncle are tailors.

I was impressed that these indigenous Ngäbe people who came from so far away as Cusapin in the Comarca Reservation were talented tailors. They lived in La Solucion, one of the worst slums imaginable, but they had their dignity and the ambition to try to rise out of that pestilent swamp. I hope they make it to the other side of the airport someday. But I still wonder about the fate of the other 99% of their neighbors. There are lots of them and plenty of unplanned babies are doomed to a life in the mangrove swamp.

Mangrove habitat.
Clyde Stephens collection.

One-room dwelling with tin roof.
Clyde Stephens collection.

20. James Norman Morris Maas: One of a Kind

JAMES MAAS WAS A perfect English gentleman and very much a scion of the British Empire, once traveling to visit the grave of his grandmother in Kolkata, India. After "seeing the world through a porthole" for 30 years as a merchant seaman, James looked for a place to retire. But certainly not in England. As he told me, "I will fight for her, and I will die for her, but I will not live in the Mother Country." United Fruit Company ships had taken him to many banana ports in Central and South America, and eventually, James became impressed with the islands and waters of Bocas del Toro, an archipelago in western Panama on the Caribbean.

JAMES'S HISTORY

We exchanged many visits with James over the years. He was small in stature and spoke softly and politely with a very proper English accent. His house was a museum of British Empire history where he modestly told incredible stories, especially about his childhood in China.

His father, Norman Maas, was trained as a naval officer and served with an artillery unit in France during World War I. In 1925, he went to Shanghai, China, as a naval architect and was the directing manager of a British shipyard. While there, he met a woman, Stella, who had left England to teach

English at a Chinese university. They returned to England in 1931 where they married. James was born a year later. In 1937, the couple returned to Shanghai, along with their son.

James's maternal grandmother also was a foreign sojourner. She went to India as a missionary in the 1850s, which explains his visit to her grave in Kolkata. It might also explain why his mother developed a special liking for the writings of Rudyard Kipling. "It's highly likely," James's godson, Simon Lang, wrote to me in an email, "that Stella was also in India at some time and got her passion for Kipling from her mum."

Norman Maas worked as an advisor to British shipping interests and was responsible for dredging sections of the Huangpu River. James showed me a large panoramic photograph with views from his father's office window that was several stories high. The scene was spectacular, overlooking the river and the cosmopolitan city with banks, factories, stores, export and import businesses, night clubs, hotels, restaurants, and plenty of bordellos. Known by some as "the wickedest city in the world," Shanghai had a mixture of every nationality from around the globe, living in extremes where anything could be found. Foreign newspapers, radio stations, restaurants and night clubs catered to just about every ethnic group in the overcrowded port city.

Non-Chinese residents lived in the huge, protected International Settlement where tens of thousands of foreigners had fine homes, including the Maas family. Each nationality had its own enclave called a concession. Within each national colony, there were luxury homes, schools, stores, social clubs, movies, country clubs and religious centers. Concessions were held by the French, German, American, Japanese, Italians, Swiss, and British. Not included were numerous White Russians in exile and destitute Jewish refugees escaping from the Nazis.

The five million Chinese who lived outside the International Settlement in Shanghai were allowed to enter neighborhoods of the foreigners to some extent. Each ex-patriot home had Chinese employees and many of them lived on the premises. For their part, the Maas family had three servants: An *amah* (nanny) for James, a cook and a gardener.

James particularly had fond memories of the time his mother took him to Tsingtao, a site north of Shanghai where a German colony existed from 1898 to 1914.[1] For the British and other colonials, Tsingtao was an exciting escape into beer, beaches and fun.

Life for the foreigners was luxurious, in contrast to the squalor of the masses. But, no matter who you were, all were threatened by malaria, cholera, smallpox, dysentery and typhoid—rich and poor. In 1914, the British and Japanese allied together against the Germans who lost to the Japanese. China later regained control of its own territory until the Japanese returned again to rule.

JAMES MEETS JIM

During one of my visits to the Casa Maas in Bocas del Toro, James was just opening a letter from England. When he finished reading, we adjourned to the waterfront patio under the cool shade of his tropical almond tree. Soon, we toasted each other with a Guinness stout. Although James was quite a private person, I could tell he wanted to relate the contents of the letter. It was from J. G. Ballard. Because I had been so isolated working on the banana plantations, I had no idea who Ballard was, but James showed me one of his books, *Empire of the*

Sun. Furthermore, he told me that Steven Spielberg had made a movie of Ballard's story. I was impressed and started asking lots of questions.[2]

James told me that he and Jim Ballard had been camp-mates as adolescent boys at a Japanese prison camp, and they still exchanged letters. All the correspondence between these two men was extremely fascinating, so I began to take notes, something that my host was not keen about. But life in China then and the experiences in the prison camp were only known by a few living survivors, compelling me to develop mental and written notes. Life Under the Japanese In 1937, Japan conquered Shanghai and life very quickly changed for everyone, including the Maas family. After initial restrictions, the foreigners were allowed some freedom to travel and resume their private and business lives. Life almost normalized except that tens of thousands of Japanese invaders were now warlords in this area of China.

The restrictions became more drastic on December 7, 1941, when the Japanese bombed Pearl Harbor. The order was given in Shanghai that all the British people had to wear red arm bands and some were locked up.

The final shock came in 1943 when all British, non-military Americans and other nationalities were put in prison camps. The Maas family was taken to the Lunghua Civilian Assembly Center where there were about 2,000 internees, of whom 300 were children. This was the point at which James and Jim became friends.

Camp internees ranged in age from six months to 88 years old. Most were British in addition to Canadians, Dutch, Australians, Belgians, New Zealanders, seven South Africans, three Russians, a Portuguese, and a Norwegian.

Thirty merchant seamen from America also were there. Fortunately, these seamen were civilians; had they been members of the military, they would have gone to one of the many infamous Japanese prisoner-of-war camps.

Up until 1937, the Lunghua Camp was a teacher-training college with seven large, concrete buildings, several wooden structures and a nice campus spread across 43 acres (17.4 hectares). When the Japanese attacked China, many of these buildings were either heavily damaged or destroyed; as a result, the college was abandoned.

In 1943, the Lunghua Camp had a Japanese commandant, Hyashi, who wrote and spoke English because he had been at the Japanese Embassy in London for several years. He was less tyrannical than other camp administrators, and he communicated with the internees while riding his bicycle around the camp with his young son.[3]

The prisoners were allowed to organize and operate a school and grow and cook food from their own gardens. In general, they were given some freedoms to survive the best way they could. Also, James' father was in charge of handing out daily work tasks that were assigned to the detainees. Both women and men washed clothes. One thing was sure; all activities stopped twice a day when there was a roll call and everyone was accounted for by name.

Food clean water, sanitation, toilets and swarms of mosquitos carrying malaria were always concerns. In the winter, the internees suffered terribly from extreme cold because there was no heat. The Maas family of three had a small cubicle where they survived until the end of the war. Although the cubicle was small, friends with an extensive family arranged to have two of their children sleep nightly at the Maas home.

Stella Maas did all kinds of chores to keep the family going, including teaching French and art. When she had spare time, she often painted watercolors of the camp under the watchful and curious eyes of the Japanese guards. Sometimes, she would set up her easel near the boundary line and paint the rice fields beyond, showing every strand of barbed wire in front of her. The paintings became part of James' inheritance, and when he proudly showed them to me in Bocas, I almost choked up.[4]

Stella also had with her 17 tomes of Rudyard Kipling. As I noted previously, she had a special affinity for Kipling. His stories and poems of India gave her a diversion despite very little privacy and minimal light for reading in their minuscule, cold Lunghua cubicle. Students in the camp school were taught English, Latin, French, mathematics, history, literature and science. Even amateur plays were presented, including some of Shakespeare's works. Volunteers gave organized lectures to anyone who would listen. There was no shortage of lecturers because the camp had many educated internees such as dentists, engineers, chemists, bankers, architects, businessmen, missionaries, medical and numerous missionary doctors, preachers and teachers.

There also was a soccer field and the inmates organized competitive sports leagues. Unknown to the Japanese, a couple of clandestine short-wave radios were built and provided some news from the outside world. The entire camp would have suffered the consequences if the radios had been found.

By 1945, the enemy was losing the war, and the Japanese guards under a new administration became more brutal to the internees. Rations were scarce and hunger was common. Many in the camp had gastrointestinal disorders and became weak.

JAPANESE SURRENDER AND TRANSITIONS

After the War After the Japanese surrendered on August 15, 1945, the Lunghua internees had no communication with the outside world except what they heard on their short-wave radios. When the Japanese guards disappeared, the campmates didn't know what to do with their freedom. The camp began to celebrate the victory but the future was uncertain. Was it really over? Where would they go and what would they do? Near the end of August, the roar of American B-29 bombers caused excitement when parachutes dropped food and supplies. This included cans of Spam, Klim powdered milk, chocolate bars, jam, and Lucky Strike cigarettes. There was a stampede to retrieve the parcels from the rice fields; it was the first time they had seen familiar foods for years . The whole camp went from famine to feast as the internee camp committee benevolently distributed food as needed. In their emaciated condition, some of the internees packed what they had and started walking toward Shanghai. Others waited to see what the liberation forces would do with them. Many had no home in England and no money; they just waited. This was the case with the Maas family; they couldn't return to their home in Shanghai because someone else had occupied it. James and some friends stirred up a little excitement by trekking into the city; a highlight of the trip was taking a dip in a swimming pool. Finally, in November of 1945, James and his mother boarded the British troop ship SS Arrawa and returned to England. His father stayed in Shanghai and resumed his work on the Huangpu River, but fell ill the following year. Consequently, he also left for England.

Because James was so young at the time, his experiences in China had only a limited impact on him. Meanwhile, a classmate who also had been at Lunghua, looked emaciated and pale compared with James who looked perfectly healthy. "Gosh," someone chortled, "he's got to be the wrong boy."[5]

By now age 13, James's friendship with Ballard continued, and the two of them attended the same prep school; when Ballard's bike was broken, he rode on the back of James' bike. In 1950, James entered Queen's College at Cambridge where he earned a Bachelor of Arts degree in 1953 and a Master of Arts degree in 1957. His studies were in chemical engineering, but he became a merchant seaman, mostly below deck, retiring after 30 years at sea.

FINDING ROMANCE IN PANAMA CITY

James occasionally flew from Bocas del Toro to Panama City, a one-hour flight on Alas Chiricanas. It was his custom to economize when it came to lodging. He became very frugal as a boy in the Japanese prison camp and in post-war Britain. James prided himself with the fact that he never paid more than five dollars a night for a room. A favorite place was in the old historic neighborhood, El Casco Viejo, which was rundown and dangerous.

When James checked into El Hotel Foyo, the desk clerk was doña Fulvia Perez, a polite and gentle lady in her senior years. She would greet him by saying "Buenos dias, señor Maas" and he would reply "Buenos días, doña Fulvia." That was all. The next time, after this same customary greeting, the dialog expanded to "Cómo está usted?" And the reply would be "Muy bien, gracias." Nothing more than that. With each check in, the conversation expanded word-by-word and phrase-by-phrase. After the couple gained enough confidence in each other, they were married, and doña Fulvia moved to his fine home in Bocas del Toro. His friends and neighbors were shocked but very happy for both of them.

THE SOCIAL CIRCUIT IN BOCAS DEL TORO

The Bocas Town social circuit consisted of a small group of mostly foreigners, sometimes hosted by the Gran Dame de Bocas, doña Lydia Thomas de Padget. She was born and raised in Bocas and lived in the old, German house of Hans Kandler with her Ecuadorian-born husband, Alfredo Padget. The gatherings were small and formal. Food and drinks were very special with the high society of Bocas. James Maas was a bachelor early on, very frugal, and he did not cook. When he hosted our group, the usual drinks included his favorite, Guiness stout, a rarity in Bocas. Dinner was rather formal and pleasant at a nice table under an elegant chandelier hanging from the cathedral ceiling. Local tropical food, not English in any way, was cooked and served by doña Isabel de Fitzgerald. After James married, doña Fulvia became part of our group.

When *monsieur* Lucien (Lulu) Ferrenbach, of Strasbourg, France, drifted into Bocas Town with his wife, *madame* Beatrice (Bibi), our social life took a different turn. We assembled in their small, dilapidated tin shack that had been pulled out of the water after the earthquake and cleaned up enough to live in. Guests gathered on the rough board porch over deep water looking north at the Caribbean Sea. Being French, the first course was always delicious paté served on Lulu's homemade French bread and washed down with copious amounts of vin rouge.

For the main course, we adjourned inside to a homemade table with two wooden benches and a couple of stools. The cuisine was always delectable and authentically French.

Both Lulu and Bibi were talented artists and had several other skills to make a living.

Silvia Chong Howard took her turn by inviting the group to her small, new house, one of the habitations built for victims of the earthquake. The best of Chinese food was served on gorgeous, hand painted plates that her father had imported from China so long ago. Rice wine added a special flavor to the oriental food.

When it was our turn to entertain, the group came in their small boats to our house at Hospital Point.

As they approached our house up the hill, each guest was welcomed with what we called a *pipa de la bienvenida*, meaning a green, succulent coconut that had an ample shot of Panama rum added through a hole in the top.

Phyllis always served the very best of tropical delights, especially breadfruit dishes, palm heart salad and pili nutsall grown on our farm. Guests loved to explore the old concrete sidewalks and see where the United Fruit Company had a major Tropical Medical Center. After 1920, the abandoned property grew back into thick jungle until we bought it in 1970.

ADIOS AMIGO

As time went on, James and Fulvia developed health problems and were forced to move to Anton, a community west of Panama City, where her family could help them. James died on February 15, 2015, and was buried there. I knew that he had asked for his version of a Viking funeral in which his body would be put in a cayuco, taken out to sea, burned, and sunk to the bottom. His lawyer, Harry Diaz, later a Panama Supreme Court magistrate, pleaded with James, telling him that he couldn't arrange a Viking Funeral, not in Panama! James then opted for cremation, which was not easy outside the capital of Panama. When I talked with Fulvia, she felt guilty for not complying, but I consoled her with the fact that James would be proud to be near her in Anton.

THE JAMES MAAS HOME AND GARDEN TODAY

With the opening of the floodgates to tourism and development in 1995, Bocas Town and the islands have been exploited in the extreme. Today, the streets are crowded with vehicles, all in a hurry. Unplanned and uncontrolled structures four floors high have blocked off all views of the sea and neighboring islands. Only a few buildings are earthquake proof. Prices have skyrocketed. Garbage, pollution and crime are major problems. Bocas del Toro is no longer the sleepy ghost town that James Maas found in 1988.

After James died, Fulvia had to sell the Casa Maas. I learned this when family members came to our house at Hospital Point early on a Sunday morning. The young couple said Fulvia wanted me to take what books I wanted from the library. They emphasized that the house had to be completely empty and vacated by that afternoon. Otherwise, all remaining contents would be sent to the dump. What a dreadful prospect.

I quickly went in my boat and landed at the Maas house to review the library collection. There was no time to sort through each book so I filled nine giant garbage bags with all of the books in the house. It was a major task loading my boat with so much bulky cargo. I had to make two trips across the bay and had barely finished when a heavy rain drenched the area. Obviously, there was no way I could keep all those books at my house. After three days of sorting, I selected only a few rare books of special interest, including the original set of 17

Rudyard Kipling volumes that Stella Maas kept at the Lunghua Prison Camp. In one of the J. G. Ballard books, I found a personal letter from the author to James dated May 12, 1991. I know that James had many other letters from Ballard but they are probably lost forever.

Luckily, I found a good home for the extra books at a bed and breakfast. It was a real challenge unloading the bulky bags of books offshore in heavy waves and getting them ashore ahead of the breakers and onto the beach. Miraculously, all the books made it without getting wet.

In 2017, as our boat passed the Maas house during our annual trip to our island home at Hospital Point, Phyllis and I noticed some changes going on in the garden and elsewhere. Later, some construction started, and I stopped by to inquire about the new owners and their plans for the house.

Fast forward to 2020, when Chiquita hired me as a consultant to orient some European executives who were visiting the banana plantations in Changuinola. That trip enabled me to enjoy a couple of weeks at our house on Isla Solarte. When I passed by the Maas property, there was no sign of his house and gardens. Instead, a three-story building had been built. My mind was confused so I met the manager and asked some questions.

As it turned out, an international organization had purchased the Maas property and called it Chabad Bocas. It is a gathering place for Jewish travelers and residents, including a synagogue and kosher restaurant named Falafel Bocas. I was invited to join them for a falafel, and I hope to accept their cordial invitation someday. In return, I will present them with a copy of this story about James Norman Morris Maas, a man whose life was extraordinary by any measure.

Panoramic view of Shanghai, 1926.
Source:The Bund, Shanghai in 1926 as seen from the US Consulate General.png Wikimedia Commons.

German Siege of Tsingtao.
Source: https://picryl.com/ media/defenders-of-tsing-tau.

J. G. Ballard.
Source: JGBallard.jpg.

Fictional movie scene from Empire of the Sun.
Jim (right) makes a friend.
Source: Movie trailer, Empire of the Sun (1987).

The Lunghua Kitchen.
Photograph: Oscar Seepol. Courtesy of Susannah Stapleton and Historical Photographs of China, University of Bristol (www.hpcbristol.net).

The school children at Lunghua. James is probably in the photo.
Photograph: Oscar Seepol. Courtesy of Susannah Stapleton and Historical Photographs of China, University of Bristol (www.hpcbristol.net).

Fulvia and James.
Simon James Lang collection.

Lulu and Bibi.
Clyde Stephens collection.

L-R: Fulvia Perez de Maas, Silvia Chong, me, Phyllis,
James Maas, Conchita Castrellon.
Clyde Stephens collecction.

James (left) and his godson Simon James Lang.
Simon James Lang collection.

The Kandler House.
Clyde Stephens collection.

Entrance to Chabad Bocas.
Photograph: Leovigildo Julian Lorenzo.

21. Earthquake on Earth Day

PEOPLE HAVE SUCH SHORT memories. They forget that Bocas del Toro sits on a seismic fault. Residents should be prepared for earthquakes by constructing earthquake-proof structures. But I would speculate that most of the new buildings in Bocas Town that were hastily built since the late 1990s would crumble if we had another big shake. That is what happens when building codes are modified or suspended to save money and time.

Originally, there were common sense guidelines to build only two floors high. Then owners and builders colluded to find ways to suspend the rules and add on a third floor. Now, many buildings in Bocas Town are four stories high on tiny lots only 15 meters wide. You cannot even see the bay between overcrowded buildings. Someone will pay the price some day and people will suffer the consequences. Some of us know that is true.

On Earth Day, April 22, 1991, at 5:00 P.M, the region was devastated by a 7.6 earthquake.[1] I was on a different island, Isla Solarte, where our new solar-powered retirement house was being built at Hospital Point, but it was not quite ready for occupancy. Fortunately, no damage was done, because I had demanded that the contractor make our new home earthquake-proof, even though he said that Bocas never had earthquakes.

Phyllis was in Bocas Town on the second floor of a comfortable, old wooden house owned by our best friend, Silvia Chong Howard, and her elderly mother, Miss Cate Howard. At 5:00 P.M., the entire structure lurched and swayed in all directions. Phyllis hung on for dear life in the upstairs hallway, bouncing around as the house collapsed to the ground. She wasn't seriously hurt but had lots of black and blue bruises.

When Phyllis later found Miss Cate in the rubble between a mangled wall and the jagged floorboards, she thought she was dead. Soon though, Miss Cate began calling Silvia's name. Silvia had been downstairs and was crushed underneath the upstairs floor. Luckily, a neighbor saw a hand waving between two loose boards; it turned out to be Silvia. Mr. Nelson, an elderly man who was under the house repairing a guitar, was pinned to the concrete floor by a heavy beam. All four of them were rescued and taken to the local hospital.

After I saw that Silvia's house was in shambles, I rushed to the hospital, but when I arrived, it was completely empty. I ran through all the halls and wards, even entering the surgery room, and found no one. Someone in the street said everyone in the hospital was removed and sent to a hilly area that was familiar to me.

It turned out that Phyllis and friends hadn't been at the hospital very long. Panic had struck, scattering the residents everywhere because of a sudden rumor that Bocas was sinking, and a large wave would soon rush over the town, so everybody should run for the hills. So I set out at full speed for the hilly area several kilometers away. There I located Silvia and her mother close by the community water tanks. Near dark and in a light rain, they were among numerous victims who were sitting on the ground in complete silence and shock. Some of them, like Silvia and Miss Cate, had serious injuries. I did not find Phyllis there; she had already gone back to Bocas Town looking for me. But James Maas, a retired English seaman, a compassionate man who barely knew who we were, when he learned of our plight, sent a message that Silvia, Miss Cate, Phyllis and I were invited to stay with him because his house was not seriously damaged. Some walls and floors were cracked, water pipes were broken and there was no electricity, but we were welcome to move in. Phyllis got the message in the street and went directly to the Maas house.

Back on the hill, I hired a small, rickety Toyota pickup truck to take us to James's house. Miss Cate was on a canvas stretcher on the ground. We placed her on the open wooden platform of the old truck and somehow maneuvered Silvia to the passenger seat in the cab. I rode in the back to console Miss Cate. The driver had to crank the engine by coasting down the steep hill and suddenly releasing the clutch. Lucky for us, his brakes did not fail as we made the sharp right turn at the bottom of the hill to avoid going straight into the sea. All four of us literally entered the Maas house with just the clothes on our backs.

Under the circumstances, James proved to be a generous host during the crisis. We all worked together to recover after the earthquake. We placed old paint buckets under the drip line of the roof to catch rainwater for cooking and spot bathing. My job was to lean over the sea wall and dip salt water into containers to clean and flush the toilets. Phyllis became a nurse taking care of Silvia and Miss Cate, both of whom had broken ribs. Silvia also had damaged teeth and a fractured jawbone. They were both bedridden and required lots of personal care. After five days with no running water or electricity,

Silvia and Miss Cate left on a special flight with other victims to Panama City for treatment. Because they were homeless, they did not return to Bocas for two years. In fact, numerous other families lost their homes and stores. Many structures along the waterfront collapsed into the water. The large government wharf fell completely into the sea. People were scared and traumatized and many left the island.

On May 13, 1991, Phyllis and I voiced our heartfelt gratitude to James for keeping us for nearly a month, moved to our new house at Hospital Point, and started retirement again. I had insisted that our house builder make our retirement house completely earthquake proof. And he did. He and I were upstairs in the new, unfinished house when the earthquake struck. Even though the new structure lurched and shook violently in all directions, there was no damage of any kind. Other buildings and people were not so fortunate.

Hundreds of houses in Almirante, Changuinola and Guabito were ruined and people died. It was a macabre sight to see fissures and bulges in the earth in banana plantations. The Company's new accounting office in Farm 8 was destroyed by a huge crack that opened up the ground around and under it. The long bridge over the Changuinola River was badly damaged and all rail traffic was halted. Railway cars loaded with a thousand boxes of Chiquita bananas fell off the tracks and overturned, some upside down. Railroad tracks twisted out of line. Airport runways were damaged with undulating bulges and cracks. Water pipes broke in houses. The big main pipes broke and flooded the streets. Electric power was cut off. It was a macabre scene. Bocas Town was already a ghost town, but now it became even more abandoned.

In Bocas Town, property values crashed. Property could

be bought at a low cost, but nobody wanted to buy. Many terrified people suffering from shock and aftershocks left the island. After their welcomes gave out with friends or relatives elsewhere, a few came back to start again but that was not easy. Those who stayed went to work to repair the damage and try to get back to normal.

Lest we forget, other major earthquakes on this Caribbean coast occurred in 1798, 1822 and 1916. Let's hope that people build wisely and are prepared for the next big shake. It might happen tomorrow.

Aerial view of Bocas del Toro, Panama.
Photograph: Clyde Stephens.

Map of Bocas del Toro, Panama.

Old photograph of Bocas del Toro, Panama.
Clyde Stephens collection.

The magnitude 7.6 earthquake in 1991.

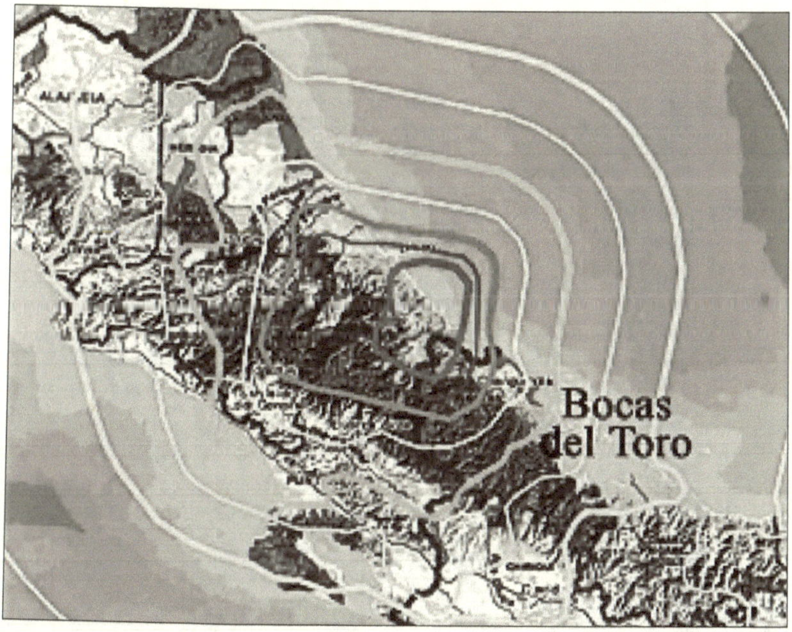

91 4 23.

Steps to nowhere. Photograph: Clyde Stephens.

Bocas Town wharf. Photograph: Clyde Stephens.

Opposite Page, top: Silvia Chong's house where Phyllis and I were staying. Bottom: Phyllis, demonstrates how she supported herself as the second floor collapsed.

The French Bakery (downstairs) and residence of Madame and Monsieur Léger, who came to Bocas after the French Canal era. The building, 50 meters north of Avenue E and 4th Street, survived the earthquake. Photograph: Clyde Stephens.

Accounting office, Chiriqui Land Company, Chiquita Brands International, Farm 8, Changuinola. Photograph: Cameron Forsyth.

San San River bridge, built in 1908.
Photograph: Cameron Forsyth.

Railcars overturned with 1,000 boxes of bananas, Guabito.
Photograph: Clyde Stephens.

Busy Bocas del Toro, 2014.
Now busier than ever.
© Ignasi Such. Used with permission.
https://www.dreamstime.com 115546028.

Our island home, since 1991.
Photograph: Clyde Stephens

Book 3

1. Close Encounters with Germans

MY FATHER GRADUATED IN 1929 with a Bachelor of Science in Agricultural Education from the University of Florida and taught vocational agriculture in high school. After several years of teaching, his agrarian roots lured him back to the soil and he purchased a farm to grow crops, cattle and raise children. In 1934, I was the first of three born on this farm in north Florida southwest of Quincy. Three more were added to the family by 1946 in other places.

In 1941, J. F. Williams came from the Florida state capital in Tallahassee and told my father that he had to sell his farm and go back to teaching agriculture. Williams was the Superintendent of Agricultural Education for the state. My father, Carl Wilson Stephens (1903-1996), told his old friend that he was happy with his family on the farm and had no intention of leaving. Williams then sternly reminded my dad that there was a war on and an urgent need to teach the boys and men how to grow food and fiber for the war effort. Williams made it clear that if my Dad did not comply, he would pull out his badge of authority and draft him on the spot. Dad had no choice. The large farm in Gadsden County on Juniper Creek, southwest of the Sawdust Community crossroads, was sold to a neighbor, Joseph Suber, who grew shade tobacco used for wrapping cigars.

My father was told that he would replace his double first cousin, Vaughn Driggers, in Apopka near Orlando way down in Central Florida. Driggers was a reserve officer and had already reported to the Army. Dad went to Apopka looking for any place to put his large family, but there was no housing of any kind to rent in the entire town.

By chance, Dad noticed an old, empty three-story mansion at the north end of Highland Avenue. When he asked the owner, Mr. Chamberlain in Orlando, if he would rent the house, the elderly tycoon from Pittsburg, told him that his country home was not for rent. My Dad then explained that he had been forced to teach vocational agriculture as part of the war effort and had no place to live. Immediately, Mr. Chamberlain changed his mind and declared that he wanted to help with the war effort, too. And yes, he would rent the house. My dad asked how much for the rent and the reply was $200. Dad said that he hardly earned that much every month. "Oh," said Mr. Chamberlain, "I mean $200 a year." They shook hands and the deal was made. The property had the mansion, a huge yard, a garden, labor house, chicken yard and a big barn with infrastructure to milk cows. The rest of the 160 acres had an orange grove, two lakes, a pasture, fields for growing crops and plenty of natural woodlands.

After I finished the 1st grade in Greensboro near where I was born, we moved to the Chamberlain mansion in Apopka in June 1941. Mama was eight months pregnant. After two trips of hauling a family of six, furniture, personal items, farm implements and four cows in a trailer pulled by a 1938 Hudson Terraplane car, we settled into our new abode—the Chamberlain mansion in Apopka. Well, almost.

My father was obligated to attend the annual vocational agriculture teachers meeting in Daytona Beach and was gone part of a week. On June 30, 1941, my mother gave birth to her fifth baby upstairs in the bedroom to the left of the stairs. Doctor Thomas McBride, the only doctor in Apopka, assisted with the birth. When Dad called from Daytona Beach, Mama asked him what to call his son. He quickly said John Philip Stephens because he had just heard an inspiring band concert with music by the great composer, John Philip Sousa, in the Daytona Beach Bandshell by the beach.

On December 7, 1941, my father heard on the radio that Japan had bombed Pearl Harbor. President Franklin Delano Roosevelt declared war on Japan the next day. Germany and Italy declared war on the United States and the USA reciprocated on December 11 . Dad and I were in the shade of a large clump of bamboo in the front yard. He explained to me what World War II was about, why we moved to Apopka, and the need for him to teach men how to grow food for the war effort.

Within a few months in 1942, the U.S. Army expropriated the city of Apopka athletic field, stadium and surrounding properties to train soldiers for war. The area, known as the "ball park," was only one block from our house—practically in our front yard. A few of these young soldiers were married and wanted to be with their wives during their last three months of training before being shipped off to war. But there was no place to put their wives in Apopka. That is, except maybe at our house. Husbands in Army uniforms started knocking on our door and pleading with my mother to please rent them a room for three months.

My parents consolidated our entire family of seven in the only bedroom with a private bathroom that was located downstairs. My mother furnished two rooms on the first floor, four rooms on the second floor and a large single room on the third floor. The two-room labor house in the back yard was also made livable. Each room had the basics: a kerosene stove, a dining table with two chairs, pots, eating utensils, linens and the most important piece of furniture—the bed. There was one toilet, bathtub and sink down the hall in the only other bathroom. The couple on the third floor had to walk down the stairs to the first floor to use the bathroom. Rent was $28 a month. We kids were assigned to collect trash from the garbage can in each rented room and burn the contents. Even though we were innocent children, we soon learned what those strange, long latex tubes were used for. We tiny boys even tried on some for size. Obviously, condoms were made for men, not boys.

All rooms were quickly booked after each three-month training cycle. Although that was a short time, my parents and us kids bonded with these young couples who came from all over the United States. It was sad when a husband went off to war in Europe or the Pacific and had to say goodbye to his wife, sometimes in our presence. She had to return home alone either by bus or train. Despite censored mail with missing sentences clipped out with scissors, we got letters from husbands telling us about battles, wounds and war. That was sad and upsetting. Wives also sent us mail telling us about their husbands and themselves. Gifts of appreciation sometimes came to our house. Once, we got a box of apples from a wife in Virginia. That was a thrill for us Florida kids who ate mostly oranges. We kept contact with some of those couples for several years after the war was over in 1945.

I learned something about New York City and Brooklyn accents. Mama told me that I could make some money by selling green, boiled peanuts to the soldier boys in the army camp. My father grew peanuts and Mama boiled them for me to sell. The first day as a peanut vendor, I walked down the sand road with several dozen paper bags of boiled peanuts, fresh out of the ground. Soon, I heard this gruff, obnoxious, loud voice, "Hey kid, what cha got der?" "Peanuts," I humbly replied. "Gimme a bag," he said. I threw a bag over the tall fence, and he stuck a nickel through the cyclone wire. That was my first sale, and I was so happy because I had earned my first five cents at age eight. I walked on down the road hoping to sell more peanuts when I heard this loud, ingratiating voice yell at me again. "Hey, kid. Come back here. I want my money back. Take these peanuts back and roast em." I didn't know what roasted peanuts were, but I knew the rest of my life that I hated an accent like that. I poked the nickel back through the fence, ran home almost in tears and told Mama what happened. "Oh, son, don't worry. Those old Yankee boys don't know how to eat peanuts." But my mother, Era Mae Bullock Stephens (1906-1977), conformed to market demand and roasted some peanuts, which we called "parched peanuts." And yes, sales went up.

We kids walked six blocks to our grammar school. The high school was on the same site but has since been converted into the Apopka City Hall, while the site of the grammar school is now bare ground. It was normal to take a shortcut and walk down the railroad track that passed by the school.

As we walked along the tracks, it wasn't unusual for us to see German prisoners of war just across the street loading and unloading railcars with cement and fertilizer. The POWs were from a camp located on the south side of Silver Lake (today, the campus of Lake-Sumter State College) in Lake County, east of Leesburg. This camp was attached to the Army Air

Corps base there, not an uncommon practice during the war years from 1943 to 1945. These prisoners were assigned work tasks in industry and agriculture that earned them 80 cents per day.

There was no doubt about these Germans being the enemy because newspapers, radio broadcasts and gossipy stories informed the public about them. We walked very cautiously on the far side of the railroad tracks so there would be less danger of being attacked by these dangerous enemies.

The truth was that these German prisoners were as harmless and docile as humans could be. They were young kids themselves, some teenagers or in their twenties, blond, blue-eyed boys who were laughing, joking, and calling to us scared kids as we walked in a hurry. I remember the guard with a gun casually slung over his shoulder, and I doubt that he was ever concerned about trouble or escapes. Many of these German captives felt lucky to have escaped the horrors of war in Europe. Now they lived in sunny, subtropical Florida, picked oranges, harvested vegetables and worked on roads. They had sports and recreational facilities, educational classes, a library, watched movies, had a canteen to spend their money, ate the same food as the guards and got medical care.

Over 9,000 prisoners of war were held in Florida during the war, with 250 of them at the Leesburg camp. The maximum number of German POWs in the continental U.S. was 371,683 in May 1945. There were POW camps in every state except Nevada and Vermont.[1] Regardless of their good treatment, the men were still prisoners and many retained their loyalty to the Third Reich, Italy or Japan.

In our school in Apopka, students told "war stories"

about Germans. When I was about eight or nine, I shocked the class by telling about a German couple, the Dietrichs, who had a small farm next to us in Apopka. They spoke English with a strong German accent. My father and I made a neighborly visit to their house to talk about his crops, chickens and cows. My wild eyes began to wander and I peeked through their windows. I told the class that I had seen arms, ammunition and hand grenades in their rooms. That really got the attention of the other students. But the teacher was doubtful and questioned my story. She was right, of course. It was all my imagination and a big lie. I have carried this burden all of my life and will, no doubt, go to hell for it.

After I entered the University of Florida in 1952, I went broke twice and couldn't pay my tuition. My father could not help because his farm in Fort Myers hadn't made money for two years in a row. I worked as a student and received a meal ticket, but this wasn't enough to cover my expenses. Those were hard times, so in September 1954, although the Korean War had ended by then, I joined the Army because the Korean G.I. Bill would pay my tuition when I finished a two-year enlistment. That would save the day for me. Fortunately, I was sent to Germany and not South Korea.

I arrived in Germany in February 1955, where the extreme cold was excruciating for a Florida boy. This was less than a decade after World War II, and I was shocked to see so much war damage still visible in some German cities. But I admired the people for their determination and hard work to rebuild their war-torn country with the help of the American Marshall Plan.

I was assigned to the 46th AAA Battalion in Fliegerhorst, a small, quiet former Nazi base in the countryside east of Hanau. My sergeant pointed out to the commanding lieutenant that the Florida boy's nose, ears and face turned purple in the frozen gun park. As a result, I became a radio operator and transmitted messages in Morse code and by voice from a warm, locked radio shack. When I had time off, I enjoyed lots of good German beer and wine in scenic old villages like Rudesheim on the Rhine River.

In Frankfurt am Main, there were still many blocks and streets with piles of rubble. In the middle of this horribly bombed historic area, I knew where there were stairs descending below the street level to Der Jazz Keller. As you went down the steps, the cloud of cigarette smoke got thicker. Then the blast of lively music really hit you. I ordered *ein stein Bier* and found a seat. Amazingly, there was a small, upright piano in the band that played American jazz—a rare musical form in Germany in the 1950s. American military guys and Germans played together for fun in a perpetual jam session. Anyone who wanted to play got their turn. English was spoken but German was the predominant language in Der Jazz Keller. I met mostly Germans in that rustic underground cellar that had survived the tons of bombs of World War II. This unique cavern served a very good purpose and helped former enemies connect with music, beer, friendship and good times. Music had always been my hidden passion. I attended performances when I could get a pass through the gate at Fliegerhorst. Within a short taxi ride in an old Mercedes Benz car, I could hear wonderful concerts in small villages that had an old church with an ancient pipe organ and a choir that performed Bach or Handel.

The best concert I remember was in Frankfurt am Main in a new opera house that had been built after World War II. To get to a performance on time was a challenge. I had to quickly change from my Army work uniform to some appropriate civilian clothes. Then I rushed out the main gate of Fliegerhorst Kaserne, took a taxi to the train station, then boarded a train pulled by a steam engine to the Frankfurt am Main Haupbahnhof—the main train station. From there, I almost ran to the opera house, about eight blocks away. I never had a reservation, but I knew that I could always get that special seat that no one else wanted. It was behind a supporting post in the balcony. I could see everything on the stage and in the orchestra pit simply by shifting to the left or right of the post. This seat was affordable for a poor U.S. Army soldier who had just gotten a raise to $112 per month. My theater ticket cost four Deutsche Marks (DM), which was one U.S. dollar and I didn't complain.

I attended the first concert performed in Germany after the war by the famous Concertgebouw Orchestra from Amsterdam, conducted by Eduard van Beinum, a renowned maestro. After they played Brahms' Symphony Number Two, the German audience gave a heartfelt applause that went on and on. I think this was in deep appreciation to the Dutch for coming to Germany so soon after the war—a decade later. The sincere applause was perhaps a subtle apology for brutally invading and terrorizing tiny Holland only a few years earlier. Based on the arrogant expressions on the faces of the Dutch musicians, all was not forgiven. But it was a start and old wounds were slowly healing between the Dutch and their nemesis, Germany. I will never forget that concert.

In that same opera house, I attended just about every opera that they presented during 1955 and the first half of 1956. Thanks to those affordable seats behind the post for one dollar, I could afford a ticket. Operas and concerts are now performed in the restored historic Alte Oper Frankfurt, and I suspect the price of a ticket is over a hundred times what I paid.

Just before I returned to the USA, I got permits to travel through Russian-controlled East Germany on the train to West Berlin. The city was still being restored after the devastating blitzkrieg in 1945, but there was an optimistic attitude to rebuild and be proud once again.

I was able to take a bus tour to East Berlin that was controlled by Soviet Russia. After we crossed the Brandenburg Gate, an East German armed guard boarded the tour bus. Nearly all of us tourists were English-speaking.

The guide gave strict orders to put away all cameras and to not take pictures. That was because the communist side was so drab and behind with restoration. Piles of rubble and bombed buildings were common sights. However, there were political showcases for propaganda where photographs were allowed. One such place was a huge, immaculate park with a giant sculpture of Mother Russia taking care of her children, the Germans, after the war. Hammer and Sickle emblems and Marxist banners were everywhere. When we got off the tour bus to see the park and take pictures, there were hundreds of uniformed army and air force men—all Russians. It was Sunday, and they were touring the beautiful showcases of East Berlin. I was 21 like many of the Russians in uniform and stood out in the crowd because I was wearing a U.S. Army uniform. That was because my civilian clothes had been shipped back to the USA. I did not feel comfortable amongst all those communist troops because of hostile feelings during the Cold War.

But before I knew it, I was shuffled to the center of a front line of Russian soldiers and airmen for a photo opportunity. A lone American soldier in uniform posing with this multitude of Russian military men was a dramatic event and cameras were clicking in all directions. People on the bus tour became very excitable. The young Russian men were also excited and wanted to exchange anything that I had as a souvenir. Like me, a few of the Russians spoke some German and we made deals to swap minor items. I traded the US Army pin on my cap for a Red Star Hammer and Sickle pin from the cap of a Russian soldier. It was forbidden and I was getting nervous, but I exchanged a few small coins from West Germany and a few other miscellaneous items in my pocket. They were happy with any memento from this Yankee soldier, probably the first American they had ever seen in their life.

With all the excitement, a Russian officer came over to see what the commotion was about. He disbanded the group and we all went our ways. It proved to me that soldiers are soldiers and people are people regardless of what uniform they are wearing. Mother Russia was behind us, hopefully looking down with her approval.

When I got back to Gainsville in July 1956, I received $110 a month from the G.I. Bill. This saved the day for me and got me a BSA degree. In addition, I earned some pocket money—55 cents an hour—by spraying roaches in the girls' dorm in the evening. When I entered the dorm, I was supposed to yell, "MAN IN THE HALL," but my voice always choked up and was hardly audible. Consequently, the girls did not hear me and were running in and out of their rooms to the showers down the hall in their underwear or nothing but a towel. Of course, I had to close my eyes but no longer than an augenblick. This earned me some money, and killing cockroaches in the girls' dorm was sort of fun.

The most important close encounter I had with a German was after I started my career with the United Fruit Company in Central America. I met and married Phyllis Steffen Witte in Honduras where she was a teacher in the banana schools for the same banana company that I worked for. Of course, my wife was an American, born and raised in Southwest Minnesota. But her genetic heritage was pure German from all sides. All of her ancestors were Prussians who spoke German in Minnesota. This was a lasting, close encounter that began in 1963. Even our children, Linda and Carl, are half German, although they were born in Panama. Their other half is a mixture of Irish, Scotch and English. Sadly, Phyllis passed away as I was in the process of writing this book.

My various close encounters with Germans have turned out to have had a significant impact on my life.

Clyde and Phyllis with Linda and Carl.
Changuinola, Panama, 1966.
Clyde Stephens collection.

Chamberlain mansion in Apopka, Florida, circa 1943.
Clyde Stephens collection.

Carl Wilson Stephens, Era Mae Bullock Stephens and family.
Apopka house, 1943. L-R: John Philip, Zola Mae, Julius Wilson,
Edith Elaine and Clyde Schubert.
Clyde Stephens collection.

German POWs at Camp Leesburg, now Lake-Sumter State College. Source: http://lssc.digital.flvc.org/islandora/object/ lssc%3A357.

Captive Germans from submarine U-118 arrive at Norfolk, Virginia, 1943. Photograph: Public domain. Wikimedia Commons. Source: [[File:U-118 Besatzung.jpg|U118_Besatzung].

Der Römer (center) after bombing in 1944, Frankfurt, Germany. Der Jazz Keller was near here in 1955- 1956. Photograph: No restrictions found. Source: luftangriffe-frankfurt-zerstoerung-der-stadt-innenstad-taltstadt-1024x683.jpg (1024×683) (skylineatlas.de).

Private Clyde S. Stephens at Der Römer, 1955. Clyde Stephens collection.

2. German Prisoners of War

THIS CHAPTER IS A sequel to another story that I wrote about Germans in Bocas del Toro, a province in western Panama on the Caribbean. The text describes Germans who came to Bocas del Toro in the late 1800s and early 1900s and became a prosperous colony of merchants, traders, importers, exporters and banana planters. A nascent banana industry was booming and attracted numerous immigrants from many countries to that verdant tropical wilderness of rivers, bays and islands. Germans came because Germany had become a colonial, commercial, scientific and military power, and its citizens were scattering around the world to serve those purposes.[2]

This sequel tells about Germans in Bocas del Toro who experienced difficult times during two World Wars. Some of them were single men and others had German spouses and German children who were born in Germany or Panama. Others married women from Panama where their children were born.

When the two wars came, the lives of these people were drastically disrupted. Being citizens with German passports, these residents of Panama had to decide where their loyalties lay. Some returned to fight for the Mother Country but most stayed in Panama to face the consequences of the times.

BOCAS DEL TORO BEFORE THE WAR

A newspaper story dated March 27, 1914, described the magnitude of converting the Almirante mangrove swamp into a sanitary landfill, almost due west of Bocas del Toro (also Bocas Town), across Almirante Bay, largely for the export of bananas.[3]

The labor force had about 500 new employees who were pumping sand, mud and coral from the bay, while buildings for housing and stores were erected daily. The reporter predicted that Almirante would soon be the commercial capital of the province, while Bocas Town would remain the official capital.

Soon after that story appeared, rumors of war, fears and opinions were recorded in several letters written by Australian-born Winifred Smith. She lived in Almirante with her American husband who was involved in the landfill project of the United Fruit Company.[4]

When the complex alliance of European adversaries instigated World War I in 1914, dredging, landfill operations and construction suddenly stopped. In her letter of August 1914, Smith tells how 400 employees of the dredging project became unemployed and had no way of returning to Jamaica and other islands where they came from. Some had families in Panama, resulting in much confusion, despair and chaos. But an Afro-Jamaican named George said that things were going to be all right because "we have the mother country behind us." This reflected the loyalty and confidence of Afro-British subjects in Bocas del Toro.

WORLD WAR I

President Woodrow Wilson said that the U.S. would remain

neutral and not intervene, despite the fact that the strategic Panama Canal would open in less than three weeks. As the war advanced, Germany became more aggressive and sank the RMS *Lusitania* that killed nearly 2,000 passengers, including 128 Americans the following May.

In one of her letters, Smith describes how the *Orinoco* arrived flying a British flag; the ship was forced to travel without lights because of the German submarine threat. The captain delivered recent newspapers from New Orleans and Colon that contained the latest war news.

Smith quotes comments by an elderly German referred to only as von Schneider, who had arrived on the isthmus with his wife in the 1890s. In exchange for recent newspapers with news about the war, von Schneider offered oranges, limes or guavas. Paradoxically, he regretted that he could not return home to fight because he had poor eyesight and was very old. He proudly proclaimed that Maubeuge in northern France had been captured by the Germans with the whole English army inside! [sic]. His rants against the United Kingdom were audacious, and he hoped there would not be one British person left. Despite this open hostility, he would ask Winifred Smith's husband for wireless reports, newspapers, and quinine to treat malaria.

Smith also mentions an Austrian, Johann Orth, who made patriotic comments about the war, and a Boat Captain, Hart, who wanted to see the French defeated because of what they had done to the church. Despite these hostile feelings, Smith said that sympathy in Almirante was mostly for the American Allies, except for a handful of Germans. "A few Germans are quite decent, and do not gloat over every victory like the others do," she wrote.

Smith penned that local French citizens were returning to France to fight. She was amazed that Afro-Antillean men from Martinique who had moved to Panama were patriotically going to war in Europe. This probably included some men from the large French Patois Town in Almirante and elsewhere in Bocas del Toro.

Ariel Rene Perez Price learned that his grandfather, Dr. Jose Antonio Price, examined Edgar Victor Georget and found him fit to fight in the French Army. Georget was born in 1895 in Punta Valiente, an indigenous area. The medical certificate was signed by Dr. Price on March 17, 1915, then stamped and signed the same day by Monsieur Raul Cochez, the French Consul in Bocas del Toro.

The twenty-year-old Georget enlisted in the French Army and never returned to Panama. His sister, Lilliane Maud Georget, was the wife of Dr. Price, although she left him and went to France in early 1918 to join the Red Cross. She never returned to Bocas del Toro either.[5] Their French father was Louis Victor Georget, a merchant, banana planter and later an administrative employee of the United Fruit Company. After the war, another son, Luis Jaques Georget, returned to Bocas del Toro from the French Foreign Legion and France. Giving life to eight children, he died in 1980.

Another young man who went to war in Europe was William Ralph Snyder, born October 1, 1893, in Bocas del Toro, and educated in England. Snyder enlisted in the military in Canada on April 13, 1915. He was the son of the American banana tycoon, Michael Theodore Snyder. His mother was a local Afro-Antillean, Seveline Mathews Shepard Snyder. Their son might have been the first casualty from Panama in The Great War. His tombstone in Boulogne Eastern Cemetery,

France, identifies him as: 426577, Private, W. R. Snyder, 3rd BN. Canadian Infantry. He died on October 3, 1916, two days after his 23rd birthday.

As part of their attempt at world conquest, the Germans tried to make a secret alliance with Mexico, just across the American border. With Kaiser Wilhelm II's army targeting England—after invading Luxembourg, Belgium, and France—it became obvious that the Americans could no longer stand by. Subsequently, nearly three years after the war started in Europe, the United States declared war on Germany and her allies on April 6, 1917. Panama did the same the next day.

PATRIOTISM

Throughout the war, patriotic activities were held in Bocas del Toro:

On November 1, 1915, a noisy band playing the British national anthem awoke the British vice consul, William H. Ponton, at 5:30 in the morning. He responded in kind by inviting the musicians upstairs to "warm up" with some British libation. They left playing "Rule Britannia" as more enthusiasts joined the parade.[6]

At the French Consulate, the band played "La Marseillaise," followed by a speech in French by Monsieur Raoul Cochez, who had two sons in the war in France. The speech was translated to English by Dr. A. O. Rawlins, a Jamaican.[7]

The Panamanian national anthem was played at the police station and at the government palace where the governor of the province, Fabio Bravo, invited the crowd upstairs to Champaign for the men and iced soda for the ladies. Through an interpreter from Spanish to English, he commended the crowd for their orderly conduct. The Reverend E. C. Notman excited the crowd by insulting and condemning the Kaiser and Germany for their cruelty. The Reverend M. C. Surgeon read a message from the King of England, then departed amid deafening cheers.

Flag sales by girls and women generated about 800 dollars in Panama silver. Even though they were told not to enter German stores, they went anyway and sold flags to Panamanian employees in the stores. One persistent little girl managed to pin the Union Jack on a German merchant as he struggled to get away.

Near the end of the war, fundraising for the Red Cross and the sale of Fourth Liberty Loan Bonds of the United States of America was reported by the Star & Herald on October 18, 1918. Columbus Day on the 12th was turned into many patriotic speeches arranged by Chairman Harry L. Peck, assistant superintendent of the Changuinola Railway.[8] Instead of long prayers at the opening and closing of each speech, organizers urged only a short benediction.

These patriotic demonstrations made German residents apprehensive and nervous about their future.

THE CANAL ZONE

The defense of the Panama Canal was a major concern, and American strategists knew that loyal Germans and their allies in Panama were a possible threat to the vital waterway. As a result, Panamanian police, nudged by American officials, began arresting all Germans and their allies on the isthmus in April 1917, including those in Bocas del Toro.

A policy statement was expressed in a letter from an official of the U.S. War Department:

> While there probably is not much palpable proof of any acts

> of hostility toward the United States upon the part of most of those interned, there is no doubt as to their attitude and their pro-German feelings. Their presence in the vicinity of the Panama Canal is not to be desired, as many of the utilities could easily be seriously injured by small amounts of dynamite.[9]

THE ROUNDUP OF ENEMY ALIENS

Germans and their allies in Panama had been under surveillance long before the United States declared war in 1917. But with the war came increasing scrutiny and arrests:

In February 1917, the Reverend S. Moss Loveridge wrote in his diary that a month or so after his family's vacation at the Aspinwall Hotel, all the guests were abruptly evacuated and the hotel was converted into a German prison camp overnight.[10]

David Zemer described how four German ships were impounded in the Panama Canal Zone in February 1917. The ships were expropriated, overhauled and given American names. The crews were eventually interned on Taboga Island.[11]

A novel eyewitness account was written by the Reverend Loveridge. In the Santa Ana neighborhood of Panama City, he bought lemonade at a German drugstore that exhibited life-size portraits of the German Kaiser and his wife. After he went next door with his drink, he recognized an American police friend in a dark corner of the room. When asked why he was hiding in the shadows, the officer cautiously replied that he was spying on the Germans. The next day, war was declared.[12]

On April 12, 1917, the *Star & Herald* reported that about 50 Germans were arrested by Panamanian authorities on their own initiative but were aided by American soldiers and the

Panama Canal Zone Police. The Germans were given that day and night to arrange their affairs before being sent to Taboga Island the next day for internment.

TABOGA ISLAND AND THE ASPINWALL

When the French sea-level canal was started in 1881, a 50-bed sanitarium for sick employees was constructed on Taboga Island some 19 kilometers off the coast of Panama City. The French Canal failed in 1889. When the Americans started building the Panama Canal in 1904, the sanitarium became the property of the Isthmian Canal Commission. The building served as a place of rest and recuperation for Canal Zone employees.[13]

When the canal was completed in 1914, the hospital was converted into the Aspinwall Hotel, named in honor of Engineer W. H. Aspinwall, who was known for his role in the construction of the Panama Railroad. The new hotel was inaugurated in 1915 and became a popular destination. Unfortunately, it closed the next year because of financial difficulties. After some renovations, the Aspinwall opened again in January 1917. Soon after, the United States officially entered World War I. The elegant structure was closed as a hotel on April 11, and opened the next day as a detention center known as the German Camp.

NOTES ON ENEMY ALIENS OF SPECIAL INTEREST

In an espionage report dated April 18, 1918, Major Edmund Butcher of the 29th U.S. Infantry Provost Marshal detail (Military Police), listed 97 people, providing their curriculum vitae and some harsh assessments of their character.[14] This report records the date that each person was interned at Taboga.[15]

On the first day of internment, April 12, 1917, a mother and son were sent to Taboga. She was Frida Bartel, 32, who was born in Polloni Pommern, Germany.

Keeper of the German American Bar in Panama City, she provided the Germans with information she obtained from American soldiers who were included among her patrons. She was considered dangerous and of loose character; the house she maintained was described as an illegal resort. Mrs. Bartel was the only woman that was sent to Taboga Island against her will.

Baron Adolf Olaf Von Gargen, 50, Frida Bartel's common law husband, was interned on April 15, 1917. The report said that Von Gargen had separated from Mrs. Bartel.

Otto Bartel, the 12-year-old illegitimate son of Bartel and Von Gargen, was voluntarily interned the same day as his mother. His mother smuggled letters out of the German Camp via Otto, who was attending school in the native village of Taboga. But when the authorities found out, he was expelled. Other parents did the same, and their children also were expelled.

Otto K. Haack (also Haak), age 43, born in Pforzhein, Germany, was the owner of The Diadem, a jewelry store on Central Avenue in Panama City.[16] He was known as a forwarding agent for German mail to and from Central and South America. Haack signed a statement that he would not cause harm to the United States or Panama and was freed on June 12, 1917, to run his business. The final statement said that he was apparently harmless but unpredictable.

Dimitri Ivan Troetsch Wilcox, a Panamanian lawyer from Chiriqui, wrote a book, *Genes alemanes en Chiriqui y Bocas del Toro*, and reviewed the capture and treatment of Germans in Panama in both World Wars.[17] He was vehemently critical of the way Americans arrested prestigious German academics who had been invited by Panama to organize, administrate and be professors in institutions of higher learning.

Troetsch writes about the close relationship that his grandfather, Carlos Troetsch, had with Dr. Richard Neumann. They exchanged visits, spoke in German, and listened to music played by the Troetsch family at their home in La Concepcion in Bugaba, Chiriqui.

Neumann came to Panama in 1911 and became a renowned professor at the Instituto Nacional de Panama. Despite his academic fame and ability to advance educational standards, he was arrested, along with other German professors. There were protests by students and staff in support of their professors, but to no avail. Neumann was taken to Taboga on November 30, 1917.

The following is the espionage report submitted by Major Edmund Butcher, which is an antithesis of Dr. Neumann's fame in Panama:

> Richard Neumann, age 34, born in Koenigsberg, Prussia, professor of Pedagogy and of languages in the Instituto Nacional de Panama. He made arrangements with other Germans to leave the Isthmus at the outbreak of the war as to avoid internment. Since that time, he has been traveling through Panama visiting with Germans and sympathizers. When apprehended in the province of Chiriqui, he was closely associated with other Germans under suspicion. He would assemble his associates in the Hotel Bolivar and meet with them until the early hours of the next morning. He is one of the most dangerous of the internees. Interned November 30, 1917.[18]

Several other German professors at the institute were Dr. Ernst Hoffmann, Otto Lutz, Eugene Lutz and his wife, Hellen Margaret Lutz, George Goetz and Agnes von Owen. Hoffmann and Eugene Lutz were interned on April 13, 1917 and his wife, Hellen Margaret Lutz, joined Dr. Neumann on Taboga on April 20.

Dimitri Troetsch also expressed revulsion regarding the arrest and deportation of the naturalist Alois Strassil Hartmann. A resident of Volcan, Chiriqui, and later the patriarch of the Hartmann-Troetsch family, he married Susana Troetsch, and they had 10 children. As a naturalist guide, Hartmann was known for his insect and bird collections. He was also a successful coffee producer.[19, 20, 21]

In contrast, the espionage report by Major Butcher on Hartmann describes him with scorn and suspicion in the extreme, as follows:

> Hartmann, Alois: Age 26; born in Hohenschladt, Morunue, Austria;[21] occupation, Naturalist. Alias Dutch Louis, Louisito, Loco [Crazy] Luis. This man, under the guise of being a hunter, travels throughout the interior of Panama and across the country to Costa Rica and associats with Germans everywhere. He has a thorough knowledge of the Interior of Panama and his reputation as a desperado is a menace to the United States. Hartmann is a person known to use a gun at the slightest provocation. Interned at Taboga, April 26, 1917."[22]

Major Butcher also reported about Carl Alexander Max Jaeger:

> Age 39; born in Frankfurt-Main, Germany; occupation, merchant. Formerly owned a saloon in Panama City. Jaeger

> was one of the party who planned to leave Panama and board a launch to avoid being interned by the American authorities. His saloon was frequented by Germans and in his conversation with American soldiers, Jaeger expressed himself as being extremely anti-American. He served in the German Army and is a dangerous character. Interned at Taboga, April 13, 1917.[23]

Perhaps the best known of the Bocas del Toro suspects was Carl Friese, who fled to Costa Rica in April 1917, aboard his boat, the *Heligolne*, a name that may identify with Heligoland, a German island in the North Sea. A friend would later say that Friese went to Costa Rica, "On account of ill health and on account of the persecution by British negroes in Bocas del Toro."

A report about Friese, dated May 24, 1918, was sent by the American Consulate in San Jose, Costa Rica, to the Secretary of State in Washington, D.C.:

> Carl Friese, German, San Jose, was on the original Enemy Trading List. He was the German Consular Agent in Bocas del Toro and the financial agent for German war propaganda. Conducts a store in Bocas that is managed by a woman who was formerly his mistress. He came to Costa Rica when the United States entered the war. He is one of the most active propagandists and agents here and communicates with the outside by messengers. Claims to be interested in rubber here but it is a blind.[24]

Friese also was targeted by the U.S. minister (ambassador) to Panama, Mr. William Jennings Price. In a letter to Washington, Price confirmed that the U.S. was seeking ways that Panama could legally justify confiscating property belonging to German aliens. A case-in-point, Price wrote, is "The rather large house of Carl Friese and Company at Bocas del Toro."[25]

Hans Kandler was an associate of Carl Friese in Bocas del Toro, and espionage reports from the American Consul in Colon and Bocas del Toro revealed more data about the detainees. Additional information was provided by British, French and Italian colleagues in Panama.

INTERNED FAMILIES

In his report, Major Butcher described some of the reasons several families were interned "voluntarily" on Taboga; they were mostly spouses and children of prisoners already detained there. Some of the reasons for detention were:

- House too close to the Panama Canal.
- Two women were destitute.
- A man wrote anti-American articles in the newspapers.
- One man operated a launch service.
- A man madc friends too easily.

A spouse interned voluntarily was given permission to come and go, but the permission was withdrawn when she was caught smuggling German letters in and out of Taboga.[26]

The Wilhelm Probst family had a farm near Boquete where the mother, Juanita Barbara Barbosa Probst, and their son, Wilhelm Jr., were born. Major Butcher's intelligence report advised that Wilhelm, the father, was probably harmless, but he had been meeting with other Germans in David. Consequently, the family was interned at the German Camp on Taboga. Butcher's report stated that Juanita's detainment was voluntary, "owing to her destitute condition." Their son, Wilhelm Jr., also "volunteered" to be interned, even though he was only eight months old. Interestingly, one could be interned voluntarily, but once detained, could not choose to leave voluntarily.[27]

When I first researched the story of Alwyn Hasso von Wedel, a photographer, postcard maker, naturalist guide, bird and plant collector, I learned details about his fate and that of the other German men in Bocas del Toro. Many of them were deported to the United States in both wars, although von Wedel was only imprisoned during World War I.

A vehement story in the *Star & Herald* newspaper revealed the hostile attitudes against the Bocas del Toro Germans, while an intelligence report in August 1918 stated the following:

> ...Bocas del Torro [sic] has been a nest for German propagandists and sympathizers. It was considered unwise to permit these people to remain where they were owing to the fact that Bocas del Torro [sic], Porto Bello, and other points along the coast might be made use of as a base for supplies of enemy submarines."[28]

ENEMY ALIENS FROM BOCAS DEL TORO ON THE BLACKLIST

Several Germans in Bocas del Toro and their businesses were placed on the Enemy Trading List, also known as the blacklist between 1917 and 1919.[29] This was based on orders issued by the War Trade Board in 1918.[30]

Attempting to subvert the blacklist had consequences, as shown in this consular report dated December 6, 1918:

> R. L. Bonilla should be placed on the confidential list because he continues to conduct the business of Herbert Leer, a Black-listed firm. Bonilla is the brother-in-law of Alberto Leer and got permission to liquidate the business interests of Alberto Leer after he was interned as a German enemy alien. Bonilla took

the opportunity to import more goods and continue to carry on the business of Herbert Leer under his own name."[31]

Nevertheless, the blacklist did not always have its intended effect. Among various reasons, blacklisting was seen as a U.S. initiative that encroached on Latin American sovereignty. The blacklist, writes historian Max Paul Friedman, "encouraged neutralism among Latin Americans whose fear of Yankee domination was greater than their suspicion of their German neighbors."[32]

Some Germans in Bocas del Toro were not deported for various reasons. (See Appendix I).

ALIENS DEPORTED TO THE U.S.

The group of 98 enemy aliens described by Major Butcher was deported from Cristobal on April 19, 1918, aboard the USAT (U.S. Army Transport) *Kilpatrick*. The ship arrived in New York Harbor on the 28th for processing at Ellis Island. (See Appendix II).[33] Hans Kandler was the only German from Bocas del Toro on this shipment.

Amazingly, the manifest also included the names of 45 American military men from the Panama Canal Zone. They rode troop class and were identified as guards for the interned Germans. This provided about one guard for every two prisoners, including the wives, mothers and children.

Hans Kandler, the Probsts, and probably the rest of this shipment of enemy aliens, was next transported by train 128 kilometers from Ellis Island to Gloucester City, New Jersey,[34] where they were held in the same facility where Frida Bartel was housed.[35]

A cover letter ddressed to Carlos Jaegar indicated how he progressed from Gloucester City to the "War Prison Barracks,

Fort Oglethorpe, Georgia." Jaegar was with Kandler on the same voyage from Panama to New York Harbor. Based on various documents, it appears that many of the detainees in Gloucester City were eventually sent by train to Fort Oglethorpe 1,126 kilometers to the southwest.

Three months later, the *Kilpatrick* transported the rest of the enemy aliens from Bocas del Toro. They departed from Colon, Panama, on July 27, 1918, and arrived in New Orleans on August 6. Following the arrival, an inspection found that one detainee had a photograph of American torpedo boats in Colon, and another man had a chart of Central American waters, along with a U.S. military-generated map of Panama. Also found was a duplicate of each internee's file with the U.S. Immigration Board of Special Inquiry.[37]

Based on various documents, it appears that all of these civilian detainees were also sent by train to Fort Oglethorpe, 805 kilometers to the northeast. (For a summary of the names, ages, professions and contacts, see Appendix III) Because of the security of the Panama Canal, the "enemy aliens" from Bocas del Toro were held at Fort Oglethorpe from August, 1918, until most or all of them were repatriated to Panama in 1919.

ABOARD THE *KILPATRICK*

In a letter, here is how the trip from Panama was described:

> ...They all had to stay in the Steerage and submit to the rough and brutal treatment by the guarding soldiers. They had to eat in a compartment which ordinarily is used for the transport of horses. The meals were inferior as to quality and they were served in a manner [that was] utterly repulsive.... The prisoners had to clean not only their quarters and the eating place but

also the toilets which also were used by the soldiers.... Most had lived for years in a tropical climate and were provided only with very light under and over clothes. War Prison Barracks, Fort Oglethorpe, Ga. Aug. 1, 1918.

FORT OGLETHORPE

By 1917, the war had suddenly generated an extremely hostile anti-German sentiment in the United States.[38] The 1910 census counted 8.6 million people in the country who were either born in Germany or whose parents were Germans. Despite the fact that most of them were Americans, over half a million were interrogated, photographed and fingerprinted. They were considered "enemy aliens" and thousands were sent to internment camps, like Fort Oglethorpe.[39]

Fort Oglethorpe is located in the northwest corner of Georgia, just south of Chattanooga, Tennessee. It was the site of American Civil War battles in 1864 where 34,000 men were killed. Beginning in 1902, the U.S. Army Calvary used this area as a base. From 1917 through 1920, Fort Oglethorpe was converted into a war prison camp that held about 4,000 enemy wounded or missing. The battlefield is now part of the Chickamauga and Chattanooga National Military Park.[40]

Three barbed-wire fences, about 3 meters high, enclosed the camp. Watch towers were located outside the barbed-wire perimeter. Each tower had a search light, telephone, a repeating shotgun, a rifle and a machine gun. To preserve order, the War Department stationed 32 commissioned officers and 452 enlisted men as guards.

Fort Oglethorpe held about 4,000 prisoners who lived in 30 barracks with very little privacy. Bath houses were outside. The detainees ate in an enormous dining hall.

German was the most common language heard in the camp, but Polish, Hungarian, Czech, Slovak, Croatian, Bulgarian, Romanian, Russian, Hebrew, Arabic and other languages from the vast complex of the Central Powers Alliance were spoken.

The fort was divided into three parts: Camp A, which was known as the "millionaires' camp," housed wealthy prisoners in private rooms. The "millionaires" paid for their own food, and for cooks and servants who were prisoners from captured German ships. Camp B held the majority of the prisoners. Camp C was for prisoners being punished for violating the rules, causing a disturbance or attempting to escape. They were restricted to half-rations. Daily life was strict and regimented. A bugle call was sounded at 5:30 AM, then a roll call at 6:30, followed by breakfast. At noon, the bugle announced lunch. There was a rest period from 1:00 to 3:00 PM. At 5:30 PM, there was another roll call followed by dinner.

During the day, there were work assignments inside and outside of the camp. The prisoners were paid in special minted aluminum currency. Prisoners had free time in the evenings to do whatever they liked.

The prison administration allowed "Oglethorpe University" to operate so prisoners could improve their education and pass the time. The rector was Wilhelm Steindorff. With so many nationalities from the European Central Powers in the camp, there was a broad diversity of languages, professions and professors who were available to teach. Courses included languages, literature, biology, mathematics, physiology, electronics, art and other courses. Several student inmates painted outstanding works of art that were later published in albums.[41]

One popular activity in the camp was playing or listening to music. The United States had interned an entire German army band with 75 members and their conductor, Willie. In 1914, Japanese forces had captured this band in the German colony of Tsingtao in China, and turned it over to the U.S. Department of Labor. With these musicians came their instruments and musical scores. The symphonic group was trained to perfection and performed twice a month to enthusiastic audiences.

Singers gave recitals of Franz Schubert's songs. Other accomplished musicians in the prison camp included Ernst Kunwald, who had conducted the Cincinnati Symphony Orchestra. Kunwald had a fiery temperament, and several people heard him make threatening comments about President Woodrow Wilson. At Fort Oglethorpe, he presented the camp with Beethoven's *Leonore Overture No. 3* and Tchaikovsky's *Fifth Symphony*. He played the piano every day and many prisoners gathered quietly around his shack to listen to the great performer-conductor. On Sunday evenings, Kunwald hosted a discussion group on the music of the great masters, particularly German speaking ones, like Wagner, Bach, Mozart, Beethoven, Brahms, and Schubert. Dr. Karl Muck, a world-renowned maestro, had conducted the Symphony Orchestra in Boston from 1906 to 1908 and then again from 1912 to 1918 and considered himself an American. This great man with the highest cultural reputation and his wife had enjoyed the adoration of Americans. Suddenly, because of the anti-German hysteria in the United States, the famous conductor was arrested in March 1918. He was sent to Fort Oglethorpe, along with 29 members of his orchestra. The fall from fame to a prison camp was mentally devastating for him.

Dr. Muck refused to participate in musical activities in the camp. Later, he finally consented to give a performance when some of the prisoners convinced him that the camp was like a German village. He selected over 100 professional musicians—all prisoners—for the rehearsals. When Dr. Muck conducted the Oglethorpe Symphony Orchestra on that memorable night, over 2,000 prisoners gathered inside the giant dining room to hear Beethoven's *Symphony No. 3*, the Eroica. One of the prisoners, Erich Posselt, wrote in his biography:

> Then Muck, a gaunt, haggard-looking, bronzed figure, stepped on the platform and raised his hands. In the moment of death-like silence preceding the first note, it was as if an electric current had run through all that unkempt audience in overalls and shirtsleeves, in heavy camp-boots and with unshaven faces. Then, jubilantly and overpoweringly, the Eroica rushed over us, welded us into one, scorched us and purged us, and shook us to the very depths of our being. It is my conviction that this performance of the Eroica was one of Muck's greatest achievements, and probably one of the greatest events in the history of music in America.[42]

It is likely that at least some or all of the enemy aliens from Bocas del Toro attended this historical performance.

One of the single-story barracks was used as an amusement room and fitted with a stage, where serious dramatic performances were given by the prisoners. On other occasions, cabaret entertainment was presented. The room also contained a library, as well as a piano and billiard table. Another activity was watching movies twice a week. This room was also used for religious services, namely Catholic, Lutheran and Jewish.

Prisoners played soccer, volleyball, handball, tennis and other sports. Leagues were organized by campmates and competition was spirited and frequently fierce. There was also boxing and wrestling. Arts and crafts, wood and metal working, drawing, writing, and painting were common ways to be creative. Prisoners could grow gardens and eat their own vegetables. As one professor said, "the well-educated men who occupy themselves with hobbies are the lucky ones, but the rest are doomed to mental stagnation."[43] Overall, medical, dental and health care were considered good. That was because the U.S. War Department demanded it—a policy for getting equal treatment for American prisoners held by the Germans.

Internees could write and receive letters but the hated censors always invaded their privacy to the extreme. Wives could visit their spouses but couples were ordered to speak only English, not German.

Despite these many activities, some prisoners were stressed into suicide, insanity and depression. A major catastrophe happened in the fall of 1918 when half of the men became infected with the "Spanish Flu" that swept through the barracks. Forty-six prisoners died during that initial outbreak that rapidly spread around the world in a pandemic that killed tens of millions.

"Spanish Flu" seems to have started in the United States, not Spain, and was spread to Europe by American soldiers packed in troop ships and on the battlefields.

The prisoners operated their own German newspaper, the Orgelsdorfer Eulenspiegel, a cynical and satirical name. News covered many subjects, but texts were frequently sarcastic with double meanings that added a sense of humor for the readers.[44]

THE END OF WORLD WAR I

Germany agreed to an armistice to end the war on November 11, 1918, and the first Panamanian detainees left Fort Oglethorpe a little over two months later.

A note typed on the manifest of the USAT *Kilpatrick* noted that 27 alien enemies were being returned to the Canal Zone. The ship sailed from New Orleans on January 28, 1919, and arrived at Cristobal on February 3.[45]

Seventeen men from Bocas del Toro were aboard; carefully noted was the amount of money each had in his possession. It varied from nothing to $77 carried by Albert Leer. Hasso von Wedel, the photographer and naturalist, had $2.[46] (See endnote for information regarding Eduardo Lange, Ernest Lange and Peter Ulhmann.)[47]

Many of the prisoners from Panama did not return to their homes until several months after the war was over. Some languished at Fort Oglethorpe for many months, causing anger, depression, frustration and suicide.

In a pleading letter written by the Committee of Aliens from Panama to the Legation of the Republic of Panama in Washington, D.C., internees emotionally and amply expressed their grief and pleas for help in returning to their homes after two years of detention.

The letter was written in Spanish at Fort Oglethorpe and dated May 16, 1919. It was sent via the Legation of Switzerland, which mediated and handled affairs of prisoners of war. The committee declared that the prisoners from Panama were in a state of deep depression and there had been two suicides in two weeks.[48]

Numerous letters were also written to judges, senators, congressmen, lawyers and the Department of Justice, but none of them ever received a meaningful answer. It turned out that a détente had developed, as expressed in a letter dated April 11, 1919, that William Jennings Price, the American Minister, sent to Washington, D.C.:

> ...I spoke to the Secretary of Foreign Affairs of Panama (His Excellency, Sr. Don Dr. Ernesto T. Lefevre) regarding Panama's wishes and he expressed himself as being quite in favor of as many being sent to Germany from the United States as possible, instead of their being returned here (Panama).[49]

When an effort was made to determine the feelings of the detainees still at Fort Oglethorpe, 19 of them wanted to go to Germany "at their own expense" and ten wanted to return to the Isthmus.

The eventual decision was that the Germans should be returned to Panama, except for a handful who specifically wanted to return to Germany.

A total of 39 Panamanian aliens were on the transport ship *Kilpatrick* that left New Orleans on June 15, 1919. The roster included the saloon keeper, Frida Bartel, and her son, in first class and the distinguished professor, Dr. Richard Neumann, who returned in second class.

Wilhelm Probst, his wife, Juanita, and two-year-old son, returned as a family of three. The wife and son were Panamanians from Chiriqui. Nearly all of these alien passengers were on Major Butcher's espionage report and were on the same ship that arrived in New York on April 28, 1918.

Hans Kandler of Bocas del Toro, Alois Strassil Hartmann of Chiriqui and Carlos Jaegar of Panama City left New Orleans on July 17, 1919, for their return to Panama.

It is not known when the last prisoners were returned to Panama but mathematics professor Eugene Josef Lutz was on the *Kilpatrick* that sailed from New Orleans to Panama on August 18, 1919.

POST-WAR ADJUSTMENTS

The first German to return home to Bocas del Toro was Eduard Lange, who arrived in mid-January, 1919. His brother Ernst Lange and other Bocas del Toro aliens were on the next ship that arrived in early February and were soon reunited with their families.[50]

No doubt, relations with former friends and associates were strained and bitter because of hostile feelings before these immigrants were deported. During their absence, there was fraud, mismanagement of assets or complete losses of some properties when locals took advantage of the imprisoned Germans. Detainee Erich Rassow wrote from New Orleans in 1918:

> ...a party of Government employees went to my farm [and house] at Cricamola River, ...and took therefrom provisions, clothes, saddles, etc. They also carried away all my thoroughbred poultry and pigs. Besides this, unknown parties practically ruined my plantation by cutting down trees, stealing Zinc [sic]. Etc.[51]

Nevertheless, the detainees picked up the pieces, reorganized their businesses and got back to work.

Carl Friese, who had escaped from Panama and exiled himself in Costa Rica to avoid internment, did not waste any time in restoring his business in Bocas del Toro. The Star & Herald on August 14, 1919, reported the following: "FRIESE COMPANY TO BRANCH OUT. Mercantile Outfit in Bocas

Making Plans for Bigger Business Than Ever. French and English Goods." The newspaper added:

> It is learned that the large wholesale and retail establishment on Front Street known as Carl Friese & Co. will be replenished with a very large and attractive stock. Goods of English and French make will be the predominating feature and rates cheap as possible. It is said that Mr. Kandler, the junior partner, will be General Manager and Mr. Friese will travel to Europe and elsewhere in the interest of the firm. Already the senior partner has shipped from Costa Rica one hundred bags of the best grade of sugar produced in that republic.
>
> With great anxiety, the general public looks for the return of Mr. Friese for only in the resuscitation of the business conducted by him and his associates do they see any redemptive ray of hope to be delivered from the present Egyptian-like oppression of high prices prevailing in the town stores for the last two years.[52]

Hans Kandler had arrived a month earlier on the *Kilpatrick* that had sailed from New Orleans on July 17, 1919, nearly six months after the other Bocas del Toro Germans came home. He quickly met with his friend Carl Friese to reorganize their partnership, as described in the *Star & Herald* article.

Hasso von Wedel returned to his photography and postcard profession.[53]

By 1923, Albert Schieck had opened an upscale bar called The Tunnel that was located in the Hotel Tunnel in Bocas Town. This was a gathering place for Germans and others with a taste for a high-class place to socialize.

Herbert Leer, a former United Fruit Company administrator and merchant in Bocas Town in 1904, was lauded as an investor and entrepreneur in 1916 in the Blue Book. He married a Costa Rican from San Jose, Lilia Bonilla, in 1915. Records do not show him being deported or returning to Panama after the war. It seems probable that he took refuge with his in-laws in Costa Rica to escape internment. He was recorded as a resident of Bocas del Toro in the 1930s.

In Panama City, Dr. Richard Neumann was given a cordial welcome back. He accepted the government's offer to resume his former position as professor and as an administrator organizing educational programs. The following year, in 1920, Panama granted him citizenship.

Alois Strassil Hartmann returned to his coffee farm in Volcan, Chiriqui, married Susana Troetsch, and they had ten children. His fame grew as a naturalist and guide, an interest that was passed on to several of his children.

Hermann Grabowsky anxiously reunited with his French wife, Gabriela, and their young baby, Elsa Mariana Grabowsky. She was born in Bocas del Toro on August 31, 1917, less than a year before her father was deported in 1918.

Wilhelm Probst and his wife lived in Boquete, David, and La Concepcion, all in Chiriqui and had 11 more children.

Relations were probably strained between German, British, American and French families in Bocas del Toro after the war, but time probably helped to heal old wounds.

PEACE, PROSPERITY AND ECONOMIC DEPRESSION IN BOCAS DEL TORO

In the 1920s, after the recovery from the Great War, German merchants, entrepreneurs and planters reestablished themselves, even though some or all of their former assets had been expropriated or pirated by dishonest custodians. Despite this disastrous period in history, it is notable that Bocas del Toro still had five consuls and vice consuls: The United States, Great Britain, Nicaragua, Portugal and Germany. Once again, Carl Friese was in charge of the latter. The banana industry of the United Fruit Company was always the major source of wealth in Bocas del Toro and much of its rich payroll was spent locally. The success of local merchants depended on this multi-million dollar source that filtered down to all levels of business. However, after Panama disease, *Fusarium oxysporum*, killed the remaining Gros Michel bananas in Changuinola by the mid-1920s, exports and payrolls declined rapidly.[54]

Cacao replaced bananas but did not require as much intense labor or management. Also, as an investment, cacao earned low revenues as compared to bananas, appropriately known as "Green Gold." Even though 4,788,080 pounds of dry cacao were exported in 1925 from the world's largest cacao plant in Almirante, it was clear that the economic future in Bocas del Toro was doomed within a few years. As a result, the United Fruit Company made plans to develop a gargantuan new virgin area on the Pacific side to replace the diseased banana plantations on the Caribbean side. The new banana port would be called Puerto Armuelles, in Chiriqui, not far from the Costa Rican border.[55, 56]

After the exodus from Bocas del Toro to Chiriqui, plus a rapid decline in the plantations, payrolls plummeted to drastic levels by the late 1920s. Banana exports ceased altogether in 1932. The number of United Fruit Company employees went from 5,000 during optimum banana production to a low number of employees working in the cacao trade. This economic crash affected all forms of business in Bocas del Toro, including that of German merchants and their employees. During the 1930s, there was only a minimal number of Company employees who worked in cacao operations. In 1934, an address book in German listed 15 German men in Bocas del Toro at that time.

This negative economic impact caused many residents to leave the province and look for a better living elsewhere. Some had nowhere to go and simply stayed where they could. Compounding this situation was the debilitating effect of the world economic depression during the decade of the 1930s. The saving grace was that people in Bocas del Toro could own small farms and grow gardens to feed their families. Many households had chickens and pigs. Fishing in the rivers and the sea was always a way of life in the province. Hunting in the nearby tropical forests put meat in the pot as well.

WORLD WAR II: ENEMY ALIENS AGAIN FROM BOCAS DEL TORO

During the 1930s, Richard and Edmund Kandler were sent to Germany for their education. They graduated from school and were inducted into the German military as Hitler's Third Reich invaded nearly all of Europe and beyond. A son of the Lange family, Carlos, also studied in Germany and served in the military.[57] Thus, young men from Bocas del Toro participated in World War II on the German side. Another roundup of Germans in Bocas del Toro began soon after Japan bombed Pearl Harbor in Hawaii on December 7, 1941, although no reports with exact dates and names were found. Documents state that within 20 minutes of the announcement of the attack, Panamanian authorities began rounding up about 300 Japanese residents,[58] plus 251 German aliens throughout the Republic. This also included Austrians, Italians, and later about 250 European Jews who had fled to Panama and other Latin American countries to escape the Nazi Holocaust.

One reason for concern in Latin America was related to me by Walter Hamer, who was a colleague of mine at the United Fruit Company. Suspicions were raised about a German employee of the UFC who was working in La Lima, Honduras. When investigated, agents found that he was a spy who had access to all the schedules of banana ships entering and leaving Honduras. He also had other sensitive information, during a raid, the agents found a radio that the spy had used to transmit messages to German submarines.

At the insistence of the United States, various countries in Latin America were arresting hundreds of enemy aliens and sending them to localities like Camp Empire, a U.S. internment facility near Balboa. Most of the men, particularly the Japanese, were harshly treated while clearing brush in high heat and oppressive humidity, watched over by occasionally brutal armed guards and vicious dogs.[59, 60]

The detainees were then shipped to the United States, entering at the port of New Orleans. Their next stop was a processing center at Camp Algiers across the Mississippi River.[61] From there, eventually 4,058 Germans, 2,264 Japanese and 288 Italians went by train to prison camps around the country, including Fort Oglethorpe, Georgia.

Although the Japanese and their families were put in separate facilities, Jews and Nazi sympathizers were initially assigned to the same quarters. This created a frightening situation for the Jews until the two groups were separated. In 1973, I met Alvaro Biasetti, a field foreman in my banana research department in Palmar Sur, Costa Rica. When I asked him about his Italian name, he said that he was born in Costa

Rica but his father was from Italy. Alvaro told me that his father had been arrested as an enemy alien in 1941 and sent to a prison camp in New Jersey. His Costa Rican wife and children did not see their father for four years until he was repatriated. Alvaro helped his mother make a living and raise his siblings during the war although he was only an adolescent.

The exact fate of Germans in Bocas del Toro is not known but there are several personal accounts that tell their story during the roundup in the province:

Carlos Reid (1893-1987) spoke of the Germans in Bocas del Toro during his interviews with Dr. Stanley Heckadon-Moreno and described their arrests: "Things went on well until another war started in 1941. Again, orders came that all Germans should be taken in; three, along with Hans Kandler, were sent in."[62]

Certainly, Hans Kandler was suspect because he had two sons who were in the German military. The names of other Germans who were arrested are not known except for Carl Friese. His tragic story was told in detail by his chauffeur, Jose Thomas:

> Once again, in December 1941, America and Germany were at war and the Panama police were sent to pick up Carl Friese and other Germans in Bocas del Toro. They were to be taken to a detention camp in the Panama Canal Zone. When Friese got permission to go home and pack his personal belongings, I drove him to his elegant house for the last time. This second detention was too humiliating for the proud and prosperous man, who had been in Panama for four decades, and he refused to be taken away. The poison he took was painfully slow and he finally died the next day, even though a doctor tried to save his life. A few months later, the famous headquarters of the Carl Friese Company burned to the ground. All was

lost except the pilings below the water line.[63]

The Probst family was deported again, but by now, the family had grown from three to fourteen members. All but one of them had been born in Panama. The family was sent to the prison camp in Crystal City, Texas.

Cleveland Nelson was 13 in 1941, when his Jamaican father worked as a mechanic for the United Fruit Company in Almirante. During hard times, his father was released from the payroll, leaving the family with no income. Cleveland continued the story as follows:

> Times were hard and money and food were scarce. Consequently, my mother became a maid for a German family. Ewald Hortz[64] worked for the United Fruit Company and was in charge of the Company's water system. When World War II was declared at the end of 1941, he was arrested, like most Germans in Central America, and hauled away to a detention camp, probably in the Panama Canal Zone. Mrs. Hortz was left alone to fend for herself and had to release my mother from her job, which made times even harder for us.

Although she was penniless, Cristel Hortz had to care for four dependents: Ruby Vargas, Irmgard, Ingrid and Ingiborg.[65] Nelson continued:

> ...[P]eople were compassionate enough to drop money in the baby carriage carrying the youngest little girl. When this desperate woman had to move from the Company house, she had no choice except to move to a farm house on the shores of the Almirante bay between the port and Bocas del Toro. Her fate during and after the war was unknown to me.

The daughter we know the most about was Irmgard, who attended the American School of the United Fruit Company in Almirante from 1937 to 1941. Her records were found in a tattered school album kept by a teacher, Mattie Grant. Irmgard was registered as German, age 4. Out of this class of five, there were five nationalities: a Panamanian, a Costa Rican, one French, one American and a German. In the school year 1939 to 1940, Irmgard was in the 1st grade and sang in an Easter concert. She appeared in a class picture with nine students in March 1941. Irmgard played a piano solo at the school Thanksgiving program on November 27, 1941. Fourteei days later, the United States declared war against Germany. Sadly, her name was not on the Christmas program. Her teacher lamented that "the war took away Irmgard," an innocent child who had been in the American School for four years.

After Irmgard's father, Ewald, was arrested in Almirante, I have learned that he was taken by the Panamanian police and turned over to the American authorities, who processed him in the Canal Zone and prepared to ship him to the United States as an enemy alien.

Under previous U.S. policy, only detainees were shipped north. As with Ewald's wife, Cristel, women were left behind and expected to care for themselves and their children, although most had no means of support. Eventually their impoverished cries caused an anti-American backlash that reverberated all the way to Washington. "...[T]hese wives and children who were left behind constitute a most dangerous focus of anti-American propaganda and ... they should be removed at the earliest possible opportunity," Secretary of State Cordell Hull wrote in 1942.[67] There was no explanation as to why the truth was considered propaganda.

The revised policy made it possible for the Hortz family to eventually reunite in the Canal Zone and travel together, arriving in New Orleans aboard the USAT *Florida* on April 8, 1942. An inspection document lists Ewald's height as 5'3", his eyes were brown, and his hair was "gray on top." Cristel was listed as 5'5" tall, brown hair, and blue eyes. No information was provided about the daughters and not even their names were given.

Surprisingly, the family didn't remain together. The *Florida*'s manifest suggested Cristel and the children went to the internment camp at Seagoville, Texas, southeast of Dallas. But their names did not appear in the census taken there in January 1943. No trace could be found of them until a document that surfaced in 1944 indicated they were in fact at Seagoville.

The detainees often passed through different camps, and one document indicated that at some point, Ewald was at Camp Kenedy, southeast of San Antonio. But three different census reports showed that most of the time, Ewald was in the Second Internment Company at the U.S. Army camp near Stringtown, Oklahoma, 201 kilometers south of Tulsa.

The camps were initially primitive and uncomfortable, but improvements were made over time. Meanwhile, there was considerable conflict at Stringtown. The German detainees there had divided themselves into two factions: "Those who wish to be repatriated," an inspector wrote, "and those who do not. The so-called loyal Germans, have threatened and intimidated all the others who are afraid that their names will be reported to the German Government as anti-Nazi." A different conflict was resolved when eight Jews from Panama felt intimidated by the Germans and were moved into separate quarters. "[T]hey are no longer threatened," the inspector wrote, "but are merely ignored."[68]

REPATRIATION

As had been envisioned by the U.S., shiploads of the detainees, some by choice, some involuntarily, were sent to the Nazis or Japan in exchange for Americans. Many of the repatriated children were born in the Western Hemisphere and did not know and did not understand the land of their fathers.

No reason has been found, but Wilhelm Probst and his family were signed up to board one of the earliest repatriation ships, the *Drottningholm*, that sailed from New York on July 15, 1942, destined for Gothenburg, Sweden, a neutral country. But a line had been drawn through the listing for "Wilhelm Probst, Sr. (wife and 11 children at Seagoville)" and the words "not going" were written in. William Jr. was listed separately, and a line was also drawn through his name as "not going." The *Drottningholm* was an exchange vessel and had 646 Germans aboard. In general, the Germans were taken to a neutral border and exchanged for Americans in Germany and other parts of Europe.

Meanwhile, the Probst family remained interned in Texas for another 18 months. Then, on February 15, 1944, 13 of the 14 family members were among 1,117 German repatriates who boarded the M.S. *Gripsholm* in New York City en route to Lisbon, Portugal. An older daughter, Ana, had married an American and chose to remain in the U.S.[69]

The Hortz family was also on the *Gripsholm.* Ewald, by now 48 years old, must have identified with the "loyal" Germans at Stringtown because he, Cristel, age 44, and their daughters were among the repatriates who boarded the *Gripsholm* in New York City on February 15. They arrived in Lisbon on February 25 and were taken to the German border, where they were exchanged for U.S. diplomats, Jews, and wounded American POWs. Each repatriated person then received $4.50 for their patriotism.[70]

The German American Internee Coalition states that in Germany, as well as in Japan, many of the repatriates suffered bombardment and deprivation...[71]

THE AFTERMATH

Carl Friese was buried in a grave just east of his former confidant, Johnny Hagenbarth, an Austrian. There is no tombstone on the bare concrete slab of this respected man of prosperity and fame in the archipelago before he died.

No records have been found about Germans that were released after the war and returned to Bocas del Toro, except for the Kandler family. Richard, who was with the Blue Division,[72] sent by Spain to aid Hitler's invasion of Russia, found his way out of the rubble of war-torn Germany. He made his way to Panama's consulate in Paris, where he obtained a new passport. Finally, he went to the U.S. Consulate and got an in-transit visa to go through New York on his way to Panama in 1946. This took time and required proof that he was from Panama.[73] Edmund followed a similar procedure. Their father, Hans, was released from internment by the Americans and returned to Bocas del Toro to pick up the pieces and begin life for the second time.

Much of the story of the interned Germans remained hidden in documents stamped secret for more than a century before researchers began to nudge them free in the 1950s. What has been revealed, writes Jacob Wasserman, is that "the expanding state has further enmeshed war, politics, civil liberties, and ethnicity. A succeeding century of wartime government-immigrant interaction began with concerts behind barbed wire during World War I in Georgia."[74]

Herbert Leer's store, (left) circa 1905.
Clyde Stephens collection.

Small houses where laborers lived were barely accessible.
Clyde Stephens collection.

The German ship Prinz August Friedrich entering Almirante Bay, Bocas del Toro, circa 1910. Clyde Stephens collection.

The sinking of the Lusitania, May 7, 1915. Engraving by Norman Wilkinson: https://commons.m.wikimedia.org/ wiki/File:Sinking_of_the _Lusitania_ London Illus News.jpg.

Frida Bartel, "Dangerous and of loose character." Application for Parole of German Alien Enemy. https://elischolar.library.yale.edu/mssa_collection, Folder 9-16-12- 4690, Box 420, Shelf 1, Compartment 16, in European War Matters. 158. Accessed at: s/8.

W. R. Snyder's enlistment oath and headstone.
Certificate: Courtesy: Ariel Rene Perez Price
Headstone: Courtesy: Emily LaRusch Snyder.

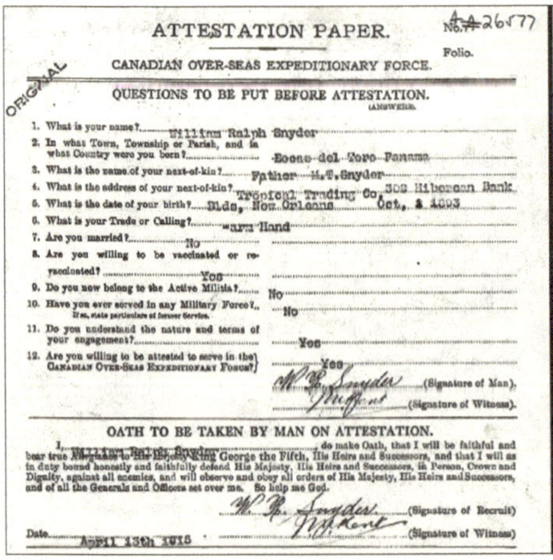

ATTESTATION PAPER. No. 426577

Folio.

ORIGINAL

CANADIAN OVER-SEAS EXPEDITIONARY FORCE.

QUESTIONS TO BE PUT BEFORE ATTESTATION.

(ANSWERS.)

1. What is your name? William Ralph Snyder
2. In what Town, Township or Parish, and in what Country were you born? Bocas del Toro Panama
3. What is the name of your next-of-kin? Father H.T. Snyder
4. What is the address of your next-of-kin? Tropical Trading Co, 302 Hibernan Bank Bldg, New Orleans
5. What is the date of your birth? Oct. 1 1893
6. What is your Trade or Calling? Farm Hand
7. Are you married? No
8. Are you willing to be vaccinated or re-vaccinated? Yes
9. Do you now belong to the Active Militia? No
10. Have you ever served in any Military Force? If so, state particulars of former Service. No
11. Do you understand the nature and terms of your engagement? Yes
12. Are you willing to be attested to serve in the Canadian Over-Seas Expeditionary Force? Yes

W. R. Snyder (Signature of Man).

(Signature of Witness).

OATH TO BE TAKEN BY MAN ON ATTESTATION.

I, William Ralph Snyder, do make Oath, that I will be faithful and bear true Allegiance to His Majesty King George the Fifth, His Heirs and Successors, and that I will as in duty bound honestly and faithfully defend His Majesty, His Heirs and Successors, in Person, Crown and Dignity, against all enemies, and will observe and obey all orders of His Majesty, His Heirs and Successors, and of all the Generals and Officers set over me. So help me God.

W. R. Snyder (Signature of Recruit)

(Signature of Witness)

Date April 13th 1916

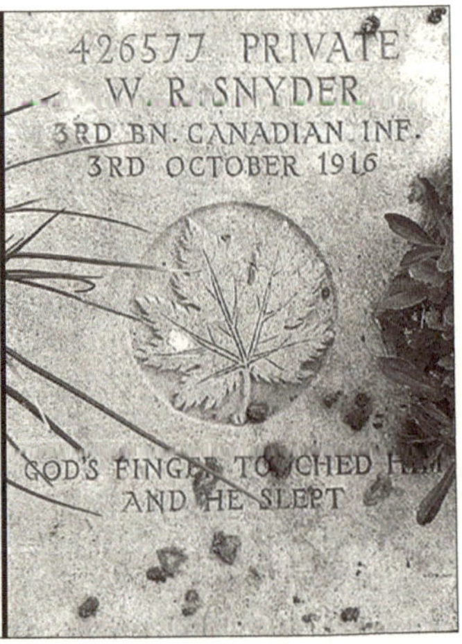

The Aspinwall on Taboga Island.
At times a hospital, a resort and a detention center.
Clyde Stephens collection.

Postcard of The Aspinwall reads: Sanitarium put up on the Island of Taboga for the benefit of Canal Imployees [sic].
(Republic of Panama.)
Clyde Stephens collection.

INTERNED MEN ARRIVE FROM BOCAS DEL TORO

A. Schicck, R. Knoeffle, Herman Grabowski, Hock, von Wedel, whose relatives were mixed up in the recently exposed work in getting passports for Huns getting out of the States before Uncle Sam started in to clean up Hunland, Herbert Leer, Mr. and Mrs. Mass and child, John Hackenbach and his mother, who worked for Carl Friese and several other enemy aliens arrived in Cristobal from Bocas del Toro. The United States and Panama governments acting together recently jailed these propagandists of hell and murder and they are now in safe hands on the Atlantic side.

Star & Herald, July 6, 1918. Clyde Stephens collection.

THE OFFICIAL BULLETIN: SATURDAY, MAY 4, 1918.

NEW ENEMY TRADING LIST AS REVISED AND ENLARGED BY THE WAR TRADE BOARD

- Friese, Carl, Dec. 5, 1917.
- Friese, Carl & Co. (Antonia C. Rylance), Mar. 15, 1918.
- Gebert, Otto, July 12, 1918.
- Kandler, Hans (Carl Friese & Co.), March 15, 1918.
- Leer, Albert, Sept. 6, 1918.
- Leer, Herbert, Oct. 4, 1918.
- Rylance, Antonia (Friese & Co.) Mar. 5, 1918.[30]

Enemy Trading List of Bocas residents. Clyde Stephens collection.

Carl Friese. Clyde Stephens collection.

Genes alemanes (German genes). Source: https://www.binal.ac.pa/isbn/archivos/deposito/dlb-np012017.pdf.

USAT Kilpatrick in New York City.
Source: Library of Congress. File: U.S. Army Transport Kilpatrick leaving NY docks with troops aboard for Boston. Broad category or subject- Ships., New York LCCN2014684125.tif - Wikimedia Commons. National Archives.

Gloucester mansion where some of the Panamanian detainees were held. The mansion might have belonged to Billy Thompson.[36] *Gloucester City, New Jersey Internment Facility. Source: German American Internee Coalition (gaic.info).*

Top: Early drawing of Fort Oglethorpe (left); men at work in the yard (right). Source: http://cookiehouse.net/enemy-alien-internment-camp-fort-oglethorpe-ga-19181920/.

Center: Fort Oglethorpe, circa 1919. The gate (left); guard with machine gun (right). Source: http://cookiehouse.net/enemy-alien-internment-camp-fort-oglethorpe-ga-19181920/.

Below: Inside a Fort Oglethorpe barrack. Source: U.S. Marshals Service, History, Internment of Enemy Aliens (usmarshals.gov).

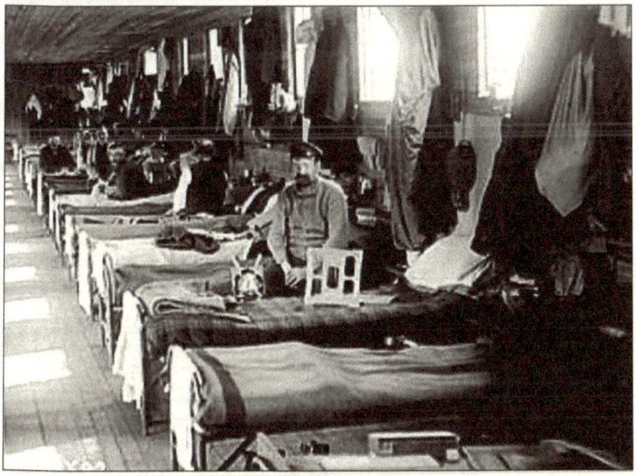

The “Millionaires” had their own chickens and vegetable garden. Source: Classy Chickens, National Archives. https://catalog.archives.gov/ search?q= 86697378. Fresh Vegetables, https:// digital collections.nypl.org/ items/ 510d47d9-b49a-a3d9-e040-e00a18064a99. The Miriam and Ira D. Wallach Division of Art, Prints and Photographs: Photography collection. The New York Public Library. 1918.

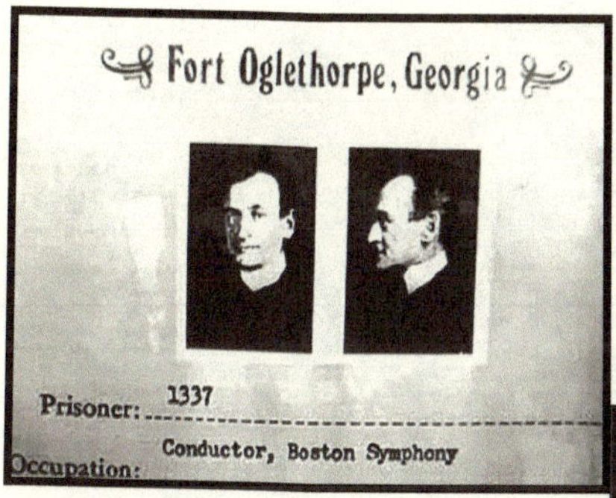

Dr. Muck, Prisoner 1337.
Source: Fort Oglethorpe. Episode 35: The 533 Boston Symphony Orchestra in World War I - HUB History: Boston history podcast.

Poster of Dr. Muck.
The Karl Muck Scandal (boydell and brewer.com).

Final Edition: Going Home.
A mythical owl (Eule) holding a mirror (Spiegel) is fleeing from the "flaming ruins" of Fort Oglethorpe. Image Courtesy: The Morris Library's Special Collections Research Center, Southern Illinois University Carbondale.

Google My Maps

COLEGIO RICHARD NEUMANN
1928

German men in Bocas del Toro, 1934

Paul Ritter*	Kaufmann/Merchant
Fritz Hagenbarth***	Kaufmann/Merchant
Hans Kandler*	Kaufman/Merchant
Heinrich Baden	Kaufmann/Merchant
Hasso von Wedel*	Kaufmann/Merchant
J. Landau	Plantagenbesitzer/Planter
Paul Haak*	Kaufmann/Merchant
Fred Filter*	Plantagenbesitzer/Planter
Herbert Leer**	Manufactured Goods
Albert Leer*	Kaufmann/Merchant
Alfred Schieck*	Kaufmann/Merchant
Ernst Lange*	Plantagenbesitzer/Planter
F. Knoefel*	Kaufmann/Merchant
Hermann Grabowski*	Allgemein/Warengeschaft#
Carl Friese**	Kaiserlich Deutscher Vize-Konsul/Imperial German Vice Consul Allgemein Warengeschaft#

* Deported during World War I and returned to Bocas del Toro.

**Escaped to exile in Costa Rica during the war.

***May have been John Emil Hagenbarth or a relative.

General merchandise store.

Taken from Adressbuch der Deutschen im Auslande 1. Band: Sudamerika 1934.

List of German men in Bocas del Toro, 1934.

Opposite Page, Top: World War II Detention Camps. Source: https://gaic.info/internment-camps/latin-americandetention-facilities/.

Opposite Page, Bottom: Colegio Richard Neumann, public high school in Panama City, Panama. Source: https:// m.facebook.com/ photo.php?bid =355151261176677&id=355148831176920&set=a.355150931176710.

South American Germans arriving at Camp Kenedy, Texas, 1940s. Source: NARA RG 59, State Department, Still Picture Branch, 59-RAG6, Camp Kenedy, 1942-43. Camp Kenedy, Texas - German American Internee Coalition (gaic.info).

South American Germans leaving for repatriation to Germany. Source: NARA RG 59, State Department, Still Picture Branch, 59-RAG6, Camp Kenedy, 1942-43. Camp Kenedy, Texas - German American Internee Coalition (gaic.info).

2325.	PROBST,	Wilhelm, Sr.	Panama
2326.	,	Juanita Barbara (wife)	"
2327.	,	Wilhelm, Jr. (son)	"
2328.	,	Frieda (daughter)	"
2329.	,	Hedwig (daughter)	"
2330.	,	Ricardo (son)	"
2331.	,	Erich (son)	"
2332.	,	Otto (son)	"
2333.	,	Gretchen (daughter)	"
2334.	,	Lottchen (daughter)	"
2335.	,	Joseph (son)	"
2336.	,	Fred Alfred (son)	"
2337.	,	Sieglinde (daughter)	"

1103.	HORTZ,	Ewald	Panama
1104.	,	Cristel (wife)	"
1105.	,	Ingiborg (daughter)	"
1106.	,	Ingrid (daughter)	"
1107.	,	Irmgard (daughter)	"
1108.	,	Ruby Vargas (daughter)	"

The Probst family and the Hortz family lists.
Source: Probst-Hortz Ship Manifest: https://gaic.info/wpcontent/uploads/2019/05/ Gripsholm.U.S.Lisbon.Feb_.1944.pdf.

Irmgard Hortz. Front row, 3rd from left.
The other children are listed in the footnotes.[66]
Clyde Stephens collection.

M.S. Gripsholm, Source: CC BY-NC-ND 4.0.
www.norwayheritage.com/p_ship.asp?sh=grips.

Fusarium oxysporum. Clyde Stephens collection.

3. The Kandler Family of Bocas del Toro

I WAS SEVEN YEARS old when World War II was declared on Germany. We children knew that the Germans were the enemy. Later in life, I learned of the atrocities and the horrors of war. Stories from both sides interested me, including rare accounts from Austrian and German friends who served in the Nazi military—some as indoctrinated children with rifles in their hands. Soon after the war, they escaped from the shambles of Germany, some of them arriving in Central America where we worked together at the United Fruit Company, socialized together and sometimes shared experiences. In telling the Kandler family story, I tried to be neutral and non-judgmental.

The Hans Kandler family is of special interest to me because I met Mrs. Kandler at their once-stately home over the water in Bocas del Toro in the late 1960s. Although the back porch was collapsing into the water, *doña* Marta Maria Siebe Kandler was very pleasant as she spoke German to my boating partner and neighbor in Changuinola, Karl Steffek. He was a United Fruit Company employee and came from Austria. *Don* Carlos, as he was known, had been an officer in the Nazi Army as a paratrooper in World War II. We enjoyed boating excursions together and that is why we motored my *cayuco* to the back porch of the Kandler house for a visit. At the time, Mrs. Kandler was a widow and living alone in that enormous, historic house.

Those were the days when the capital of Bocas del Toro was a quiet, peaceful ghost town with no traffic on the streets.

After the house was sold in the 1980s, sensitive and historic documents of the Kandler family were found hidden in the attic by the new owners, Ligia Thomas de Paget and her husband, Alfredo Paget. Copies of some of those documents were passed on to me by Larry Mahan, a friend and historian.

After Phyllis and I retired in Bocas del Toro in 1991, we were entertained by the Pagets in their historic Kandler house that they had meticulously renovated.

Doña Ligia grew up in Bocas Town with the two Kandler sons, Richard and Edmund. Both boys were educated in Costa Rica and Germany in the 1920s and 1930s and served in the Third Reich military forces in World War II. Over the years, I acquired more data about Hans Kandler and his family. In 2021, his granddaughter, Maria Marta Kandler, of San Jose, Costa Rica, graciously provided family pictures, documents and many personal details about the family. Much data on ship travel and the wars was found on the Internet.

Hans Kandler was born in Hamburg, Germany, on November 30, 1881. On March 29, 1902, at age 20, he sailed on the Hamburg-Amerika Line from Hamburg aboard the SS *Calibria* and arrived in Colon. At the time, the Isthmus of Panama was still a department of Colombia. From Colon, Kandler went to Bocas del Toro and became a merchant.

When the German Imperial Navy made a port call in Bocas del Toro in March 1904, Kandler was required as a German citizen to go aboard the cruiser SMS *Falke* and take a medical exam. He was found fit for service, but did not heed the request to return to Germany and serve in the military.

By January 1, 1908, Kandler was a partner of Carl Friese &

Company. That same year, he took a vacation in New Orleans, New York and the White Mountains of New Hampshire. Carl Friese was the German vice-consul in Bocas del Toro, but when he traveled to Germany for six months in 1911, Kandler assumed the position. In one document, Kandler states that he was the vice-consul from April 9, 1910, and held that position when Carl Friese traveled to Germany for several months at a time.

Hans Kandler returned to Germany via Ellis Island in 1912 on the SS *Prinz Joachim*. After a visit, probably to see his mother, he departed from Antwerp, Belgium, on the SS *Vaderland* on September 7, 1912. His contact in Germany was listed as his mother, Dora Lier Kandler of Marienburg. Kandler's final destination was Bocas del Toro.

In August 1917, three years after World War I had started in Europe, there were seven consuls or consular agents in Bocas del Toro. Hans F. W. Kandler was listed as the vice-consul for Norway. This consular agency served Norwegian banana and cargo ships arriving and departing at the United Fruit Company banana ports in Almirante and Bocas Town, the capital of Bocas del Toro. It was an anomaly that a German like Hans Kandler was the consular agent handling ships of the American banana company, even after the United States had declared war on Germany four months earlier on April 6, 1917.

Since the beginning of the war in 1914, there was strong anti-German sentiment by the British, French, American and Afro-Antillean islanders who had migrated to Bocas del Toro to work in the banana industry. After each patriotic protest and war-bond campaign, Germans and Austrians grew more apprehensive and concerned about their future in Panama.

Hans Kandler was placed on the Enemy Trading List, also known as the Black List. This was based on orders issued by the American War Trade Board in 1918. He was shown as follows: "Kandler, Hans F. W. (of Carl Freise & Company) Bocas del Toro, March 15, 1918."

Kandler's partnership with Friese was no secret and is confirmed more explicitly in a formal statement by Carl Friese's mistress, who later became his wife. The following quote comes from the American Consular Agent in Bocas del Toro, dated December 19, 1918:

> The name should read: Antonia C. de Rylance; not Antonia Ryland. I have her signature before me on a Costa Rican passport which she surrendered to me today upon her arrival from San Jose to see Mr. Friese. She told me that she was an equal partner with Friese and Kandler in the business here.

She was a young widow, who had become Carl Friese's partner, both romantically and apparently in business. Many Bocas people thought she was from Chiriqui, and she may have been from there, but she had a Costa Rican passport.

After three years of neutrality, the United States officially entered World War I on April 6, 1917. Because of his German nationality, Hans Kandler was arrested by the Panamanian Police in the Carl Friese store on November 14, 1917, and delivered to the United States Army in the Canal Zone. On November 30, he was interned at the former Aspinwall Hotel that had been quickly converted to an enemy-alien holding center known as the German Camp. This scenic facility was located on Taboga Island in the Pacific Ocean about 20 kilometers off the shore of Panama City. An espionage report in 1918 by Major Butcher, 29th Infantry, Provost Marshal of the U.S. Army in the Panama Canal Zone, gives us a biased glimpse of what the American intelligence investigator thought about Hans Kandler:

> Age 36, born in Hamburg, Germany, merchant; formerly worked for Carl Freise [sic], the German Consul in Bocas del Toro. This man has been in charge of Freise's business since the declaration of war, the latter having fled across the border into Costa Rica so as to avoid internment. He was formerly assistant to the German Consul; is very pro-German and very dangerous. Sometime ago, Kandler requested permission to visit his firm for the purpose of straightening out his affairs and also as to close the books of Freise and Company. He was given two weeks leave and sent to Bocas del Toro in the custody of a Panamanian Police official whom he bribed into allowing him to go at liberty. The Panamanian policeman fled across the border into Costa Rica but was apprehended there and returned to Panama. Kandler was at large for several days and his actions during that period indicate that he was not in Bocas del Toro for the purpose for which he stated at the time of his request. Interned at Taboga, November 30, 1917.

Hans Kandler was the only German prisoner from Bocas del Toro in the group of 98 enemy aliens that were described by Major Butcher in his cover letter dated April 18, 1918. This entire group was deported from Cristobal on April 19, 1918, on the US Army Transport *Kilpatrick* that arrived in New York Harbor on April 28 for processing at Ellis Island.

The War Department sent a passenger list with names in alphabetical order via a cablegram dated April 11, 1918. Kandler was number 40 out of 98 interned aliens. Alois Hartmann was number 33 followed by Dr. Richard Neumann, number 58.

They were in second class and due to arrive in Hoboken, New Jersey. Amazingly, another page listed 46 American military men from the Panama Canal Zone.

On the handwritten manifest list of passengers for the United States immigration officer at the port of arrival in New York Harbor, Hans Kandler was number 22. Pertinent data about Kandler was recorded as follows:

> Age 36, single, merchant, German, 1-2608, last residence was Bocas del Toro, Panama, $2,400 on-hand, contact in USA: Edmund Kandler (brother) at 152 Sherman Ave. New York; height 5' 10", ruddy complexion, black hair, gray eyes, birthplace Hamburg, Germany.

A footnote indicated he would be processed on April 29, 1918. Kandler was sworn in at Ellis Island with his "Application for Permission to Enter the United States" and declared the following general information:

> German; would like to return to Bocas del Toro to attend to his business; was born in Hamburg, Germany, on November 30, 1881; has resided in Bocas del Toro, Panama, as a merchant and is a partner of the firm of Carl Friese & Co. since January 1, 1908. He was in transit through New York in September 1912. The purpose of his current trip was to reside with his brother in New York and await the end of the war.

At the bottom of the application was a photograph and the signature of Kandler at age 36.

The next day, on April 30, Hans Kandler was sworn in before a Board of Special Inquiry held at Ellis Island, New York

Harbor, N.Y. He gave the customary personal information: He never had the intention of becoming an American citizen. He wanted to return to his place of business in Panama. Otherwise, he wanted to live with his brother, a clerk, and sister-in-law. His brother had previously been a seaman but was released because he was a German.

Kandler further stated that he never served in the German military but would be drafted into service if he returned to Germany. He never had any position under the German government but acted as vice-consul in Bocas del Toro, for six months in 1911 or 1913. The last time Kandler was in Germany was in 1902, when he sailed from Hamburg to Colon, and had lived in Bocas del Toro ever since, except for an occasional vacation.

During Kandler's inquiry in New York, he was asked about his arrest on November 30, 1917. The interrogator noted that he had been released for two weeks in February 1918 to arrange his affairs. He was in business with Mr. Carl Friese, formerly the German vice-consul in Bocas del Toro. At the time of the inquiry at Ellis Island, Kandler stated that: "Mr. Carl Friese is now in San Jose, Costa Rica. I saw him last April 1917, when he left (Bocas del Toro)." When asked why Friese went to Costa Rica, Kandler stated that: "On account of ill health and on account of the persecution by British negroes in Bocas del Toro."

Another interrogator asked Kandler if he was under an official guard during his release for two weeks, and his answer was: "Yes, a lieutenant of police. During the time, he went to see his father. I got $30 and was paying his expenses on the trip."

When the ruling of the Board of Special Inquiry was released, it read as follows: "The board is of the opinion that the alien is admissible, under the Immigration law, but being a native of Germany, he is excluded under the President's Proclamation of April 6, 1917. Unanimously excluded." Signed by the Secretary.

GLOUCESTER CITY, NEW JERSEY,
DETENTION CENTER

Hans Kandler and probably the rest of that shipment of enemy aliens from Panama were transported by train some 128 kilometers to Gloucester City, New Jersey, where they were held in a detention center. Philadelphia is located across the Delaware River in Pennsylvania, but the Philadelphia Immigration Station was in Gloucester City on the east bank of the river.

On May 31, 1918, Hans Kandler swore to and signed an eight-page form with 45 questions that he filled out for the Department of Justice. The title of the document reads: "Questionnaire to be filled out by alien enemy detained at Philadelphia Immigration Station, Gloucester City, New Jersey." His answers are the same as those he had given in previous interrogations. The following are summaries of new questions and answers that were not recorded before.

1. He had not served in the German military but was examined in Bocas del Toro in March 1904 on board the German cruiser SMS *Falke* and found fit for service. Kandler did not follow the subsequent request to go to Germany and serve.

2. At the request of his partner, Carl Friese, who was the German vice-consul in Bocas del Toro, Kandler assumed the position while Friese was absent for five months in 1911 or 1913, but Kandler states that he was vice-consul from April 9, 1910, until the day of his internment.

3. He had $2,400 in cash with him that was deposited with the cashier of the Philadelphia Immigration Station, Gloucester City, New Jersey. His share in the firm of Carl Friese & Company was about $50,000.

4. He had not applied for citizenship in the United States but had applied for citizenship in Panama in February, 1917. The application was held up until the end of the war.

In a note to the authorities, Hans Kandler clarified his position:

> My brother, Edmund C. Kandler of 152 Sherman Ave., New York, has already written to the Department of Justice offering to furnish a bondsman and supervisor for me if the Department would release me on parole. I am asking my brother to see if he can arrange employment for me with the firm for which he works or somewhere in New York. I would like to live with him in New York.

On July 3, 1918, the United States Marshal in Philadelphia, issued the following notice to officials: "You are authorized to arrest under Regulation Twelve President's Proclamation: Hans Kandler. Detained Gloucester City."

Kandler was sent with other Panamanian enemy aliens to Fort Oglethorpe, Georgia, just south of Chattanooga, Tennessee. The train trip was about 1,120 kilometers long.

The German Empire surrendered on November 11, 1918, but repatriation of enemy aliens in prison camps in the United States was slow. One group of 17 German prisoners from Bocas del Toro sailed on the USAT *Kilpatrick* from New Orleans and arrived in Panama on February 3, 1919. Hans Kandler departed from New Orleans nearly six months later on July 17, 1919.

Soon after his arrival and release in Panama, he quickly met with his friend Carl Friese to reorganize their partnership.

At age 39, in 1921, Hans Friedrich Wilhelm Kandler married Marta Maria Siebe Beer, age 27, in San Jose, Costa Rica. She was born in Costa Rica of German immigrants who arrived there in the late 1800s. Her father was Johann Louis Siebe Bötcher who was born in Hamburg, Germany. He was a jeweler in Costa Rica. In 1880, he married Adolphine Ernestina Maria Beer Witting, who was of German descent, born in Costa Rica. The Beer family produced and exported coffee.

Because the Kandlers were Lutheran by heritage, the civil marriage was officiated by a Protestant official of the Church of England in San Jose, Costa Rica, part of the Diocese of British Honduras. The certificate states that the father of Hans Kandler was J. A. Edmund Kandler, a merchant. The father of Marta Maria Siebe was Luis Siebe, a jeweler.

The Kandler couple started their married life at his home in the island town of Bocas del Toro, where their first son, Carl Richard, was born on July 8, 1922. The second son, Edmund, known also as "Eddie," was named after his grandfather, J. A. Edmund Kandler, who lived in Germany, and an uncle, Edmund Kandler, who lived in New York City in 1917. Edmund was born in San Jose, Costa Rica, on March 31, 1924.

These proud new parents wanted to present their two baby sons to friends and relatives in Germany. The family of four boarded a banana ship, the HMS *Camito*, in Cristobal, made port calls in Limon, Costa Rica, and Kingston, Jamaica, and then crossed the Atlantic. The ship arrived in Avonmouth, England, on June 29, 1924. From there they traveled to Germany. Richard was two years old and Edmund was an infant of three months. After a holiday of nearly four months, the

Kandlers returned on the *Rhodopis* of the Hamburg-Amerika Linie on October 20, 1924, from Hamburg. Their final destination was Cristobal.

Four years later, the Kandler family went to Germany for another extended holiday. After their visit, they boarded the *Orinoco* of the Hamburg-Amerika Line at Hamburg on October 6, 1928, destined for Limon. Along the way, the ship made port calls at Antwerp, Southampton, and Cherbourg. They traveled first-class on this new ship that was built in 1928. Their ages were listed as: Hans Kandler, 46; Martha, 34; Richard, 6; Edmund, 4.

EDUCATION OF RICHARD AND EDMUND

The Kandlers enjoyed a pleasant and prosperous life in Bocas del Toro during the 1920s and 1930s. Hans Kandler and his partner, Carl Friese, had a diversified and profitable export-import business and various stores. They lived in fine homes, owned farms and traded with locals for agricultural and native products. This affluence enabled the parents to send their sons to German schools in Costa Rica and Germany despite the world economic depression. Although the boys learned English and Spanish in the streets of Bocas del Toro, they never went to school on the island. There is evidence that Hans Kandler was reluctant to send his sons away from home. But their mother, Marta, had great hopes and ambitions for her sons to excel in good German schools—not found in Bocas del Toro.

From 1928 to 1933, Edmund and Richard attended Die Humboldt Schule, a prestigious German school in San Jose. In March 1934, finally overcoming their father's reluctance, the two Kandler boys, at ages 10 and 12, along with German Costa Rican students, some with their mothers, sailed on the

Tacoma of the Norddeutscher Lloyd Line from Limon for Bremen, Germany. This information comes from their Costa Rican colleague, Carl Walter Steinvorth, who said in German:

> *Dem stimmte seinem Vater zu, vielleicht widerstrebend, und so ergab es sich, das Richard K. im Maerz 1934 mit noch anderen jungen Deutschen und deren Muettern auf dem Schiff "Tacoma" des Norddeutschen Lloyds nach Bremen, Deutschland, abfuhren.*[2] (His father agreed to this, perhaps reluctantly, and so it happened that in March 1934 Richard K. set off with other young Germans and their mothers on the North German Lloyd ship "*Tacoma*" to Bremen, Germany.)

From Bremen, they journeyed to Bad Blankenburg in the Harz mountains in the state of Thüringen in the center of the country. This is where they studied under the tutelage of Karl Reithel and his wife who were Oberlehrers, or schoolmasters. The couple had no children and adopted the Kandler boys like their own sons. Hans and Marta Kandler came from Panama in 1936 to visit their sons.

In 1938, the parents returned to Germany when their sons were 14 and 16. Hans Kandler gave his family an extended recreational and educational tour of his country. Beginning at the port of arrival in Hamburg, they visited the imposing Lutheran Church of Saint Michael—Sankt Michaelis Kirche. Their picture album showed other places that they visited: Kiel, Berlin, Bockschmiede in Thuringia and the German Alps on the Swiss and Austrian borders. This was probably the last time the family would be together until 1946 in Panama.

Richard and Edmund continued their studies at a school in the nearby town of Rudolstadt. On January 29, 1941, at age

18, Richard passed the *abitur*, an in-depth final oral and written exam that was required to graduate from the high school level. This also qualified him to enter a university. Because of his good grades, he planned to advance his studies in diplomacy and foreign relations.

WORLD WAR II AND ITS IMPACTS ON THE KANDLERS

World War II changed the lives of the Kandlers and other Germans in Bocas del Toro and elsewhere in Panama. After Japan bombed Pearl Harbor on December 7, 1941, President Franklin Delano Roosevelt declared war on Japan, Germany and Italy. That is when the Panama Police once again arrested Hans Kandler and turned him over to American officials in the Canal Zone. Documents on his internment in prison camps in the United States and his return to Panama after the war have not been found.

What is well documented was the suicide by poison of Kandler's close friend and business associate for 40 years, Carl Friese, who refused to be arrested when the police came for him in December 1941. To add further grief to the Kandlers, Carl Friese's prestigious store burned to the water line a few months after he died. Only charred remains of the foundations were seen in the deep water along the shore of Calle Uno between Avenida E and F. That was the end of a renowned commercial dynasty that had been in Bocas del Toro for four decades.

Carl Friese was buried beside an Austrian associate, Johann "Johnny" Emil Hagenbarth, in the Bocas del Toro cemetery, but no tombstone marks his grave. Only a blank concrete slab covers the burial site, and I suspect the location is known only by me. The slab was shown to me by Jose Thomas, Carl Friese's former chauffeur and aide. Carl's widow, Antonia C.

de Rylance, bequeathed their Chevrolet convertible to Jose Thomas for loyal and faithful service. He drove the vehicle as a taxi in Bocas Town.

RICHARD AND EDMUND KANDLER
DURING WORLD WAR II

During the rise of Adolf Hitler in the 1930s, some of the German Costa Rican students wisely left Germany in 1939 and returned home to their families in Central America. However, the two Kandler boys continued their studies in Germany.

This was a revolutionary time when frenzied patriotism attracted millions of young Germans who were lured into the service of the Nazi Party. Others who were not so dedicated had no choice but to join the cause or leave.

When Richard and Edmund Kandler finished their studies in 1941, they attempted to return to Costa Rica but it was too late. The two brothers enlisted in the German military forces. One photograph shows them together. Edmund is in a navy uniform and Richard wore an army uniform.

Growing up in Panama and Costa Rica, Richard became fluent in three languages: German, Spanish and English. Consequently, he was sent to Frankfurt to be certified as an interpreter. He was assigned to the Blue Division whose volunteer members were from Spain and spoke only Spanish. They were a complex, maverick military group.

Adolf Hitler supported General Francisco Franco during the Spanish Civil War in the 1930s and sent German military forces to Spain to fight against the Republicans, communists, socialists, anti-Catholics and their allies. When Hitler declared war on Stalin, communist Russia and the Soviet Union in June 1941, it was payback time.

Franco was reluctant to commit Spain as a full partner and combatant ally with Germany. One reason was because Spain was still recovering from a devastating civil war. Nevertheless, to show good faith with the Germans, he agreed to send the Blue Division that would be comprised of civilian Spanish volunteers. More than 18,000 men came forward by July 1941 and went by railroad for training at Grafenwöhr in southeast Germany. Upon arrival, they swore allegiance to Hitler and the Third Reich and were then inducted into the 250th Infantry Division of Germany. Tens of thousands of other volunteers from Spain came later.

Somewhere between Spain and Germany and the Eastern Front between Europe and Russia, Richard Kandler joined this group of volunteers. He acted as a link between the undisciplined but fearless volunteers from Spain and the stern, disciplined German soldiers with a Prussian military heritage. One result was conflicts between the two nationalities in the Division, which was sent to the Eastern Front into Russia, where brutal battles and extreme cold caused high casualties and suffering.

The disparities between the Germans and the Spanish are described by Walter Hamer, who worked on a housing project in Barcelona, Spain, from 1965 to 1969. Walter associated with a Spanish architect, Samuel Maña, who had been a volunteer in the Blue Division. The Spaniard's comments on the war are as follows:

> The Spaniards were treated very badly by the Germans and were always given the most dangerous battle assignments. Cold weather gear was unequally distributed between the Germans and the Spanish soldiers. As a result, I suffered permanent lung damage from the bitter cold.[3]

Despite this inequality, the esprit de corps of the Spanish Volunteers was high and they had some notable victories. However, two horrible freezing Russian winters and brutal warfare caused heavy losses on both sides near Leningrad. Kandler was in the battle of Krasny Bor in February 1943, where the Spanish volunteers and Germans were far outnumbered by the Russians. This was the beginning of the end for the Blue Division that had to retreat back to the west. The Spaniards were ordered to return to Spain and that was the end of the historic Blue Division.

For his valor, Richard Kandler received two medals from Spain and the Iron Cross, Second Station from Germany. The Iron Cross had been revived by Adolf Hitler to recognize "exceptional bravery and/or leadership in the face of the enemy."[4] Richard was not yet 21. In February 1943, the young soldier from Bocas del Toro was treated in a military field hospital.

Another record shows Richard Kandler in Silesia, a former Prussian province in northeast Germany. During a retreat on March 20, 1945, shrapnel from an explosion entered his jaw and lodged near the third cervical vertebra. He was unconscious for a week. When he awoke from the coma, he remembers that the nurses treated him very well. But there was a language problem—he spoke only in Spanish. Eventually, Richard discovered that he had been transferred to Prague, Czechoslovakia. The explosion resulted in a severe cerebral concussion, loss of balance, paralysis of facial muscles on the right side and mental depression. Somehow, he partially recovered from this horror of war but its effects left him with what today is called post-traumatic stress disorder or PTSD.

During treatments and convalescence, he had plenty of time to think about what he had experienced in war. One conflict preyed upon his mind: When the bugle suddenly sounded early one morning, he jumped up in his bed and hit his head very hard on a beam above. This left the soldier incoherent and dazed. He could not rush to the battlefield with the rest of the platoon and stayed in the barracks to recover. Richard Kandler was the only survivor because none of his comrades returned after the battle.

Another horrible memory was when he was in a room with disabled soldiers hanging from contraptions holding them together. They had lost some or all of their limbs and depended on the mercy of others to help them. Richard remembers holding cigarettes to the mouths of these suffering men so they could smoke.

Despite his medical condition, in April 1945, Kandler was transferred to a camp where the Germans were holding Russian prisoners. He rode in a railcar with disabled men where there was no guard. This provided the opportunity for him and a buddy to jump off the train and go into hiding. They slept in barns with hay and farm animals and hoped they would not get caught. They probably knew the war was ending. Germany finally surrendered on May 7, 1945. Like millions of displaced former German soldiers, civilians and other nationalities wandering about Europe after the war, Richard walked through the rubble and dealt with the chaos that followed. His main thought was how to get across the Atlantic Ocean and back home to Bocas del Toro. This would prove to be a major challenge.

(See Appendix V for similar experiences of Fred Koch and Bernie Firchow, both Germans, who eventually became Chiquita banana employees.)

GETTING BACK HOME AFTER THE WAR

Otto Probst of Panama City confirmed in 2021 that Richard and Edmund "Eddie" Kandler from Bocas del Toro, were in a Displaced Persons Refugee Camp in Kevelaer, a town in northwest Germany near the Dutch border. Although the Probst family had been born and raised in Chiriqui, the entire family of 14 was arrested and sent to an enemy alien camp in Crystal City, Texas, in 1942. This was because their father was a German-speaking Prussian. In 1944, 13 of the Probst family were repatriated to Nazi Germany and swapped for American prisoners. After the war was over in May 1945, the Probst family was interned in the Displaced Persons Refugee Camp in Kevelaer, where they met the two Kandler men. This was the beginning of a friendship between two Panama families whose fathers were German. Later, Otto Probst worked for the Kandlers in Bocas del Toro in 1948 at age 15.

From the Kevelaer refugee camp, both the Kandlers and Probsts applied for further processing in Munich in southern Germany—trying to find a way to go home to Panama. It is not known if they maintained contact in Munich but there are documents showing how Richard Kandler returned to Panama. It is likely that Eddie Kandler and the Probsts followed similar procedures.

In the 1980s, hidden documents were found in the attic of the former Kandler house in Bocas del Toro and helped to explain how Richard Kandler returned to Panama. His daughter, Maria Marta Kandler, discovered additional papers in her possession in Costa Rica that showed that her father was in Munich and Paris in 1946. Data from these sources enabled us to retrace Richard's footsteps in 1946, 17 months after the war was over.

During that first chaotic year, beginning in May 1945, Richard probably walked and wandered with hundreds of thousands of destitute and hungry displaced persons trying to find their way home. This led the multitudes to southern Germany where the United Nations Relief and Rehabilitation Administration (UNRRA) was organized in Munich in 1945. The organization had 12,000 employees, and they helped displaced people with the basic necessities such as food, shelter, medicine, clothing and fuel for survival in war-torn Europe. Bob Bruce gave me an account of his experience in Munich after the war.

In 1946, to lodge the masses of war victims, the Americans and other Allies established a transit and resettlement center in the Funkkaserne, a former Wehrmacht barracks in the Munich district of Freimann. In this camp, displaced persons were processed for returning to their country of origin or to emigrate to a new home. Richard Kandler was a temporary resident of *Funk Kaserne* in Munich-Freimann. The following records the paper trail left by Richard Kandler from Munich:

Just after turning 24 years old on July 8, 1946, he traveled to Paris, France, where the Consul of Panama issued him Panamanian Passport Number 2 on July 24. It is not known how he identified himself to prove his Panamanian nationality. The passport was valid for two years and described him as 1.74 meters tall, white, a bachelor, Panamanian by birth, born in Bocas del Toro on July 8, 1922. Jorge Tulio Royo signed the passport that was "gratis" and noted that Kandler was returning to Panama via France. On this same date, the Panamanian consul in Paris wrote a letter to the French Authorities saying that Kandler was authorized to travel via Paris to Panama.

Richard Kandler returned from Paris to Munich, where the Panamanian consul's letter was authenticated by a translator who signed and stamped it at the bottom. The stamp read: Funk Caserne, UNRRA. He was staying at the United Nations Relief and Rehabilitation Administration barracks with thousands of others waiting their turn.

On September 24, 1946, in Munich, the American consul-general wrote a letter to: Mr. Carl Richard Kandler, Funk Caserne, Munich-Freiman: "Sir: This is to confirm that the Consulate General has been instructed by the Department of State to give aid, in so far as circumstances permit, toward your repatriation to Panama." This proved that the American authorities approved his return to his home in Panama.

Kandler left Munich and went to Paris again. On December 3, 1946, the vice-consul of the United States of America in Paris issued him a transit visa to pass through New York on his way to Panama. In addition to the usual information, the application noted his temporary address as: Hôtel du Ministère, 31 Rue de Surene, Paris. His present legal address was Bocas del Toro, Panama, but the address in New York City was in care of the Consulate General of Panama. The time of arrival in New York was "about December 13, 1946," and his departure from La Guardia, the New York international airport, was December 18, 1946. This indicated that he would be flying home to Panama. Other data on the form declared that he was 1.7 meters tall, weighed 68 kilos and was blond with blue eyes. It was audacious that the former German soldier declared himself as a "student" during the previous five years.

Richard must have arrived in Panama about December 18, 1946, just in time to celebrate Christmas with his family. He had not seen them since 1938. It had taken him 20 months after the end of World War II to get back home. Nothing is known about the arrival of Edmund after the war.

POST-WAR YEARS IN BOCAS DEL TORO

Without a doubt, the Kandler sons felt great relief and satisfaction in rejoining their parents and friends on Isla Colon. Richard had bonded once again with his close friend Martin "Tino" Sparks by enjoying the pleasures of boating, swimming and fishing.

Their father, Hans Kandler, spent all the war years in an enemy alien prison camp in the United States. When he returned to Bocas del Toro, once again he had to pick up the remains of the Carl Friese and Kandler enterprises, just like after World War I.

Richard and Edmund joined their father as partners in the import-export business that became the largest in Bocas del Toro.

The Kandlers owned the leading hardware and construction material agency in the province. Their sawmills at Split Hill, Miramar, Guabito and Isla Colon produced lumber for local buyers and markets in Panama City. During an interview, Jose "Tito" Thomas, Jr., told me that he had worked at the Split Hill sawmill for the Kandlers in the 1950s to earn money for school expenses.

The Kandlers also owned farms. Cacao, the source of chocolate, was harvested on their large and remote properties up the Sixaola River at Grassy Banks, Sibube and Tiger Hill. Cured cacao beans were put in sacks and hauled by a mule to the river bank.

In the 1920s, when bananas were grown on these rich alluvial soils, the Almirante railroad extended to these farms all the way to the end of Las Delicias. In the 1930s, floods washed out the rail line. With no railroad, cacao was transported in large *cayucos* down the Sixaola River and loaded on railcars that transported the sacks to Almirante for export.

At Tiger Hill on the bank of the Sixaola River, the Kandlers had a cantina, store, trading post for cacao and a few austere guest rooms for those who stayed too late. This locale was described to me by Karl Steffek, my Austrian neighbor at Farm 8 in Changuinola. He knew the Kandlers, spoke German with them and drank beer in a hammock on their porch overlooking the Sixaola River and Costa Rica on the far bank. Cool river breezes from the high Talamanca mountains made the lowland tropics feel pleasant. Customers from both sides of the international border exchanged stories and the latest news from their countries.

For Karl Steffek to get to the Tiger Hill cantina, he had to go from Guabito through thousands of hectares of shaded cacao trees to the end of the United Fruit Company railroad in a motor car on rails. From there, he walked a short distance to the cantina. This was a welcome break from his job that involved management of cacao for the banana company in this outback part of the world. Originally, in the early 1900s, these fertile loam soils produced tens of millions of bunches of Gros Michel bananas before the Panama disease fungus wiped out the plantations. As each banana farm died, cacao trees were planted to replace the "Green Gold"—as bananas were known then. The Kandlers acquired control of the three most remote cacao farms of the United Fruit Company and found a way to make *Theobroma cacao* a profitable business. They also leased other cacao farms owned by the United Fruit Company along the Sixaola River.

Because Richard and Edmund married Costa Ricans, their pregnant wives went from Bocas del Toro to San Jose to have their babies. When the right time came, the mother and new-born child returned to Bocas del Toro. Richard's wife was

Vera Sancho Ocampo, and they had the following children: Ricardo, 1954; Maria Marta, 1955; Hans, 1956; Johanna, 1958; Carl, 1964. Maria Marta remembers a happy childhood in Bocas del Toro. Richard spoke to his wife and children in Spanish but sometimes scolded them in German. He was an avid reader and passed that trait on to his children. He read books mostly in English but was fluent in three languages. His English had a Caribbean accent that he had learned in Bocas del Toro from the Afro-Antilleans in the province.

On September 10, 1957, Hans Kandler suddenly died at age 75 in Bocas Town after living there for half a century. He went peacefully in his rocking chair on the porch overlooking the crystalline waters of Bocas del Toro. The family took his body to San Jose, for burial in the Foreigners Cemetery.

Sometime in the 1950s, Richard and Edmund Kandler acquired control and management of the original Carl Friese banking agency, which was an official concession issued by the Panamanian government. This position enabled them to broaden their knowledge about finance and investments in the province. But around 1961, something went terribly wrong. Carlos Reid describes it in *A Creole from Bocas del Toro*, by Stanley Heckadon-Moreno: "...there was a cry of bankruptcy. The government clamped down on everything. Hundreds of poor people lost all their savings. Many hearts were broken. Many became mentally deranged."

Richard and Edmund suddenly left Panama with their families in 1961 and never returned. Maria Marta Kandler was only six years old then and told me how distraught and traumatized the family was to abruptly leave their home and happy life in Bocas del Toro. She said that the public thought that her father and uncle "took the money and ran" but she emphasized how untrue that was. They had practically nothing of value in Costa Rica. The National Bank of Panama foreclosed on their properties and sold all the Kandler assets except the house that was in *doña* Marta's name. In Costa Rica, Richard earned a living as a translator and was involved in real estate, amongst other things. Sometime after their arrival in Costa Rica, Richard and Edmund became estranged from each other. Only in their older years did the brothers start talking to each other again. The bereaved widow, doña Marta Kandler, continued to live alone in the stately old Kandler house over the water for several more years. Simon Balid told me that Mrs. Kandler would take the United Fruit Company passenger boat from Bocas to Almirante to see a doctor in the banana company hospital. She was a good friend of the Balid family and stayed overnight in their home upstairs over their store—*La Casa Rosada* (The Pink House).

When Mrs. Kandler's health declined, her sons moved her to San Jose, to be closer to the family. To make this move, friends took her to the international border in Guabito, where she exited Panama for the last time. After walking across the long railroad bridge over the Sixaola River, she entered Costa Rica. The local grass airfield was several kilometers down the tracks and she got there by riding a flat, hand-pumped railcar with the other passengers. Captain Eduardo Vanolli flew her in his single engine Cessna 180 along the Caribbean beach. The flight took 30 minutes to Limon. From there, she transferred to a LACSA Airline C-47 that flew to San Jose. Marta Maria Siebe Kandler had returned to the land of her birth after living in Bocas del Toro for half a century.

Richard Kandler applied for a war pension for his military service with the Third Reich. This meant that numerous detailed forms had to be presented to the German Embassy in Costa Rica. His longtime German Costa Rican friend, Carl Walter Steinvorth, acted as a witness and swore that the information was true. Steinvorth had been in the home of Karl Reithel, the Oberlehrer in Bad Blankenburg, where the Kandler boys went to school. In 1939, Steinvorth's parents prudently got him back home to Costa Rica when Hitler was rampaging across Europe. Despite the absence during the war, Richard and Carl Walter remained life-long friends after Richard returned in 1946. Like many combat veterans, Richard remained haunted by the demons and horrors of war the rest of his life but he rarely talked about them.

It took years to process the application for the pension but it was finally approved. Those detailed pension records were inherited by Richard's daughter, Maria Marta Kandler, of San Jose. She generously shared these documents and photographs with me for this story.

Hans Kandler (1881-1957), Marta Maria Siebe Kandler (1894-1980), Richard Kandler (1922-1996) and Edmund Kandler (1924-1991) were buried in the Foreigners Cemetery in San Jose. Their family story was a mixture of bitter and sweet.

This long-forgotten chronicle is an example of the challenges that German immigrants faced on both sides of the ocean during those sometimes happy but often turbulent times. The Kandler family left Bocas del Toro in 1961, but the Kandler house is still at the south end of Third Street, standing above the crystaline waters of Bocas Bay.

SMS Falke, 1892.
Source: Wikipedia.org.

Marriage certificate of Hans and Marta Maria
in San Jose, Costa Rica, 1926.
Courtesy: Maria Kandler.

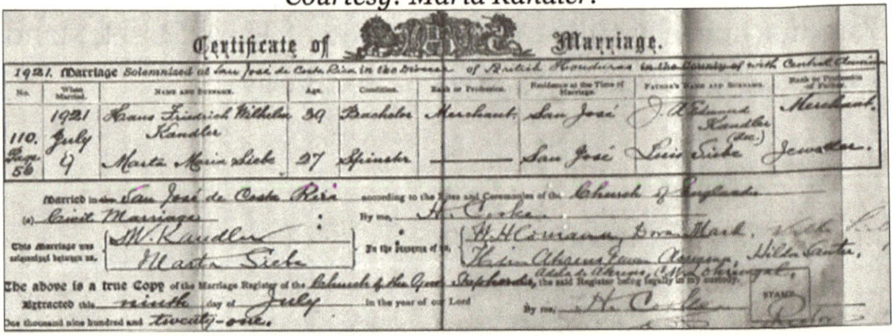

Certificate of Marriage.

1921. Marriage Solemnized at San José de Costa Rica in the Diocese of British Honduras in the County of with Central America

No.	When Married.	Name and Surname.	Age.	Condition.	Rank or Profession.	Residence at the Time of Marriage.	Father's Name and Surname.	Rank or Profession of Father.
110. Page 56	1921 July 9	Hans Friedrich Wilhelm Kandler	30	Bachelor	Merchant.	San José	J. A. Edmund Kandler (dec.)	Merchant.
		Marta Maria Siebe	27	Spinster	—	San José	Luis Siebe	Jeweller.

Married in San José de Costa Rica according to the Rites and Ceremonies of the Church of England
(a) Civil Marriage By me, H. Cooke.

This Marriage was solemnized between us, H. W. Kandler, Marta Siebe; In the presence of us, H. H. Cowan, [illegible], [illegible], Hilda [illegible], Adela de Alvarez, [illegible].

The above is a true Copy of the Marriage Register of the Church of the Good Shepherd, the said Register being legally in my custody.
Extracted this ninth day of July in the year of our Lord One thousand nine hundred and twenty-one. By me, H. Cooke.

STAMP

Hans Kandler and Marta Maria Siebe, circa 1921.
Courtesy: Maria Kandler.

Hans Friedrich Wilhelm Kandler, 36. Questionnaire filled out at Philadelphia Immigration Station. April 29, 1918. Source: https:// www.fold3.com/image/60859648?terms=kandler,hans.

Original Kandler house was remodeled by the 1930s.
Courtesy: Maria Kandler.

Richard and Edmund riding on Third Street, 1928.
Courtesy: Maria Kandler.

Carl Friese owned the building (far left) over the water at the south end of Third Street before it became the Kandler home. Note the Carl Friese sign on the roof: Import & Export Co. Ltd. The building served as an office, warehouse and store downstairs and a home upstairs. The United Fruit Company office (right) shows William H. Ponton, on the steps. Another similar banana company building is next to the general office building. The United Fruit Company originally used the building at the center rear as a store downstairs and a home upstairs for the Louis Victor Georget family. This picture was taken in 1922 and shows the decline of the town after the United Fruit Company moved operations to the new port of Almirante in 1909. Clyde Stephens collection.

Edmund (left) and Richard (right), with their host family in Germany, 1934. Courtesy: Maria Kandler.

The Kandler family in Germany at Weinhaus Eberitzsch in Bad Blankenburg, 1936. Richard (left) is 14, and Edmund (right) is 12. Hans (center) is 55 and Maria (behind Edmund) is 43. Courtesy: Maria Kandler.

Richard Kandler (far right) with classmates, 1936.
Courtesy: Maria Kandler.

Richard in army uniform and Edmund in navy uniform.
Courtesy: Maria Kandler.

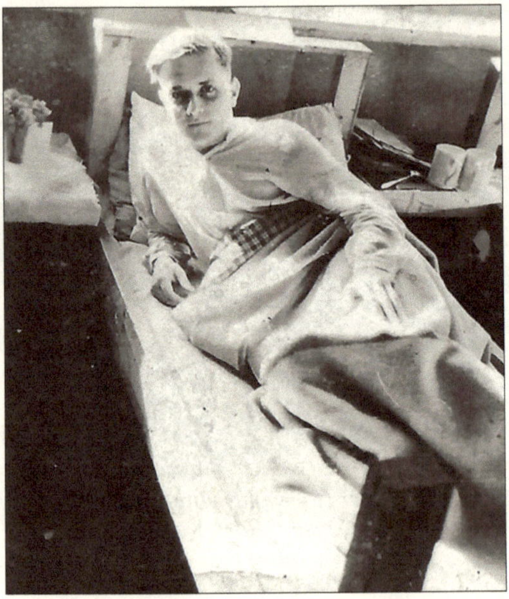

Richard in a field hospital, February 1943.
Courtesy: Maria Kandler.

Richard survived extreme sub-zero winters,
especially in Russia. Courtesy: Maria Kandler.

Richard in a trench outside of a bunker.
Courtesy: Maria Kandler.

German friends meeting with the Kandlers in Bocas del Toro, 1947, to celebrate Richard's return. Hans and Marta (left), Richard (far right), Lotte Bauer-Schlichtergoll, Richard's cousin from Costa Rica (4th from left). Courtesy: Maria Kandler.

Richard with friend Martin "Tino" Sparks, 1949.
Courtesy: Maria Kandler.

Eddie, Hans and Richard Kandler.
Courtesy: Maria Kandler.

Kandler sawmill in Miramar, 1950s.
Courtesy: Maria Kandler.

Richard Kandler with his wife Vera Sancho Ocampo
in Costa Rica. Courtesy: Maria Kandler.

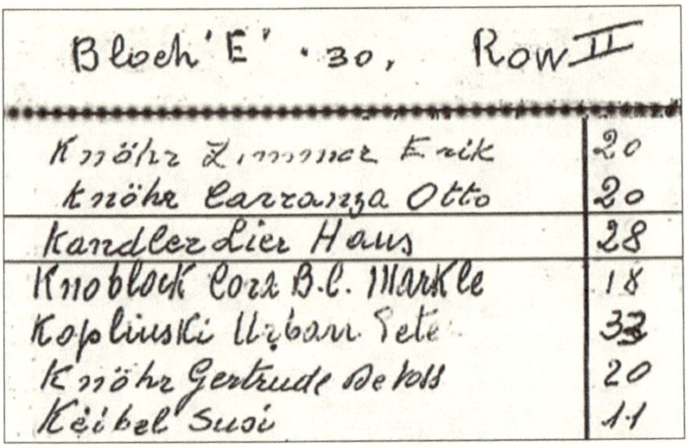

Block 'E' · 30, Row II

Knöhr Zimmer Erik	20
Knöhr Carranza Otto	20
Kandler Lier Hans	28
Knoblock Corr B.C. Markle	18
Koplinski Urban Pete	33
Knöhr Gertrude de Voss	20
Keibel Susi	11

Hans Kandler is buried in Block E., Row II
Foreigners Cemetery, San Jose.
Source: https:// www.scribd.com/document/102659602/Lista-de-personas-sepultadasen-el-Cementerio-de-Extranjeros-de-San-Jose-Costa-Rica.

The Foreigners Cemetery, San Jose, Costa Rica.
Courtesy: David Ballard.

The Kandler House, Bocas del Toro, Panama.
Courtesy: Maria Kandler.

4. Alwyn Hasso von Wedel 1873-1957

BIRD AND PLANT COLLECTOR
ON THE CARIBBEAN COAST OF PANAMA

This story has been published by Olson and Stephens in a scientific journal[1] and by Dr. Stanley Heckadon-Moreno in *Epocas*[2]. Nevertheless, I wanted my readers to be able to know something about the life of this extraordinary man who contributed so much to science and photography in Bocas del Toro. I first wrote about von Wedel in *Banana People* after I interviewed his daughter, Emma Felicita von Wedel Lopez, in the 1990s.[3] In 2015, in Changuinola, Elsie Howard revealed details about her personal relationship with von Wedel beginning in the 1930s in Bocas Town[4]. Smithsonian ornithologist Dr. Alexander Wetmore reported important files about von Wedel, including correspondence from Webster W. Chase, a close friend of von Wedel. United Fruit Company friends in Almirante, Tom and Alice Louise Dunn, spoke highly of von Wedel and his work. These references led to a long and complicated story.

EARLY BACKGROUND

Alwyn Hasso von Wedel was born on July 27, 1872 or 1873, in Dordrecht, Cape Colony (now Eastern Cape province), South Africa. His mother was a South African of English descent and his father was a Prussian officer stationed in southern Africa.

As a boy, he was sent to school in England and afterwards to a military school in Germany. In 1888, at age 16, he was a shipping clerk in New York but later became a purser on a banana ship that took him to Central America. After wandering around, he arrived in Bocas del Toro in 1898, where there was a prosperous German colony and a booming banana industry. Von Wedel worked as a clerk for the United Fruit Company and took up photography as a hobby. His intense interest in pictures got him a job with the banana company to record investments and work in progress with photographs. His fame led to a partnership with Dockrell. Together, Dockrell & von Wedel produced 25 postcards from 1911 to 1916.

WORLD WAR I

On July 6, 1917, the Star & Herald newspaper reported that von Wedel and other Germans from Bocas del Toro were detained in Cristobal. The tone of the story was vitriolic and in the extreme, as follows: "The United States and Panama governments acting together recently jailed these propagandists of hell and murder and they are now in safe hands on the Atlantic side."[5] One reason that von Wedel was arrested is because censors had intercepted three letters that he had written to gun suppliers requesting that these firms mail to him their latest catalogues and best prices for export.[6] Obviously, such queries about guns made the animal collector very suspect. Unfortunately, censors did not know that he needed guns to shoot birds to sell to museums. He also needed guns to kill wild game to eat at his remote camps in the forests. Von Wedel was listed as an "enemy intern" aboard the USAT *Kilpatrick* that sailed from Cristobal on July 27, 1918, and arrived in New Orleans on August 8. Evidence indicates he was with numerous other

Panama Germans, who were interned at Fort Oglethorpe in north Georgia during the war.

Von Wedel and 26 other "alien enemies" left New Orleans on the USAT *Kilpatrick* and arrived in Cristobal, Panama Canal Zone, on February 3, 1919. A note on the manifest of the ship listed how much money each person had on them upon arrival in Panama. Hasso von Wedel had two US dollars. Seventeen of the men were from Bocas del Toro. Each one went back to their families and businesses and started life over again.

Hasso von Wedel made a trip to Avonmouth (Bristol, England) on the SS *Bayano* that departed from Colon on July 9, 1922. His destination was Germany. The motive for this trip is not known.

In the late 1920s, von Wedel was trained by several prominent ornithologists to collect and preserve bird skins for museums and other collectors. Under the tutelage of Fredrich Hedge Kennard, et al., von Wedel delivered over 1,000 bird skins from Bocas del Toro that represented 272 bird species between 1926 and 1929. The bulk of these bird skins went to the Museum of Comparative Zoology at Harvard University. In the Brandt collection in Cincinnati, 399 specimens were from von Wedel. These birds were collected from Puerto Obaldia in San Blas from 1933 to 1934 and from Bocas del Toro from 1933 to 1939.

During some of five years in the remote wilds of San Blas near the Colombian border, von Wedel was partially enabled with transport from Colon by the United Fruit Company who had small banana plantations at Perme. Their little rail lines went to the interior where von Wedel collected. Communications and living conditions there were minimal and von Wedel suffered many hardships. Some of his patrons were frequently unappreciative and caused him even more grief during these times of economic depression. He finally gave up collecting in San Blas around 1934 and returned to Bocas del Toro. Von Wedel collected birds up the Cricamola River with a young assistant, Martin Sparks, in 1938. This same year, he began to collect plants with the botanist Robert E. Woodson of the Missouri Botanical Garden. Von Wedel was credited with collecting over 3,000 botanical numbed specimens. Ten new species were named after him.

Meanwhile, he had moved from Bocas del Toro to Almirante where he continued to produce postcards and Christmas cards. He could have offered other biological specimens to collectors but time did not allow him to expand his services.

WORLD WAR II

Once again, Germans in Bocas del Toro and the rest of Panama were arrested and sent to prison camps in the United States for repatriation swaps to the Nazi government in exchange for Jews and American prisoners held by the German government. This began on December 7, 1941, when Japan bombed Pearl Harbor. Strangely enough, Hasso von Wedel was not taken like the rest of German-speaking people in Bocas del Toro. There was a good reason he became exempted.

When the Japanese invaded the Philippines, they controlled the world supply of Manila hemp. The U.S. Government made a contract with the United Fruit Company to convert their cacao farms to Manila hemp to produce rope in Changuinola. The United Fruit Company selected Hasso von Wedel to be their official photographer to record development and operations of the Manila hemp project. Photograph albums were attached to budget reports that were submitted to the Company headquarters in Boston. This position entitled von Wedel to live in a Company house in Base Line in Changuinola. I visited that house in 2020.

Von Wedel and Webster Chase were working colleagues, close friends and drinking buddies. Chase left some correspondence about von Wedel. After the war was over in 1945, the traditional banana company continued to produce Manila hemp until the Philippines recovered their traditional industry. In 1950, banana men started planting bananas once again and the hemp operation closed in 1954. Von Wedel died on October 7, 1957, in the United Fruit Company hospital in Almirante and was buried in a special crypt made by his friend Webster Chase in the Base Line Cemetery in Changuinola. My ssistant, Leovigildo Julian Lorenzo, helped me search for hours for the tomb in 2015, but to no avail. The burial ground is overpopulated, swampy and overgrown. It is believed that von Wedel's crypt was repossessed by some one else.

MYSTERY OF HASSO VON WEDEL'S ARTIFACTS AND TREASURERS

When I interviewed von Wedel's daughter in the 1990s in Bocas del Toro in her wooden home on stilts, she told me that all of her father's belongings disappeared after he died. Since then, I have tried to figure out what happened to these valuable field notebooks, hundreds of postcards and photographs, and his photographic equipment.

The logical person to inherit von Wedel's historic files and belongings was Webster W. Chase. Chase was sent to Limon, then to Villa Neilly, Costa Rica, to help build a new carton box plant in Coto to pack bananas. When he returned to Farm 8 in Changuinola, he was my neighbor and I knew he had a photo lab. After he died in 1964, his wife, Lorna Chase, had to leave the Company house. She moved to a tiny apartment in Limon. Later, Mrs. Chase was moved by one of her daughters to an assisted-living home in David, Panama.

In conclusion, with each move after Webster Chase's death in 1964, more and more of von Wedel's treasures were disposed of; there was no place to store them and probably no interest in finding a place for them. What a loss for all of us. In the end, I was the lucky one to buy numerous von Wedel postcards for my own personal album. Thus, some of his phenomenal work has been preserved.

When Storrs Olson asked me to collaborate for our article in the *Archives of Natural History*, he suggested we would be creating a lasting memorial to a heretofore unrecognized naturalist. I believe that we and others have now made that memorial a reality.

Bocas del Toro and San Blas, where von Wedel lived, clerked, photographed and collected animal and plant species.

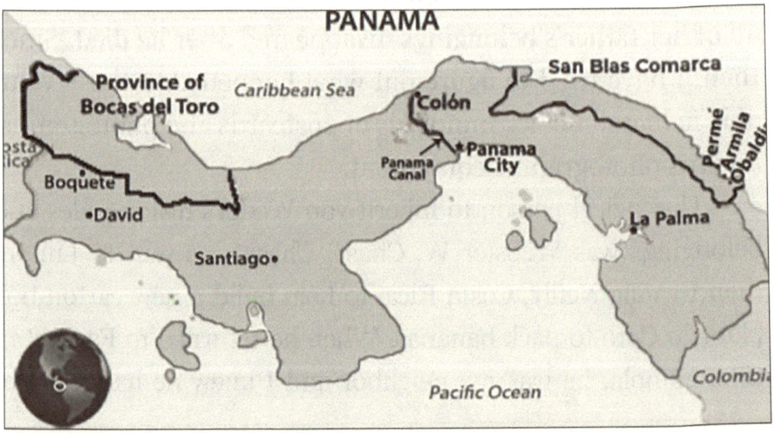

Six nationalities of United Fruit Company employees in Bocas del Toro, circa 1906. Von Wedel is in the front row, center. Clyde Stephens collection.

Hospital Point postcard by Dockrell & von Wedel.

Custom House and United Fruit Co. Offices, Bocas del Toro, R. P.

Von Wedel postcard #4, south end of Third Street, Bocas del Toro. Clyde Stephens collection.

This favorite scene by von Wedel was falsely claimed by several writers as being located in various countries. Clyde Stephens collection.

Bringing out and stacking Bananas for shipment.

Von Wedel, circa 1930s.
Courtesy: Elsie Howard.

Von Wedel and Ngäbe boy.
Courtesy: Elsie Howard.

Von Wedel house, Base Line, Changuinola, 2020. Courtesy: Miriam Oquendo Fajardo.

Crypt (left) that Webster Chase prepared. Photograph taken by Alexander Wetmore on January 17, 1958, and labelled "Tomb of Hasso von Wedel, "at left of gate. Changuinola, Prov. Bocas del Toro, Panama. (Smithsonian Institution Archives RU 7006, Box 184, Album 3, Image Number 6583 or 6584.

A final picture before von Wedel died in 1957.
Courtesy: Fr. John Kennedy.

5. The Probst Family

WHEN I WROTE THE chapter entitled "German Prisoners of War from Bocas del Toro," I was astounded to find so many tragic disruptions in the lives of thousands of German, Japanese and Italian families living in Panama and other Latin American countries. Several families in Panama were used as examples in that chronicle. The Probst family had an extraordinary story that needed to be told separately. Much data was found on the Internet about the family being arrested, detained and shipped to "enemy alien" prison camps in the United States. Their repatriation to Germany during World War II was also recorded on ship manifests in 1944. However, personal accounts were missing, and that is when I began a search in Panama to see if any of the family came back to the country of their birth after the war.

Queries about the Probst family were sent to several Panamanian friends. One of them, Captain Luis "Ibu" Alvarado, made telephone calls and wrote e-mails that led to new contacts. Dr. Stanley Heckadon-Moreno remembered a beautiful girl from high school in David named Siegelinde Probst. She was finally located through a former classmate, Jeannette Moreno, Stanley's cousin. Later, Siegelinde Probst provided her brother Josef Probst's telephone number. Someone from the San Jose Island Resort, where Otto Probst had been a partner and major shareholder, also responded. The person lamented that Otto had died several years ago. Then Frank Amador Selles quickly wrote that Otto Probst was still alive. He knew because Otto had married his aunt, Dora Selles, from Bocas del Toro. This was confirmed by Dr. Jorge Selles, Otto's brother-in-law. The trail was getting hotter.

After many inquiries, e-mails and long phone calls with so many volunteers, a cordial interview was held in Panama City at the Albrook Inn on October 3, 2021, with Siegelinde and her brother Josef. They provided incredible details that made this chapter possible. Ibu and Stanley attended the gathering and took notes. Josef showed us the only known copy of a book in Spanish that his father, Wilhelm Probst, originally wrote in German: *Vida y lucha de un hombre solitario* (Life and Struggles of a Solitary Man), printed in Mexico in 1975. Many details in this chapter come from that rare book.

Just when I thought I had enough information, Jeanette Probst provided many more details from her father, Otto Probst. In May 2022, Gretchen Probst Bentley astounded me with many facts about the family in prison camps and their life in Nazi Germany as repatriates. Thanks to these four Probst siblings and a team of friends, some of the incredible saga of the Probst family has been recorded.

WILHELM PROBST AS A TEENAGER

Wilhelm was born on September 14, 1894, about 188 kilometers east of Berlin in Meseritz in Posen in old Prussia. At age 13, he went to work as an apprentice to learn farming and earn money. However, the farmer treated him as though he were a man and ordered him to do unreasonably difficult labor. On one occasion, young Wilhelm impulsively grabbed a pitchfork, told the brutal man to stay away from him, and ran into the woods. The farmer responded by reporting the violent threat to the police.

While in the forest, Wilhelm followed a construction crew that was building a high-voltage power line. The young lad was curious and very precocious. He proved that he was capable of performing certain work tasks and was hired to learn the trade. His boss eventually trained him to install high-powered lines.

After a year or so, he took a job at Limbach, far to the south in Bavaria near the Alps. During a terrible winter day that stopped work for everyone, Wilhelm was at a tavern playing billiards when a suspicious man came in and asked about him. Then he was escorted to the police station and put in jail. His boss tried to pay a fine to release him, but he was told that Wilhelm had to serve his sentence in jail based on what had happened in Meseritz. He was put in a solitary cell and made to suffer. Six weeks later, Wilhelm was released and eventually regained his composure. He decided to join a former boss named Pandow in Dessau, Sachsen-Anhalt. Pandow was happy to see him again and noted that at 16 Wilhelm had developed into a big, strong man. Soon, he was advancing in skills, reputation and opportunities with more pay. With this success, he forgot the nightmare of jail in Limbach and became a happy and satisfied person again. It was during this period that he and his two brothers, Richard and Otto, were photographed together in Prussia.

Wilhelm's reputation and expertise grew, as well as his earnings. So much, in fact, that he was taken out of a work gang and given his own crew of 15 men to supervise. They put in power lines from Quellendorf to Saale. The people were nice and he liked the region.

Next, he went to Alsace which was part of the former French Republic that had been occupied by the German Empire since 1871. Wilhelm knew there were strong resentments by the French against the Germans from the east side of the Rhine River, but he liked the people of Alsace and they liked him, sharing with him their wonderful wines.

He and his workers installed power lines in the Colmar region of Alsace. It was here where Wilhelm enjoyed his youth more than any other place in Germany. By the time he was 18, he had money in several bank accounts.

With youthful enthusiasm and financial security, he got a strong case of wanderlust and decided to travel to the Americas. Before he left Prussia, he made sure that his parents in Meseritz were taken care of financially.

ADVENTURES IN THE AMERICAS

After arriving in New York in 1913, he made his way to Chicago. From there, he went to Texas and became a cowboy.

Legends about the Mexican revolutionary, Pancho Villa, fascinated him, and he thought about crossing the border to join the revolutionaries. However, a new adventure lured him to California where he planned to catch a ship to Alaska to mine gold. Alas, the boat had already sailed. From California, Wilhelm boarded a German ship as a seaman and went to Chile. When he attempted to leave the ship, the captain denied him permission, so he abandoned the ship and swam ashore.

When World War I broke out on July 28, 1914, Wilhelm vowed to return to Europe and join the Prussian and German forces of Kaiser Wilhelm II. He visited the German Consulate in Chile, but was told it was not possible for him to be repatriated. Next, he visited the German Consulate in La Paz, Bolivia, but was again turned away. Besides these disappointments, he suffered from severe nose bleeds while he was in the towering Andes Mountains. He had no choice except to return to the lowlands and look for a ship back to Germany.

The only way to get back to Chile was to walk across part of the Atacama Desert, the driest place on earth. When he finally reached Valparaiso, Wilhelm stowed away on a ship going to Panama. At sea, he encountered numerous French patriots who were returning to France to join the war effort against Germany. They sang French anthems and patriotic banters with great esprit de corps. This irritated Wilhelm as a patriotic Prussian, and he countered with nationalistic songs that he sang in German. The British captain of the ship threatened to throw him overboard unless he shut up. To assure that Wilhelm was out of sight, the captain ordered him to go below deck and shovel coal. Upon arrival outside the recently opened Panama Canal, Wilhelm was ordered to get off the ship and stay on one of the off-shore islands. Before long, he swam ashore and after a series of adventures and misadventures, found work on a train that traveled between Colon and Panama City. But there were conflicts, and he was forcefully exiled back to the island he had escaped. A couple of men befriended him with money and passage on a boat back to Panama City. Soon after, he started walking west to Chiriqui.

A NEW LIFE IN CHIRIQUI

Wilhelm traveled 500 kilometers on horse and wagon trails to western Panama, crossing the large rivers that had no bridges. He liked what he saw along the way. In David, he met a Frenchman who hired him to collect coconuts from the outer islands of Chiriqui and load them on a sailboat. When a severe storm became a threat at sea, the coconuts were thrown overboard to save the sailboat and themselves.

Wilhelm found his way to Boquete in the mountains and started a farm. There he met and married a young woman, Juanita Barbara Barbosa, a *campesina*, half Amerindian and half Latina. She was one of eight women left of an indigenous tribe known as the Doraz, or Dorasque, who lived along the Caldera River and Dolega. Having intermarried with the local population, they ceased to exist as a distinct group, but they left behind a written language.

WORLD WAR I

The United States did not officially enter the First World War until April 6, 1917, three years after it started in July 1914. A month later, in August, the Panama Canal was ceremoniously opened. But the U.S. was extremely concerned about the Canal's security because of the large colonies of Germans in Bocas del Toro, Colon, Chiriqui and Panama City. Some of these expatriates kept a low profile and were worried about the future of their own lives and businesses in Panama. Others were more openly pro-German and that is what worried the Canal's defenders. The antagonisms that were roiling Europe were also reflected in Panama, impacting alien Germans, French and French West Indians. The Germans were immigrants who had benefited from the rise of Germany as a colonial power. Thousands of French patriots had come from France and the French West Indies when Ferdinand de Lesseps attempted to build a sea-level canal in the 1880s. The United Fruit Company also brought in a large French-speaking labor force to work in the banana plantations in Bocas del Toro. Also drawn by

Canal and plantation job opportunities, untold thousands of English-speaking West Indians were very pro-British. Both groups vowed to defend their colonial mother countries and return to Europe to fight in the war. Almost all of them were born and raised in the Caribbean or in Panama.

It was in this atmosphere that the American military in Panama started spying on the Germans and their allies. Not surprisingly, there were some open and public displays of German loyalty, patriotism and support of the Kaiser's war machine. However, espionage statements were decidedly biased and unfair to many innocent German and Austrian residents. Unfortunately, this led to a massive roundup of all the Germans and their allies living in Panama, including their Panamanian spouses and children.

In an espionage report dated April 18, 1918, Major Butcher of the 29th U.S. Infantry Provost Marshal, issued the following comments about the Wilhelm Probst family living in Chiriqui:

> William[1] Probst, age 23, born in Merseritz [sic], Posen, Germany. Occupation, farmer. Lived in Llano Bonito in Boquete. Visited very frequently in David where he met with Max Newman and other Germans. Apparently harmless but his internment was recommended by an agent of the Intelligence Office who made a trip to the Interior. Based on his report, Probst was interned. Interned at Taboga, November 30, 1917.
>
> Juanita Probst, age 19, born in Boquete, Chiriqui, Panama. Wife of William Probst; voluntarily interned at Isla Taboga owing to her destitute condition. Interned January 24, 1918.
>
> Wilhelm Probst, Jr., 8 months of age, born in Llano Bonito, Chiriqui Province. Child of William and Juanita Probst. Voluntarily interned January 24, 1918.

> The prisoners were taken to the former Aspinwall Hotel on Taboga Island, 17 kilometers from Panama City in the Pacific Ocean. The elegant structure became a detention center known as the 'German Camp.'

The Probst family of three were among 98 so-called enemy aliens living in Panama who were assembled in Cristobal and deported on April 19, 1918, aboard the USAT *Kilpatrick.* The ship arrived in New York Harbor on April 28 for processing at Ellis Island. From there, the detainees were sent by train to Gloucester City, New Jersey, for more processing at the Philadelphia Immigration Station. Next, Wilhelm and other adult male enemy aliens from Panama were shipped by train 1,100 kilometers to Fort Oglethorpe in North Georgia where there were about 4,000 prisoners. While there, Wilhelm contracted the Spanish Flu during the epidemic that spread to Europe from the United States during World War I.

A report from the Swiss Legation listed the women and children as enemy aliens from Panama who were held in Gloucester City, and who were to be returned to Panama, including Juanita and William Probst, Jr. This indicates that Juanita and her son were detained there in late 1918 and early 1919 and did not immediately go with Wilhelm to Fort Oglethorpe.

Another letter had the following subject line: "List of Internees at Fort Oglethorpe Returning to Panama First Class." The list included Wilhelm. Juanita and their son were also on the list. They had recently been sent from Gloucester City to Fort Oglethorpe by train to join Wilhelm and prepare for transportation to New Orleans and Panama.

The Probst family returned to Panama on the *Kilpatrick* that sailed from New Orleans on June 15, 1919, seven months after the German surrender. The baby, Wilhelm, Jr., was now listed as two years of age. They were on the same voyage as the respected academic, Dr. Richard Neumann. Originally, the three Probsts were assigned to first class, but the passenger list of the *Kilpatrick* shows them in second class.

STARTING ANEW AFTER WORLD WAR I

After returning to Chiriqui, they found a way to start their lives again. In time, they built a ranch in the high mountains above Boquete. This remote farm required hard work in adverse conditions for the family to survive.

By 1928, Wilhelm, Jr. was 11 years of age and had learned to shoot a pistol and a rifle. His mother had become accustomed to living in the outback wilderness and managing the family and household. Under these circumstances, Wilhelm felt confident enough to leave his wife and five children alone on the farm so he could accept a job near David. He rode his horse down the long, mountain trail and reported for work with a German company from Hamburg. Wilhelm was hired to install posts and high-tension cables from a generator powered by a new German-built hydroelectric plant outside of David.

When the project was finished, he returned to his ranch and family. After several months, he was hired by the United Fruit Company which was converting virgin rain forests into banana plantations near the Costa Rican border. His job was to install telephone lines in the new banana port of Puerto Armuelles, along the railroad tracks and in the new banana plantations all the way to Progreso. He also erected new electric power lines for the banana company. This was in 1928.

As Wilhelm, Jr. matured into a young man, he also became an employee of the United Fruit Company in Puerto Armuelles and worked in the electric department. It was noted that the sons learned about electricity from their father.

On one of these sojourns to the lowlands, Wilhelm, Sr. caught a terrible fever and almost died. The description of the malady in his book suggests that he may have had cholera caused by unsanitary food and water conditions. Malaria was also suspect. The doctor advised him to remain there for professional treatments and emphasized that his illness was acute. However, Wilhelm declared that if he was going to die, he wanted to be with his family at his ranch in the mountains. The doctor told him that vultures would eat him along the trail if he tried to return home, and that he would never again see his wife.

Wilhelm writes that the walk home normally took three days but he was on the trail for eight days. There were times when he missed the trail altogether and wandered randomly in the direction of his farm. When he finally got home, he had fits of high fever that caused delirium and hallucinations. He writes that in this altered state of mind, he pictured his mother on her deathbed even though it had been 20 years since he had actually seen her in person. But when the family in Germany told Wilhelm about her death, it was at the same time as his visions of seeing and hearing her. He never fully recovered and was never the same man again. It took six months before he could do light chores. One activity was hunting with his dogs and killing three pumas that had eaten some of his hogs.

Eventually, the family left their isolated farm in the mountains and moved closer to Boquete, where the children could go to school. This is where Josef was born. Later, the family moved respectively to Caldera, Concepcion and Dolega, where Gretchen was born. The last two babies, Alfred and Siegelinde, were born in a Panama City hospital. Wilhelm assisted with the birth of all of their babies in their humble home except for the last two. The father quipped humorously that "children were cheaper by the dozen."

WORLD WAR II

Germans in Panama came under scrutiny once again when Adolf Hitler began his rampage across Europe in 1938. One tragic incident known as "Cotito" showed how savage certain individuals in Panama's National Guard could be against German-speaking residents in Chiriqui. A bronze plaque in German, Spanish and English in Cotito summarizes the tragedy:

> In a common grave 200 meters from this point, on the farm founded by Karl and Margarethe Schmieder, lie the remains of ten of the twelve Swiss and German pacifist victims gunned down in a brutal police massacre in the early morning hours of July 7, 1941, during the first presidency of Arnulfo Arias Madrid. In their honor, that of the two victims buried in Volcan, and the eight wounded, their descendants and friends have erected this memorial in the year 2014.[1]

During the period of this massacre, Wilhelm was taken to jail in Concepcion. His angry wife walked to the police station with two pistols and angrily announced that she was not leaving until her husband was released. A message was sent to Panama City reporting her demands. President Arnulfo Arias ordered the police to release Wilhelm and not harass the family any longer. Juanita made her point because she had an aggressive Prussian personality similar to that of her Prussian husband.

After Japan attacked Pearl Harbor on December 7, 1941, President Franklin Delano Roosevelt declared war on Japan, Germany and Italy. Almost immediately, Japanese, German and Italian residents of Panama and elsewhere in Latin America were rounded up and harshly detained in the Balboa Internment Camp and a transit holding facility at Camp Empire in the Panama Canal Zone.

In his book, *Nazis & Good Neighbors* (2003), Max Paul Friedman describes the abusive treatment of civilian prisoners:

> Run by the United States military, Camp Empire temporarily housed civilian internees, including women and children, from all over Latin America. The male prisoners, most of who ordinarily ran businesses, oversaw farms and ranches, were salesmen or teachers, were treated as brutally here as if they were convicted criminals. They were forced to engage in strenuous physical labor in intense heat and high humidity, cleaning tangles of brush with *machetes*, while armed guards with vicious dogs stood watch. Even the worst of [the other] U.S. internment camps proved more comfortable.[2]

In the beginning, the prisoners were mostly men. Later, spouses and children also were detained as "enemy aliens." During their detention in Balboa, Juanita went into labor and was taken to the Santo Tomas Hospital in Panama City. She gave birth to the Probsts' twelfth child, Siegelinde, on March 25, 1942.

Nine days later, new-born Siegelinde, her mother and 11 of her 12 children were forcibly shipped from Cristobal on the USAT *Florida*, arriving in New Orleans on April 8, 1942. Juanita and her children were then sent to the Seagoville Detention Center, an enemy alien camp located southeast of Dallas, Texas.

Wilhelm, Sr. and his oldest son, Wilhelm, Jr., who was 25, had been sent previously to the enemy alien prison camp in Springtown, Oklahoma. The father and son were separated from the rest of the family for many months.

INTERNMENT CAMPS

The Probsts were among 4,788 enemy aliens from Latin America who were interned in several "enemy alien" prison camps in the United States during World War II.

The entire Probst family of 14 was scheduled to be deported on a neutral Swedish ship, the SS *Drottningholm*, due to sail on July 15, 1942, from New York to Gothenburg with 646 Germans aboard. They were being sent to Germany in exchange for Americans and other Allies being held by the Nazis. But for unknown reasons, plans changed for the Probsts. They were listed on the *Drottningholm*'s manifest list as "not going."[3]

Also listed as "not going" was the family of Ewald Hortz who had been employed by the United Fruit Company at the banana port of Almirante. The four Hortz children are the same ones who attended the UFC American School in Almirante.

ASHEVILLE, NORTH CAROLINA

When the Probst family did not sail on the SS *Drottningholm* to Sweden on July 15, 1942, Juanita and her 11 youngest children were sent by train from Seagoville, Texas, to Asheville in North Carolina. They were lodged in the Grove Park Inn, a luxurious tourist resort for rich people. The U.S. government had contracted with the hotel for rooms to hold prisoners during the war years. Josef remembers the hotel as being very classy.

Wilhelm and his oldest son are not in a photograph taken at the hotel, because they were still being held at the

Stringtown, Oklahoma, prison camp. In Panama City in March 2022, Otto Probst told his daughter Jeanette that the family was treated well in Asheville and was given good clothes to wear—as illustrated in the family portrait at the hotel. He remembered that the Probst family in Panama was very poor and never had formal clothes like dress pants, coats and ties or nice dresses.

Gretchen said the government provided German school teachers to teach the children of enemy alien prisoners. The family was at Grove Park Inn for several months and received good medical treatment. During this time, Richard, age 15, had surgery on his leg because of bone cancer. He is in the picture at the hotel.

CRYSTAL CITY, TEXAS

In late 1942, Juanita and her children were then sent by train to Crystal City, a desert wilderness near the Mexican border southwest of San Antonio, Texas. Wilhelm and his older son were also sent to Crystal City to join the family. This would be their prison-camp home for over a year until their repatriation to Adolf Hitler's Germany in February 1944.

As the largest Texas internment camp, the Alien Enemy Detention Facility in Crystal City eventually held almost 4,000 individuals and had more than 500 buildings on 290 acres (117 hectares). In order to comply with international law and promote as positive an environment as possible, the INS (Immigration and Naturalization Service: Enemy Alien Internment) designed the camp much like a small community. There were houses, food and clothing stores, schools from grades one through 12, auditoriums, warehouses, administration offices and a 70-bed hospital. Facilities also included places of worship, a post office, bakery, barber shop, beauty shop, a Japanese Sumo wrestling ring, and a German biergarten. The internees there printed four camp newspapers: Los Andes (Spanish), the Jiji Kai (Japanese), the Crystal City Times (English) and Das Lager (German). A popular place was the large swimming pool. Anna, Heddy and Frieda Probst went to sewing classes in the big sewing room where Japanese and German women attended. Gretchen was confirmed as a Catholic in the camp chapel.

The Probst family lived in House Number Q-65 that had a private kitchen. Gretchen remembers that the building was very hot in the summer. Bath houses and toilets were outside and used by other neighbors. Werner R. Ulrich, a former internee at Crystal City, has prepared thorough research and intricate drawings of the camp, which can be seen at this site: Crystal-City-Internment-Camp-07012018.pdf (gaic.info).

A romance developed between the oldest daughter, Anna Probst, and a military man at the Crystal City camp. He was transferred to another post for duty but the couple kept in contact by correspondence. His name was Hart and he was seriously pursuing the young woman from Panama.

REPATRIATION AND LIFE IN WAR-TORN GERMANY

In early 1944, the Probst family of 14 began a new stage of their repatriation to Hitler's Germany. They were sent from Crystal City by train to New York to board a ship for Europe. During this time, Hart had diligently expedited marriage papers and the couple was married a day before the Probst family left on the ship from New York in February 1944. Anna was saved from deportation to Nazi Germany—but not the rest of the family.

The marriage between Hart and Anna did not last very long. It was said that he was in the FBI or some related duty and was shot and killed. Later, Anna married Ralph Walker, an older man who was not in good health. Anna worked in a tire factory in Ohio during the war.

The Probst family of 13 and the Ewald Hortz family from the Almirante banana port were deported on February 15, 1944, aboard the SS *Gripsholm* from New York City to Lisbon, Portugal. This ship flew the neutral flag of Sweden and served to exchange enemy aliens, prisoners, members of the diplomatic corps, Jews in Europe and others on opposite sides of the war. The ship's manifest list indicated that there were three passenger categories: German consular officials and their families (28); French diplomatic officials and their families (18); German nationals (1,117).

The German nationals on this voyage had been arrested in Central America in the following countries: Guatemala, El Salvador, Honduras, Nicaragua, Costa Rica, and Panama. From South America, "enemy aliens" came from Colombia, Ecuador, Peru and Bolivia. From the Caribbean: The Dominican Republic and Haiti. Many children and spouses were citizens of those countries. A considerable number of prisoners had resided in the United States of America, where some of their spouses and children had been born and held U.S. citizenship. The youngest passenger on the voyage was a girl who was three days old. The oldest passenger was an elderly lady of 71 who had been forced from her home in Guatemala.

The prison camps where these passengers resided before the voyage were Crystal City and Seagoville, which accommodated families. Stringtown in Oklahoma, Camp Kenedy and Fort Stanton housed mostly male prisoners without families.

After their arrival in neutral Lisbon, the family's train passed through General Francisco Franco's Spain, an ally of

Adolf Hitler. They continued through southeast German-occupied France and probably to the northern border of Switzerland. It is likely that Wilhelm, Juanita and their 11 Panamanian children were swapped at the Swiss-German border for an equal number of American prisoners, Jews and other prisoners held by the Nazis. To save the Probst men and young boys from being inducted into the Nazi military service, a previous agreement had been made by the American negotiators that repatriates would not be inducted.

LIFE DURING THE THIRD REICH

After being swapped at the Swiss-German border, the family was sent to Munich where they stayed for some time. This is where some of the family began to separate. Otto Paschek, who was a nephew of Wilhelm, came from Meseritz in northern Germany to Munich to find his relatives from Panama. He returned to Meseritz with two of his Probst cousins, Gretchen, 8, and Frieda, 19. They stayed with their Aunt Berta Paschek.

Wilhelm, Jr., 25, remained in Munich because he met a German woman named Tita. The couple had a baby named Silvia. They stayed in Munich until the baby was three years old.

From Munich, the rest of the family was given lodging and a medical exam by the German government in Nürnberg. When their train was bombed, a German officer instinctively fell on top of young Josef Probst to protect his life. The rear of the train was badly damaged. Close calls from bombing and seeing massive destruction traumatized the family. After some time in Nürnberg, Wilhelm decided to take the family to his birthplace in Meseritz. His sister, Berta Paschek, took in the whole family temporarily because there was no place to go. This family house had a strong Prussian heritage with old swords and other military memorabilia decorating the walls. Some of the family moved in with other relatives until bombing caused damage to one of the houses. The Probst family rented an apartment and accommodated themselves the best they could. Rent was paid with money earned by the men and older boys who found jobs. Wilhelm was hired as an electrician by Siemens—the giant German corporation.

ENDING THE WAR

As the end of the war approached in 1945, everything became chaotic. The Americans, French and British were pushing fast from the west. Thousands of defeated German soldiers, tanks, guns and trucks came by the Probst house day and night, heading east in fast retreat. Many men had already deserted, abandoned their rifles and were walking nervously in despair. The end was near.

A very short time later, traffic reversed in the other direction—toward the west. The Russians were coming triumphantly from the east by the thousands, day and night, with their armies, trucks, tanks and guns. The conquerors were jubilant and vengeful. But the Probst family was terrified at what they were seeing. Some of them watched the horror from small windows at ground-level in the cellar. They saw Russians killing German soldiers and civilians in the street, then tanks ran over the dead or dying bodies. When they heard about the Russian conquerors raping German women, Frieda and Heddy hid in the attic out of fear.

During this turbulent transition, some of the Probst men were arrested and imprisoned several times by the conquering Russians. They were released a day or so after each detention. However, one arrest lasted several months. Wilhelm,

Eric and Ricardo were imprisoned on the second floor of a jail in Meseritz where food and water were minimal. Juanita and her children took food and water from their rented apartment to the prison and hoisted it up to the second floor with a small pulley. Juanita begged the Russians to release her husband and sons but to no avail. Her children acted as interpreters because she did not speak German or Russian.

According to Gretchen, it was 12-year-old Otto who became the father figure in the house and held the family together while Wilhelm, Ricardo and Eric were in jail. Otto organized ways to find food, water and supplies to bring home to the destitute mother and children. They were seeing the horrors of war and trying to survive. One of those horrors was experienced during the previous winter when the Germans and Russians were fighting in their area. To get water, the Probsts had to go to a river, cut blocks of ice, and once back in their apartment, melt them. Everyone had seen dead bodies of Germans and Russians floating in the same river before it froze into ice. They had no choice but to use the water.

After the Russian conquest, Joseph Stalin quickly dictated that the Polish border would be expanded to the north, south and west. As a result, Germans who had been there for centuries suddenly found themselves without a country. Millions of Germans in several expropriated parts of old Prussia, Czechoslovakia and Germany left suddenly in 1945 and 1946, running for their lives—mostly toward the west. Polish people, their language and culture replaced the Germans in Posen and Silesia. They changed the names of German towns to Polish names and the entire area became part of the new country of Poland up to the Oder River. The realignment of boundaries was done quickly under Stalin's orders.. This included the birthplace of the Probst families in Posen—now Poznan, Poland. Meseritz became Międzyrzecz.

Near the end of the war, something happened that would profoundly benefit the Probst family. It began when an allied fighter plane was shot down by German aircraft. The parachute saved the pilot's life, but Germans were waiting for him when he landed and imprisoned him. Later, it became the responsibility of an invading Russian officer to determine the prisoner's nationality. When the officer learned that Wilhelm Probst spoke English, along with Spanish and German, he asked him to determine whether the prisoner had an American or a British accent. Probst proclaimed that the prisoner's accent was American. Fortuitously, the Russians released the American pilot—thanks to the identification by Wilhelm. This pleased the Russian officer who soon struck up an unexpected friendship with the Prussian.

The war in Europe ended on May 8, 1945. Soon thereafter, Wilhelm decided it was time to get his family of 13 out of Meseritz and go west to Berlin even though the city had been bombed into rubble. His Russian officer friend, who had attended Oxford University in England, helped him buy a wagon and two horses. The wagon came with food and provisions for the family to survive as they drove west. A cow that gave milk was tied with a rope and followed behind the wagon. The officer also gave Wilhelm a Russian flag that could be displayed on the wagon for safe passage and a military travel permit to allow the family to move west through the many checkpoints on the way to Berlin. As a final gesture, the Russian officer had a homemade Panamanian flag fabricated to put on the wagon. The Probst family was well-supplied and protected as compared to the other destitute and displaced groups going in all directions after the war.

When the Probsts arrived at the Oder River to make the crossing from Poland to the German side, a large gang of Polish men surrounded their wagon and started to pillage their supplies. Fortunately, just as Russian troops were coming to their rescue, Wilhelm hit the horses with his whip and made the animals lunge forward. The marauding men jumped aside and the family continued to Berlin, about 100 kilometers to the west. The trip from Meseritz to Berlin took about two weeks.

Berlin, the once-proud capital of the Third Reich, was in ruins and was destitute. When the Probst family arrived, they first went to the British Sector. The British then sent them to the American Sector. Eventually, the Americans put them in a vast camp of displaced persons in Tempelhof behind barbed wire. The family was given barracks to live in and food. Despite these conditions, the Probst family felt appreciative of the food and medical care they got. The children ran all over the place, climbed through the barbed wire and explored the rubble and ruins around Tempelhof. Life was getting better.

Sometime during this period in 1945, Wilhelm sold the two horses, the milk cow and the wagon, although he was sure that the animals would be killed and eaten immediately. The Americans then sent the family by train from Berlin to a displaced-persons facility in Kevelaer, 500 kilometers to the west, close to the Dutch border in northwest Germany.

KEVELAER CAMP FOR DISPLACED PERSONS

For centuries, Kevelaer had been a venerated Catholic city that attracted masses of pilgrims to its holy shrines. All that changed during the Third Reich under the Nazis. Forced-laborers from numerous countries, especially from Ukraine,

Poland and other eastern European countries, were kept in an immense labor camp there.

In March 1945, British and Canadians captured the town with Sherman tanks and liberated the forced-labor prisoners and residents from the Nazi occupation. Many of the former prisoners started walking east toward their homes but others loitered, pillaged and vengefully took what they could from local residents. Brutal revenge was common by assaulting German civilians and former German soldiers trying to find their homes and families. The British, French and Americans were challenged to reestablish law and order and civility.

To accommodate displaced persons like the Probsts, the same facilities of the Nazi forced-labor camp at Kevelaer were converted into living quarters for the cold, hungry and homeless refugees. Upon arrival, the standard procedure for every displaced person was to be deloused and bathed. They slept in a building that processed tanned hides, which were put on a cushion of hay scattered on the floor. Others had just the hay for a bed. There was no privacy because 200 other displaced persons shared the sleeping area, the toilets and the unsanitary bath area. Food was minimal and the main daily dish was a watery *osenschwanzesuppe* (oxtail soup). The British administrators were challenged to keep order and provide the basic necessities for so many displaced persons from so many countries.

Students were taught in German by nuns in a Catholic school where discipline was very strict. Medical problems were treated half-an-hour southeast of Kevelaer in Hamborn, at a good hospital managed by Catholic nuns. Ricardo Probst had his bone cancer treated there. When any of the Probst family came to the hospital to visit a patient, the patient would hide left-over or spoiled food and give it to the visitors to smuggle back to the family in Kevelaer. According to Otto, Kevelaer is where the Probst family met Richard and Edmund Kandler from Bocas del Toro. This encounter between two German families born and raised in Panama would later result in the Kandlers inviting young Otto to work for them. Kevelaer was the home of the Probst family for about six months. Afterwards, they went south to Munich for processing to return to Panama.

PROCESSING DISPLACED PERSONS

There were millions of destitute, displaced persons walking in all directions across Europe trying to find their way back to their home country. Therefore the United States helped organize the UNRRA (United Nations Relief and Rehabilitation Administration) in Munich in 1945. The organization had 12,000 employees who aided displaced people with the basic necessities such as food, water, shelter, medicine, clothing and fuel in winter for survival in war-torn Europe. When the displaced foreign-born refugees arrived in Munich, UNRRA provided its services, the same as Richard Kandler received when he was processed in Munich in 1946. They were sent to a former Wehrmacht barracks in Munich-Freimann. This camp housed thousands of displaced persons while they were being processed to return to their country of origin, including the Probst family. It seems probable that Richard and Edmund Kandler also made contact with the Probst family in Munich because they had previously met them in Kevelaer.

Besides staying at the barracks, the Probst family was housed in the infamous Nazi concentration camp of Dachau near Munich, because there was room available after the war. The family also shared a farm house outside of Munich at Günzenhausen with a German farmer and his family who grew hops. The Probst family worked for their keep and were paid with fruits, vegetables, eggs and other farm products. This was a pleasant experience for the family, compared to what they had been through during and after the war.

THEIR FATES AFTER THE WAR

Most of the Probst family had stayed together during the war but now they began to go their separate ways.

The first to leave Europe for Panama in 1946 was Frieda, 21, and Hedwig (Heddy), 20. Money was scarce, and it was decided that the two adult sisters would get the first available funds to return home by ship.

Next, Juanita and the seven youngest of her 12 children prepared for departure, probably also in late 1946. The children were: Erich, 15; Otto, 13; Lotchen, 12; Gretchen, 10; Josef, 8; Alfred (Fred), 7; and Siegelinde, 5. They traveled by train to Paris where the Panamanian Consulate gave them new Panamanian passports and visas.

From Paris, the family rode a train through Spain to Genoa, Italy, where they boarded a comfortable Italian cruise ship, the SS *Jagiełło*, destined for Panama. On the way, they stopped at Madeira in the Cape Verde Islands, Barranquilla in Colombia and finally docked at Cristobal in the Panama Canal Zone. After spending some time in Panama City, the family went to their birthplace near Boquete in Chiriqui.

Wilhelm, Sr., stayed in Germany with the Siemens company, helping to restoring electrical systems., because Panama would not give him papers to return. Siemens later sent him to Afghanistan for several years to install high-tension lines. He sent money back to Panama to help support his family.

Wilhelm, Jr., by now 27, could not get papers to return to Panama. He stayed in Munich with his German wife, Tita, and their daughter, Silvia, until the baby was three. The family eventually settled in Venezuela where he worked and later died. Silvia is a medical doctor in Caracas.

Richard, 17, also could not get papers to return to Panama and remained in Germany. In addition, he had been diagnosed as terminally ill with tuberculosis and was isolated in a hospital. A Peruvian medical doctor volunteered to care for him and nursed him back to life. After his recovery, he went to Mexico, worked and received more bone cancer treatments. Later, Richard settled in Bogota, Colombia. The extreme cold in Bogota was not good for his health and he moved to lower elevations. He married a Colombian, had a family and died in Colombia.

After the war, Josef visited his sister, Anna, in North Carolina after he attended a school for four months in Minnesota. Anna never had children and she died in North Carolina.

Lotchen and Gretchen married Americans and lived in the United States. Lotchen passed away in the U.S. Erich went to California, married, had a family and died there.

Frieda married Carlos Lauterbach in Panama and had three sons: Otto, Eric and Richard.

Hedwig (Heddy) lived in Panama with her husband, Andres Mojica, of Spanish origin. Their children, Ivonne and Heddy, both live in Germany; and Antonio Mojica in Panama. She had wealth from investments in coffee and a home in El Valle, where Gretchen visited her father when he was staying with Heddy.

In 2022, five of the 12 children were still alive: Gretchen in North Carolina, Alfred (Fred) in David, Chiriqui, and Otto, Josef and Siegelinde in Panama City. It is believed that most of the 12 children, except Anna, acquired German passports because it made it easier to travel.

Some of the second-generation Probsts born in Panama also obtained German passports. Even more amazing, several of the third generation born in Panama now travel with German passports.

WANDERLUST CONTINUES

After the war, having already survived dangerous experiences, the younger Probst members were well prepared for more adventures and wanderlust in Panama and other parts of the world. By then, it might have become a compulsion to move on and succeed.

Otto Probst was born in 1933. His daughter, Jeanette, a dentist in Panama City, revealed that her father met Richard and Edmund Kandler after the war in Kevelaer at the UNRRA camp near the Dutch border. The Kandler brothers recommended young Otto to their father, Hans Kandler, who had been a prisoner in American enemy alien camps in the United States in both World War I and II.

In 1947, Otto was 15 years old when he flew to the island town of Bocas del Toro to begin work with the Kandler family. Upon arrival, he went to the office to present himself for work. The first person he met was the secretary, Dora "Dulce" Selles, a local woman whom he married later on in Panama City. Dora had been a school teacher in Bocas del Toro. Young Otto worked at the Kandler sawmill at Split Hill (*Loma Partida*), supplied logs for the sawmill and worked with cacao operations on Isla Colon. He also assisted with the Kandler cattle herd.

After several years in Bocas del Toro with the Kandlers, Otto moved to the wilds of Darien at the eastern end of the

Isthmus of Panama on the Pacific side in remote Patiño. He was in charge of a timber and sawmill operation of the *Cooperativa Agrícola de Patiño*, owned by Arturo Miller and Camilo Quelquejeu of Panama City. Mr. Quelquejeu also operated an airline with one small Cessna that flew from Panama City to Patiño and Las Palmas.

Dr. Jorge Selles, who was born and raised in Bocas del Toro, was Otto's brother-in-law and remembers visiting his sister, Dora, at the sawmill in Patiño. The Probst family lived in the area's only house. They had no neighbors and there was no store or school. Their lifeline to the outside world was the Cessna that landed on a short, grass airstrip. Dr. Selles made three visits to Patiño during breaks from medical school in Mexico City during 1963 and 1964. Otto had three daughters. When the expectant mother was close to giving birth, she would fly in the Cessna to Panama City, give birth and return a few days later with the newborn infant. Jeanette, their first child, was born in 1959. When she reached school age, she went to Panama City to enter the 1st grade in 1965.

Jeanette Probst remembers when her grandfather Wilhelm visited their isolated house in Patiño. While he was there on an extended visit, Wilhelm planted a vegetable garden in the patio by their house.

After 14 years in one of the most remote and wildest tropical areas in the world, Otto moved his family from Patiño to Panama City. His new job was provided by Earl Silas Tupper (1907-1983), the American founder of Tupperware. In 1965, Mr. Tupper had just bought San Jose Island in the Perlas Archipelago, 92 kilometers southeast of Panama City. Despite the fact that the island had been used by the American military to test all sorts of armaments and chemical warfare during

World War II, Tupper had big plans. Otto was involved in these plans to build a tourist resort. He built roads, improved the airfield and expanded the infrastructure.

Because the American military would not assure Tupper that his island was safe, he lost interest and sold the island in 1983. The money he received was donated to the Smithsonian Tropical Research Institute in Panama City where the Tupper Auditorium was built. The new buyer was a consortium named Desarrollo de San Jose that was owned by Otto and Josef Probst, George Novey III, and 50 others. Eventually, the Novey family gained complete control of this consortium. Today, it is known as Hacienda del Mar.

Because of her love for Isla San Jose, the ashes of Dora Selles de Probst were deposited in the sea near the resort. In 2022, Otto was living with his daughter, Jeanette.

Josef Probst was born in 1937 in Boquete. Several of his stories illustrate some of the family's wanderlust. In the 1930s, his older brother, Wilhelm, Jr., walked over the Continental Divide to Bocas del Toro, a major trek. In 1950, Josef and his younger brother, Fred, were growing tomatoes in Boquete, but wanted to explore more opportunities on the Caribbean side of the mountains. When Fred was 10 and Josef was 13, they also walked across the Continental Divide to Bocas del Toro. This journey took several days through virgin rainforests in cold, rainy mountains and then down to sea level.

Josef eventually became an employee of Hertz Car Rentals in the United States and saw 26 states during his time there. Because he spoke English, Spanish and German, Hertz later assigned him to Europe where he worked in several countries.

Gretchen Probst Bentley was in Fort Bragg, North Carolina, near Fayetteville, because her sister Lotte lived there. Lotte had married Army Sergeant Bouchet in Panama and was later transferred to North Carolina. Gretchen was a dance instructor for the Arthur Murray Dance Center, where she met her future husband. They did not marry until she returned from a hiatus in Japan. They had a son. The couple helped Anna Probst and her husband, Ralph Walker, move from Ohio to the Fayetteville area to be near Gretchen and their sister Lotte.

Gretchen took care of her parents for many years at her home in North Carolina. Both had health problems and she helped them with expenses and medical care. Wilhelm was a diabetic and had blood pressure and heart problems. Her parents traveled back and forth to Panama. During this time, Wilhelm's book was translated from German to Spanish. Josef arranged to get 10 copies printed in Mexico.

In May 2022, I interviewed Gretchen by telephone. She is 87, drives her own car and does her own shopping. Gretchen provided a wealth of Probst information that helped fill in the blanks of this chapter.

Siegelinde Probst graduated from high school from Colegio Felix Olivares in David, Panama, in 1958. She was the first woman to graduate from the University of Panama with a degree in engineering. Later, she continued her engineering studies in Munich, Germany. In the 1960s, she worked with the great geologist, Robert Stewart, during studies for widening the Panama Canal. Her work then took her to the wilds of Darien in Panama and Colombia in the early 1970s to study the prospect of another canal route. She retired when the Panama Canal was transferred to Panama at midnight on December 31, 1999. Siegelinde lives in Panama City and writes a poem every day. Contributions from Siegelinde and Josef Probst helped make this chapter possible.

Wilhelm Probst could not get permission to return to Panama for many years even though his wife and children petitioned Panama's president. Meanwhile, he continued to work with Siemens until he retired due to a back injury on the job in Nicaragua. He and his wife received a social security pension check from the German government. After retirement, he visited his son, Richard, in Bogota, Colombia.

No one remembers when Wilhelm finally returned to Panama and stayed. One testimony provides a clue; in 1958, Stanley Heckadon-Moreno was in high school in David at Colegio Felix Olivares and remembers Siegelinde Probst, who was a senior. Her father sold yogurt from a push cart to the students who called his product Bulgar yogurt—alluding to the Prussian as a Bulgarian. Wilhelm Probst was about 63 years old at that time. He visited different children after his return to Panama.

Wilhelm and Juanita Probst left Chiriqui and went to Panama City. Wilhelm died October 8, 1973, and Juanita died February 28, 1981. Both are buried next to each other in the Jardin de Paz cemetery.

Probst brothers: Richard, Otto and Wilhelm, circa 1911, Griefswald, Prussia. Courtesy: Josef Probst.

Wilhelm Probst, 16. Courtesy: Josef Probst.

Wilhelm in Texas.
Courtesy: Josef Probst.

Cover of Wilhelm Probst's book. Courtesy: Josef Probst.

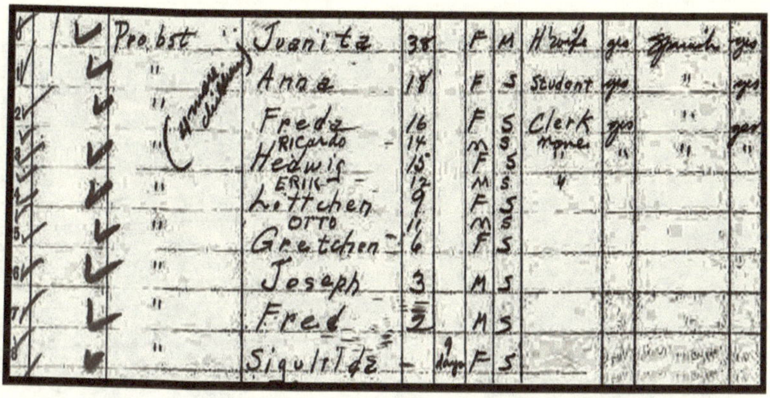

Probst	Juanita	38		F	M	H'wife	yes	Spanish	yes
"	Anna	18		F	S	Student	yes	"	yes
"	Freda	16		F	S	Clerk	yes	"	yes
"	RICARDO	14		M	S				
"	Hedwig	15		F	S	none	"	"	"
"	ERIK	12		M	S	"			
"	Lottchen	9		F	S				
"	OTTO	11		M	S				
"	Gretchen	6		F	S				
"	Joseph	3		M	S				
"	Fred	2		M	S				
"	Siqulilde	–	9 days	F	S				

Probst family on ship's manifest. Source: https://www.ancestry.com.

Grove Park Inn.Source: Miscellaneous Postcard Collection, North Carolina State Archives, Raleigh, NC, USA.

Twelve members of Probst family in Asheville, N.C., 1942. Back row, L-R: Eric, Frieda, Richard, Anna, Lotchen, Juanita holding Siegelinde, Heddy, Otto. Front row: Gretchen, Alfred, Josef.Courtesy: Jeanette Probst.

Dorasque writing. No restrictions found.
Source: DORACES3.JPG (1600×1200) (bp.blogspot.com).

Crystal City Internment Camp southwest of San Antonio, Texas. The Probst family lived in the cottage shown within the black square. Crystal City Internment Camp: University of Texas, Institute of Texas Cultures, San Antonio. Crystal City, Texas Family Internment Camp - German American Internee Coalition (gaic.info).

Crystal City family housing, federal high school for Japanese and German students. Source: Carroll Brincefield. https://www.thc.texas.gov/crystalcity.

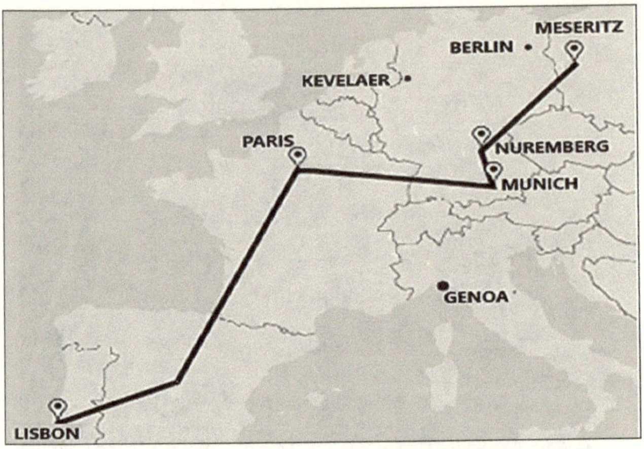

In 1944, the Probsts were exchanged for Nazi prisoners.
Map: Collin Hill and Ken Huff.

The route back to Panama.
Map: Collin Hill and Ken Huff.

Ten members of the Probst family in Kevelaer after the war.
Courtesy: Jeanette Probst.

The only known complete Probst family photograph, Crystal City, Texas, prison camp. Back row, L-R: Otto, Frieda, Richard, Anna, Wilhelm, Jr., Hedwig, Erick. Front row: Gretchen, Alfred, Juanita, Wilhelm, Sr., Siegelinde, Josef, Lotchen.

Clyde Stephens at the mass burial site of 10 German-speaking victims, located at the entrance to Orfidio Sicilia's coffee farmhouse on the Volcan to Sereno River road.

Meeting at Albrook Inn, Panama City, Panama, October 3, 2021. L-R: Josef Probst, Clyde Stephens, Luis "Ibu" Alvarado, Siegelinde Probst, Stanley Heckadon-Moreno.

6. The Stargardters
A Cosmopolitan Family[1]

IN MARCH 1973, THE United Fruit Company transferred me south from Changuinola, Panama, on the Caribbean, to Golfito, Costa Rica, on the Pacific. You have to look at a map to see that the international boundary runs north and south. I was in charge of the Banana Research Department and our family lived in the banana company port town where 235 inches (5,969 mm) of rain fell that year.

My wife Phyllis was one of the American School teachers and told me about a family of students who traveled every school day from the banana plantations in Coto near the border of Panama. It was a long ride on the new clay and gravel road across the lush forested hills. The family name was Stargardter. The father was a German who had married a woman from Colombia. They had five children.

Our own children, Linda and Carl, became friends with the Stargardter children and over the years, our paths crossed in Costa Rica and Puerto Armuelles, Panama. This story is better told if I fast-forward to the year 2020.

GERMAN ROOTS

In an e-mail to Francisco Stargardter in London, I asked him about his family background. He connected me with his

94-year-old uncle, Hans Stargardter, the family historian who lives in Bloomfield, Connecticut. After several interviews with the uncle by telephone, I was astounded by their incredible family heritage.

At Francisco's recommendation, I bought a book called *Microcosm, Portrait of a Central European City.*[2] This 585-page tome aptly describes the complicated ethnic and geopolitical changes over the many centuries in Breslau, once the renowned capital city of Silesia. The Stargardters originated in Strehlen, a town south of Breslau. Silesia and Breslau were absorbed by Poland under the direction of Joseph Stalin and the conquering Soviet Union in 1945. Today, the once venerated German city of Breslau is now called Wroctaw and Strehlen is Strzelin.

During this traumatic human shift in Silesia in 1945 and 1946, millions of Germans ran for their lives, brutally forced from their homeland. They mostly rushed toward the west, many of them on the railroads, herded for days in overcrowded, locked box cars. In turn, they were replaced by Polish and other national and ethnic groups.

That state of affairs is the way some of the Stargardters found their German homeland when they retraced their heritage in 1984. The former home, farm and businesses were now in Poland.

CHILDHOOD MEMORIES

Alfred Stargardter and Käthe Bücka married in Strehlen on April 30, 1926. Their first son, Hans, was born on April 16, 1927 in a hospital in Breslau, a drive or train ride of about 45 minutes north of Strehlen. The next two boys, Rudolf and Friedrich were born in Strehlen.

Käthe's family had been in the Strehlen area since the early 1800s. Around 1890, Käthe's maternal grandfather gifted his daughter, Rosa Feibusch, a large farm house for her wedding. Her husband was Dr. Bücka, a pharmacist. The family lived upstairs. The large building was really two homes: One for the Stargardters and the other was where Käthe's parents lived. Downstairs, there were offices for two brick factories and an office for the farm that Rosa Freibusch Bücka owned.

The enclosed patio had a nice garden with flowers and vegetables. There were trails in the patio to ride bicycles. For fun, the boys would run up brick stairs to play on the top of a wide stone wall. Their wild child play scared their grandmother, Rosa, and she stopped their fun by putting broken glass on the top of the wall.

The family owned adjacent land where they grew wheat, oats and beets. Crop rotation was practiced. Horse power was used but they also had a lot of machinery. Some family members did farmwork along with hired help.

On the farm, there was an abandoned granite quarry with a pool of clean water where the family enjoyed their own private swimming hole.

The family was wealthy because they owned the two brick factories. The original brick factory was started by their ancestor in the early 1800s, the ownership originating from the mother's side of the family. The factories of the 1900s were modern and profitable. Because of this affluence, Alfred Stargardter owned a nice Opel car.

Even though they had electricity, the grandmother was a traditionalist and retained 20 lamps in a chandelier in her room, just in case.

The boys walked to the local public school in Strehlen.

As Germans, the Stargardter family blended in and socialized with their neighbors and the community. However, all that was about to end.

THE RISE OF ADOLF HITLER

In 1933, Adolf Hitler consolidated his power that would lead to brutal conquests and the Holocaust. In *Microcosm, Portrait of a Central European City*, Hans Stargardter tells what happened to his family and describes the impact of *Kristallnacht* in Strehlen:

> We lived in Strehlen, a small town of about 15,000 people, 47 kilometers south of Breslau. We were quite well to do. My great-great-grandfather had come to the town as a peddler in the early 1800s. He became wealthy and the money stayed in the family. My parents had a brick factory and a farm. We lived in a 200-year-old farm house.
>
> It was 1936 when I can remember the first incidents of anti-Semitism. As a youngster, I didn't really believe it. There was a boy from my class that was not allowed to see me anymore. He had to join the *Jungvolk*, the Hitler Youth.
>
> In 1938, it became nasty. Crystal Night was November 9, 1938. They hit us about 6:00 in the morning, the black shirts and stormtroopers. They sent the stormtroopers to another little town so the homeboys would not be inhibited. They came just before dawn. They took my mother and father out of bed, left us children alone with grandmother. Of course, we were terribly upset. All the three of us boys could hear was the noise of breaking glass, of turning over furniture and breaking things. About 7:30 or 8:00, I ran downstairs and asked one of the men, "When are my parents coming back?" And he said, "*Geh'rauf oder ich*

gebe dir 'ne Ohrfeige!" (Get upstairs or I'll give you a smack). Then they went to the next Jewish home. We were the first ones to be hit, lucky because they became drunk later.

They took our car with them. We went downstairs and saw that the groundfloor was completely demolished. Every window was broken. They had taken a knife through every picture except the pictures of Frederick the Great that they didn't touch.

That same afternoon, an honest policeman came and took my father away. He told my father: "Mr. Stargardter, I suggest you take a lot of money because you will be gone for a long time." My father did take a lot of money. It turns out that he stayed for about 10 weeks in Buchenwald.

Later that day, the stormtroopers vandalized the synagogue and poured cow urine all over the Torah and inside the building. It was not burned, however. As a matter of fact, there was no burning in Strehlen.

The people in Strehlen, the day after the house had been vandalized, everyone came to look at the house, like looking at a car accident. Most just shook their heads and said, "My God, what has happened here?" There were very few who shook their fists and yelled, "Damned Jews!" or something like that. Most people were completely indifferent or they were afraid to say anything. They didn't have enough guts to stand up and say, "This is wrong."

My father told me that when they were admitted into the camp, they walked through a line of attendants, or whatever those animals were, and they were hit with sticks. But the worst thing that happened to him was that he got scarlet fever. He came back in nine or ten weeks because he had been decorated as a soldier in World War I with the Iron Cross. In those days, it was a reason for releasing Jews from concentration camps; I do not think it lasted long.

My mother at the time was in the process of selling the properties. We were required to sell everything within a certain amount of time. It was November 1938, and we stayed another six months or so. We could travel anywhere so far as I know. There was no requirement for permits; there weren't any badges.

Some stores and restaurants had signs: JEWS STAY OUT. We left Strehlen two months before the war. I wasn't there for the worst. I remember my father saying, "They're not going to touch us. The government is doing business with us." The brick factories were selling bricks to the German Air Force to make underground hangars. They weren't going to touch us. That was the attitude. And, of course, the people who waited the longest never got out.

Mother lived through all this and I don't really think that she was bitter. She was more German than Jewish right to her death in 1981. Two of her best friends were Germans. Tante Anita was a German woman in Strehlen, an employee of the German Railroad. She came to our house when we left, even though she had been threatened that she could lose her job.

We were forced to sell our property, and in August 1939, my parents, my grandmother, my two brothers and I left Germany on a German ship with something like $10.00 apiece. The final sale of our properties was conducted by an uncle after we left. In 1940, he wrote a letter to Guatemala stating that he managed to send a large sum of money to a bank in Switzerland. In spite of many efforts to retrieve it, the money is still there.[3]

Jessica Stargardter wrote a thesis based on interviews with her grandfather, Hans, titled: *Hans Stargardter, My Experiences in Hitler's Germany* (updated in 1997). This publication related more facts.[4]

The Nazis forced Jews to leave Germany. They were not allowed to take money or valuables of any kind. Properties and businesses were expropriated by extortion and death threats. One night the boys were sent to bed early and they knew something strange was happening. Their father, Alfred, was drilling holes in the legs of a table, putting diamonds in the holes and sealing them up. The table was put on a ship that was destined for Guatemala but the ship got only as far as Vigo, Spain, because the war started. The ship returned to Hamburg, Germany. Someone, somewhere, may still have that table with the legs full of diamonds.

Hans also remembers when Jews were forbidden to have precious metals. He saw a mountain of gold and silver on the dining room table waiting to be turned over to the Nazi government. They got a receipt for the accumulation of treasures from four generations. The receipt was lost but was useless anyway. Hans tells us that,

> My friends and playmates were forbidden to see me anymore because they joined Jungvolk, being Cub Scout age. Young teenagers joined the Hitler *Jugend.* When I was in the fifth grade, the teacher sent me home one day. He had to teach the superiority of the Aryan race and the inferiority of Jews to the class. That saved me from embarrassment. There were only about 40 Jewish families in the town of some 12,000 people.[5]

One unforgettable time was the invasion of Czechoslovakia. Men, trucks, guns and tanks roared by the Startardter house all day and night on their way to conquer their neighbors. When Jews were forbidden to attend public schools, an uncle in Berlin took Hans into his home for an education at a Jewish school.

> Our family only spoke German. Jews who spoke Yiddish were looked down upon. We only heard Hebrew during prayers and the reading of the Torah in our small synagogue whose members practiced Liberal Judaism. You might say we were casual.[6]

According to a book by Fritz Moses, their synagogue was saved by Alfred, who discretely sold the building to the mayor of Strehlen:

> We were so provincially German that we did not know any better. For five generations, my family had been in Germany. We were more German than Jewish. We got out just in time. All of my cousins, uncles and aunts who did not get out were not alive when the war ended. Some died in Bergen Belsen, Dachau, Theresienstadt or Auschwitz.[7]

MORITZ "MAX" STARGARDTER'S ESCAPE TO CHILE

Max was born on August 5, 1885, in Kulm, Prussia. Alfred was a younger brother born in Elbing on October 10, 1894. When Nazi aggression was at its worst, Max, his wife, Lotte, and their children barely escaped from Germany, taking refuge in Chile. Their grandson, Andres Bustos, described their plight in a book he wrote in Spanish, *MAX, Prusiano y Alemán, pero Judío*. (MAX, Prussian and German, but Jewish).[8] The author relates how his grandparents and their two daughters, Eva and Ulla, had to deal with extreme harassment by Nazi officials in their hometown of Elbing.

Repression, extortion, threats of death camps, intimidation and all the horrors of negotiation forced Max and his brother, Paul, who was a business partner, to finally sign their extensive and renowned business assets and personal properties over to the Nazis. They expected 15% of the market value but never got anything, not even a receipt. After signing in good faith, they were told to get out of Germany in 14 days or they would be shipped to death camps. The two daughters were smuggled to England but miraculously reunited with their parents later in Italy. Several countries refused to give them visas because they were Jews. One exception was the consulate of Chile in Genoa, Italy. But even Chile demanded an extortionate amount of money to enter its country. Max had no money because the Nazis did not allow Jews to take money or valuables when they departed Germany.

But Max and his family were saved thanks to Selma Bissinger who lived in the U.S. and had married one of Lotte's cousins; the cousin died and remarried. She and her new, wealthy husband agreed to pay the high costs of visas and transportation. Still, when the family boarded the MS *Orazio*, in Genoa, they were in for a surprise; the captain ushered them to first class. The Bissingers had ensured that the family would have the best accommodations available.

The family arrived in Valparaiso on June 13, 1939, penniless but happy to start their new lives as Chileans. Although they encountered traditional forms of anti-Semitism in Chile, life was good and their nightmare of hell under the Nazis was over. The four of them lived the rest of their lives in Chile.

Unfortunately, Lotte's mother remained in Germany because she thought it was safe to stay in her comfortable home in Berlin. She became a victim of the Holocaust, like so many other family members.

ALFRED: A NEW LIFE IN GUATEMALA

Alfred Stargardter also got his family out of Nazi Germany before it was too late. All identification documents had been altered in 1938 to comply with the command that all Jewish women add "Sara" and the Jewish men add "Israel" to their names. Passports had a large J stamped on the document to show that the person was a Jew. Each Jew was allowed 10 Deutsche Marks which amounted to practically nothing. Baggage was minimal and no valuables were allowed to leave Germany.

The family arrived in Hamburg in August 1939 and boarded the *Heredia,* a United Fruit Company ship. There were many other Jewish refugees on board. They made a port call in Santa Marta, Colombia, then disembarked in Puerto Barrios, Guatemala. There were no special problems with immigration officials to enter the country, but there was a problem when they had no money.

Normally, a train took passengers to Guatemala City in the highlands, but Hans never remembers riding the train, suggesting they could not afford the price of a ticket. Instead, Hans's family traveled via rustic back roads on slow buses and even on horse-drawn carriages. It was a long, uncomfortable trip. In Guatemala City, some of the wealthy Jews helped the family find a place to live, earn a living and go to school. The local schools did not meet their German standards so a Jewish school was built. The head *schulmeister* (schoolmaster) was a German Jewish refugee. Classes were taught in English and Spanish but not in German. Hans, Rudolph and Friedrich (Fritz) attended this school. The Stargardter boys learned Spanish in the streets and were speaking quite fluently in a year.

Alfred was uncertain about how to provide for his family, but he decided to do what he knew best—produce bricks. He hand-built a large oven and produced 40,000 bricks a year. It was hard work as Hans remembered while helping his father.

HANS GROWS UP IN GUATEMALA

Besides helping his father make bricks, Hans worked in an ornamental ironworks shop when he was 13 and 14 years old. He had a break at noon and walked half an hour back to their house for lunch. One day, the servant of a 90-year-old German Jew stopped him in the street and invited him for lunch. After that, he ate a kosher meal every day with *Herr* Engle.

He matured as a teenager and became a receptionist clerk at the Hotel Victoria in Guatemala City. Everybody asked him questions about everything. He found the answers and broadened his knowledge about many things. This caught the attention of a representative from Florida Southern College who offered Hans a work-and-study scholarship. He accepted and went to Lakeland, Florida, in 1945 at age 18. He worked as a landscaper on the campus to earn his keep. After four years, he graduated in 1949 with a degree in engineering. Hans met his future wife, Hennie, the first day at college while playing tennis. They honeymooned in Guatemala, the only time he ever returned there.

Hans was employed most of his life as a mechanical engineer and scientist with Pratt & Whitney, designing turbines for jet engines. He had about 200 patents to his name. In 2021, Hans was 94 and gave me two telephone interviews from his home in Bloomfield, near Hartford, Connecticut. Unfortunately, he died at his home in April 2022.

Bloomfield is the same city where his mother, Käthe Bücka Stargardter, died on May 17, 1981.

RUDOLF BEGINS A NEW LIFE

Rudolf was listed on page three of his mother's passport, stamped April 4, 1939, as Rüdolf Israel, age 11. On page seven, the birth certificate data shows him as Rüdolf Jakob Stargardter. In Guatemala, he attended the Jewish school. He married a Guatemalan, Maria del Carmen Morales, in 1950. After his father died of a heart attack in 1952, he helped take care of his widowed mother. Later "Rudy" and his wife moved to New Orleans and raised three children. He was involved with the Guatemalan consul and later became an American citizen.

Rudy was involved in the import-export business, supplying spare parts and industrial supplies in Central America. He and his wife died in New Orleans.

FRITZ

Fritz's life is portrayed in the chapter that follows. Like me, Fritz spent his entire professional life as a *bananero* in Latin America.

ALFRED STARGARDTER

Alfred died in 1952 after 13 years in Guatemala. He is buried in the Jewish Cemetery and has a tombstone inscribed in Hebrew and Latin writing and with a Star of David. His mother-in-law, Rosa Bücka Feibusch, also died in Guatemala and was buried there in 1956.

FATES OF THE FAMILIES

Incredibly contrasting futures awaited Isidor and Fanny Startgardter's immediate descendants:

EMIGRÉS

- Paul and Margoth Stargardter and their daughter, Inge, managed to get to the Philippines, where Margoth died from tuberculosis. Later, Paul and Inge emigrated to the U.S., where Inge and her three daughters still live.

- Alfred Stargardter was passed over by the Nazis because he had earned an Iron Cross in WW I. He and his wife, Käthe, managed to emigrate to Guatemala. Käthe appears in her German passport as Katherine Sara.
- Max and Lotte Stargardter moved to Chile.

HOLOCAUST

- Gustav and Lene, one of Lotte's sisters, escaped to Paris, from where they hoped to emigrate. But they were arrested and lost their lives in a concentration camp.
- Franz and Boby Stargardter were detained in Holland; they died in the Bergen-Belsen camp. They had no children.
- Alma and Max Heymann succumbed in a concentration camp, along with their daughter, Ruth. However, their son, Horst, was living in Israel after the war.
- Greta and Bruno Danzinger did not survive the Holocaust.

KINDERTRANSPORT[9]

- Gerhard Segel, Gustav and Lene's son, was sent to England via a Kindertransport, and was later reported to have joined the Foreign Legion.
- Before the Danzingers were put to death, their children, Kurt and Peter, were sent to London on a Kindertransport.
- Max and Lotte Stargardter's two daughters, Eva and Ulla, arrived in London via a Kindertransport.

FORCED CHOICE

- When World War II started, two of the cousins who

were in Great Britain were told they had a choice; they could either join the British Army or move to Australia. The older cousin, Horst Heymann, chose the army. A younger cousin chose Australia and was deported, creating what was known as the Melbourne Branch. Michael Heymann says, "They did very well down there."[10]

THE FOLLOWING GENERATIONS

- Hans is in Connecticut and much of his family is in the Northeast.
- Rudolf settled in New Orleans, and that is home-base for his children and grandchildren.
- Fritz was in Colombia, Costa Rica, Panama, then in Asia and elsewhere before retiring and dying in Santa Marta, Colombia.
- Fritz and Rafaela's five children live in various countries. One is in Mexico, one is in Colombia, two are in Panama, and one is in England.

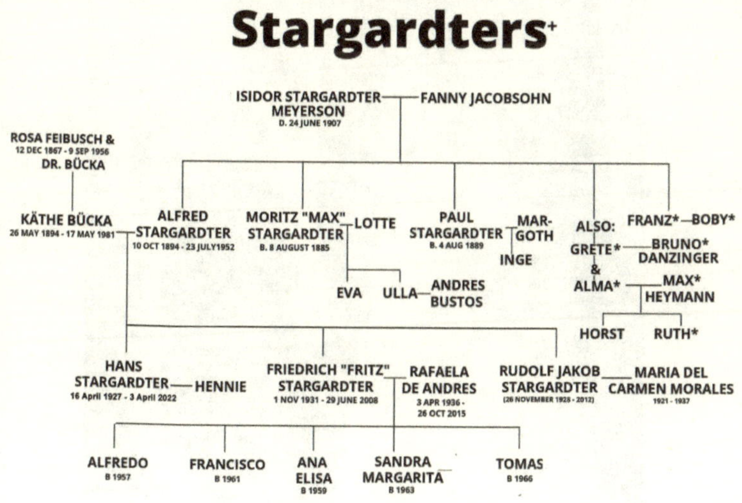

*Died in Nazi extermination camps.

*Only names used in the story are listed here.

Alfred Stargardter and Käthe Brücke's wedding photograph in Strehlen, April 30, 1926. The scratches on Alfred's suit in the photo were caused by Hitler's knife-wielding Brown Shirts who vandalized their home on Kristallnacht in 1938. Tomas Stargardter collection.

Horst Heymann in his British Army uniform in Palestine. He changed his name to Chaman Heymann. His sister Ruth and his parents, Alma Stargardter Heymann and Max Heymann, perished in Auschwitz in 1943. Tomas Stargardter collection.

L-R: Hans, Rudolf, Fritz and their mother Käthe in Strehlen, circa 1935-36. The older boy (2nd from left) is their first cousin Horst Heymann. Tomas Stargardter collection.

L-R: Rudolf, Hans, Horst (Chaman) and Fritz, 1995. Tomas Stargardter collection.

Alfred Stargardter (2nd from right) in the artillery on the Eastern Front during World War I. He was a decorated war veteran who received the Iron Cross. This military distinction saved him from extermination in Buchenwald.
Tomas Stargardter collection.

Alfred Stargardter, World War I.
Tomas Stargardter collection.

At the North Sea shore resort in Germany.
Käthe and Alfred with Rudolf, Fritz and Hans.
Tomas Stargardter collection.

The four Stargardter brothers in the 1930s.
L-R: Max, Franz, Paul and Alfred. Their parents were Isidor Stargardter Meyersohn and Fanny Jacobsohn. Max and family escaped to Chile. Alfred and family escaped to Guatemala. Paul got his family to the Philippines. The other three siblings, Franz, Alma and Greta Margaret and some of their families, perished in the Holocaust. Tomas Stargardter collection.

Katherine "Sara" Stargardter's passport, April 1939. A "J" for Jew was stamped in the upper-left corner, and "Sara" was inserted to identify her as a Jew. Her three sons are recorded on the same passport, each with the word "Israel" added to indicate they were Jewish. Hans Stargardter collection.

Boarding the SS Heredia in Hamburg, August 1939. R-L: Rosa Bücka Feibusch, Hans, Käthe Bücka Stargardter (hidden behind Hans), Hans, Rudolf, Friedrich "Fritz" and Alfred. Hans Stargardter collecction.

Kindertransport children arriving in London.
Tomas Stargardter collection.

Sculpture by Hans Stargardter, circa 2008. Hammered and welded brass, abstract form mounted on a wood base. Tomas Stargardter collection.

7. Friedrich "Fritz" Stargardter
Future *Bananero*

IN GERMAN, FRITZ IS the nickname for Friedrich, and that is the name that Friedrich used all of his life. However, on his German passports, he was always Friedrich Stargardter. When the family left Germany, his mother's passport listed him as "Friedrich, Israel (the word "Israel" was a Nazi contrivance to identify Jewish males), age 8." His birth certificate shows his name as Friedrich Stargardter, born in Strehlen, November 1, 1931.

After the family left Germany for Guatemala, Fritz learned Spanish rapidly in the streets. He attended the Jewish school in the capital, and after finishing at the high school level, he acquired the same work and study scholarship at Florida Southern College that Hans had received. Fritz worked with plants and on the campus landscaping.

Fritz was quite athletic and was on the college rowing team. During the early 1950s, founder Glenn Kirkpatrick selected Fritz to be on FSC's intercollegiate waterski team. On May 16, 1952, the campus newspaper, *The Southern*, wrote a novel story entitled "Florida Southern Student Invents Weird Contraption." It describes how Fritz attached a pair of old waterskis to an abandoned bicycle frame that was pulled by a power boat on Lake Hollingsworth, south of the college campus. This was a precursor of many innovations in the future.

After graduating with a bachelor's degree in citrus and horticulture science on September 3, 1952, he remained in the area until the following year. More than likely, he performed in waterski shows at the famed Cypress Gardens in Winter Haven. One of his acts included the very daring pyramid formation.

In 1953, Fritz was shown skiing during a rehearsal with Esther Williams, star of the movie *Easy to Love* that was filmed at Cypress Gardens.

OFF TO COLOMBIA

Fritz was employed by the United Fruit Company sometime after graduation, in 1953, to work in banana operations in Colombia. The new recruit traveled to the New York office and probably left on a banana ship at Pier 3, North River, for Santa Marta. One of his assignments was to run the dining hall in the banana district of Sevilla. In a photograph taken in the dining hall, there are three young Germans at the table. One was Wolfgang Schulz, who told me the following about the languages spoken by the group:

> Clyde, I only spoke to Fritz or Fred Koch in English. There were always people around us, and we were conscious of how impolite it was to talk in a language they couldn't understand. It had been only 11 years after World War II, and we wanted to blend in as 'American' as we could and be part of the crowd.
>
> Fritz never spoke to me in German either, and he never mentioned his background nor education in Florida. Also, Fred Koch never mentioned his past in Germany. No one ever asked me about my background, either. Everybody was judged by the way you got along with people and the quality of your work. It was a great life.

Love and romance began at the swimming pool at El Prado, the Company compound in Sevilla. Rafaela de Andreis studied and lived at a Catholic school in Santa Marta, and on weekends, she would go by train to Sevilla to see her family and Fritz.

The fame and fun of waterskiing was still on Fritz's mind, and he introduced the sport to Santa Marta, something completely new on the north coast of Colombia. One account tells about Rafaela driving his Jeep along the edge of a drainage canal in Sevilla, pulling him on skis in the waterway.

Both of Rafaela's Colombian parents had Italian and Spanish heritage and both were first-cousins. They were Pedro de Andreis Noguera and Elisa Abello de Andreis. They raised their children in the banana plantations where he was employed by the United Fruit Company all of his career. He retired as a superintendent of agriculture. Three of their daughters married Company employees, one of whom was Fritz Stargardter, on April 21, 1957. Rafaela was Catholic and their five children were raised accordingly. Alfredo, 1957; Ana Elisa, 1959; and Sandra Margarita, 1963; were born at the United Fruit Company hospital in Santa Marta. Francisco, 1961, came into the world at the Company hospital in Sevilla. Tomas arrived in 1966 at the Company hospital in Golfito, Costa Rica. The entire family could be identified as Banana People.

FRITZ MOVES UP THE LADDER

Fritz had several assignments in Colombia that groomed him as a "Banana Man" and prepared him for management. One job was controlling Sigatoka, a devastating fungal banana leaf disease. In November 1960, as superintendent of the Sigatoka control, he declared that applying 20 gallons of Bordeaux fungicide per acre, sprayed by air, would control the disease. This was revolutionary. About the same time, Fritz and others were involved with soil surveys that determined where to plant bananas in the outback areas of Turbo in the Gulf of Uraba, in the southwest area of the Caribbean Sea. Once the surveys were done, Fritz was among the rugged and experienced employees who were selected to develop this immense area for banana production. In the beginning, they lived in a basic work camp constructed in a clearing in the rainforest. Turbo was the last of the great banana developments that the United Fruit Company created from a vast, virgin rainforest with no infrastructure of any kind. There were no roads, bridges, airports, shipping ports, drainage canals or towns with a labor force. Also absent was the traditional Company town with a hospital, commissary for food and supplies, water system, machine shop, school or houses for employees. No traditional recreational amenities like club houses or tennis courts.

In 1961, after living conditions improved and nice houses were built in the isolated Los Almendros compound, Fritz moved his wife and young children from El Prado in Sevilla to the Turbo area. They flew in a single engine Otter fully loaded with their basic household needs. The long flight along the Caribbean coast to Turbo took several hours.

Alfredo remembers that a Company vehicle met them at the airport. On the way to the compound, they came to a river crossing where a flash flood had washed out the bridge. The Stargardter family and their cargo were loaded onto a provisional barge that had been mounted on floating oil drums and were pulled across the flooded river by a cable hitched to a truck on the opposite bank. This was their introduction to their new home in Los Almendros.

Alfredo described their new habitat as an adventure for the family. The compound was in the middle of a clearing of the rainforest where humidity was high and it rained constantly. They were surrounded by tropical forests of giant trees and streams. Wildlife of all kinds was abundant. These were new and exciting sights that they had never seen before.

Conditions and amenities were very basic in the compound. There was no traditional Company store. Groceries and supplies had to be bought in a small frontier town with one muddy street. The ride on a mud and gravel road took about 40 minutes to get to Apartado for purchases.

Despite the natural beauty of their environment where the kids roamed and played, there were tropical diseases to worry about. In 1963, Rafaela contracted malaria when she was pregnant with Sandra. In the midday heat, she was wrapped in a blanket, shivering with chills and fever. Since there were no medical facilities in the area, she was flown to be with her parents in Sevilla for treatment and recuperation. To add to the challenges, she was driven to Santa Marta to have the baby. Just before arriving, the vehicle ran out of gas. The newborn baby tested positive for malaria, but fortunately there were no complications.

Los Almendros had no school, and Alfredo was flown at age six to Sevilla to live with his grandparents so he could attend the United Fruit Company school in El Prado. During long holidays and vacation time, he would fly back to the Turbo area to be with the family. Returning to school in Sevilla was always upsetting for him.

After about three years in Turbo, the Stargardters moved back to Sevilla. Once again, education became a problem because the American School in Sevilla had closed. Alfredo and Francisco attended the Company school in Santa Marta and lived with their grandparents who had recently retired to that port city. On the weekends, the two boys would happily return to Sevilla to be with their parents and siblings.

Once more, Fritz was involved in Sigatoka control. However, the banana plantations in the once expansive Santa Marta division were in severe decline. Civil and governmental conflicts caused even more instability. Soon after, around 1966, after growing bananas there since the 1890s, the United Fruit Company closed the enterprise and turned the assets over to the Colombian government.

That year, Fritz moved his family from Colombia to Coto, Costa Rica, where he became the manager of a new banana box manufacturing plant in Corredores near the satellite town of Villa Neily. Cardboard boxes were made by the millions to pack Valery bananas grown on the Coto and Palmar plantations.

Sticking the corrugated cardboard boxes together required tons of corn starch glue that had to be imported. Since Fritz was an innovator, he built a tower-like structure that converted rejected green bananas into glue. That saved money and used a local banana by-product that normally was dumped in the garbage. But there was a problem. When the auditors discovered that no funds had been appropriated for this glue factory, Fritz had to find an explanation. True to his style, he found a way to solve the accounting problem. The new Pan-American Highway was only a kilometer from their house and was still under construction. Alfredo and Francisco were transported daily to the American School in Golfito, an hour of travel each way in a truck. The other siblings did the same as they became of school age.

Rafaela was pregnant when they arrived in Coto and there were some complications. Fritz arranged for his mother, Käthe Bücka Stargardter, to come from Connecticut to Coto to help. She spoke to the family in a heavy, German-accented Spanish that she had learned during their years as Jewish refugees in Guatemala. Tomas was born in 1966 at the Compañia Bananera de Costa Rica hospital in Golfito.

In 1977, Fritz was transferred back to Santa Marta as general manager to improve, modernize and increase new plantations owned by private Colombian growers who contracted with Chiquita.

This new business approach was to set up independent Colombians as banana farmers who would sell their fruit to the United Fruit Company. The Colombian government and banana growers were responsible for providing amenities, infrastructure and social benefits for employees in their farms and the community. The United Fruit Company did not own or operate any banana farms, but it provided technical advice and services to assure good production and quality. This included management of Sigatoka control.

In 1980, Fritz returned to Costa Rica as general manager of the Compañia Bananera de Costa Rica operations in Coto, Palmar and Golfito, where the manager's office was located. They lived in the manager's house at the base of the steep, forested mountain where lots of rain fell most of the year. It was a gorgeous setting and a wonderful community to live in—with all the perks and amenities. Unlike Honduras and Colombia where security had been a problem, the Stargardter children could walk to the Mount Adams American School on the edge of the golf course and the Golfito airport.

Fritz was very social at all levels and he also promoted good labor relations. However, that was difficult because the banana labor union leaned far to the left and was constantly breaking labor contracts and provoking conflicts. Labor leader Isaak Marchena and his staff had their headquarters in Villa Neily. They had a big red communist flag with the hammer and sickle proudly displayed outside the front door of their office. This impossible situation foretold the future of the huge banana operation that began in the late 1930s. Thousands of employees and their futures were at stake.

By 1980, all five Stargardter children had finished their first eight years of schooling provided by the United Fruit Company, now called Chiquita Brands International. As each reached the age of 14, it was time to "fly the nest," go abroad to high school and then continue at a university. The following is an outline of their studies:

Alfredo: Thomas More Prep in Kansas; Kansas University, BS degree in social anthropology.

Francisco: Thomas More Prep in Kansas; Kansas University, BS degree in political science.

Ana Elisa: Karl C. Parrish High School, Barranquilla Colombia; Kansas University, political science.

Sandra: Karl C. Parrish High School, Barranquilla, Colombia; Kansas University, education.

Tomas: Subiaco Academy, Arkansas; Kansas University, journalism.

Four of them received master's degrees. Ana Elisa studied German by living with a German family, then at Goetheschule in Giessen and at Johannes Gutenberg Universitat in Germersheim.

At age 18, all five of the Stargardter sons and daughters

chose to get German passports. Although each one had a passport from the country of their birth—Colombia and Costa Rica—it was more convenient to travel with a German passport. Ironically, Ana Elisa is the only one that is fluent in the German language.

When the five children married, the in-laws made the family even more international: Their spouses were Colombian, English, Italian, Panamanian and Nicaraguan. These families today are dispersed in five countries. Two of them worked for Chiquita Brands International in the banana business: Alfredo worked for ten years in finance, and Ana Elisa was in charge of quality control in Italy. Tomas is a journalist in Mexico. Francisco is a major fruit and vegetable broker in England and working on a doctorate at Oxford. Sandra is a teacher with the International School in Panama. Ana Elisa is a professor of languages in La Universidad de Magdalena in Santa Marta, Colombia.

These are the rewarding results of the large investment that Fritz and Rafaela Stargardter proudly made in educating their children; probably the most educated and cosmopolitan family in the Chiquita Banana domain.

At one time, the United Fruit Company initiated some downsizing and donated several large, valuable buildings and facilities in Golfito to a University of Costa Rica and Kansas University partnership. Fritz brokered this arrangement because of his close affiliation with Kansas University where all five of his children graduated. Later, these ample facilities enabled academic studies in many disciplines that greatly benefited Costa Rica and tropical science.

After nine years in Golfito, in February 1982, Phyllis and I were transferred to Puerto Armuelles across the border in Panama. Fritz and friends gave us a wonderful farewell party, which included a nice dinner-dance at the club and some emotional speeches. We were presented with a large ornamental oxcart that served as a bar, something very beautiful and certainly useful.

FRITZ GOES TO PANAMA

In 1983, Fritz was transferred as general manager for the United Fruit Company in Puerto Armuelles, where Phyllis and I had been sent. This Company banana division began in 1927 and was developed from the ground up. At one time, the Chiriqui Land Company, the United Fruit Company's subsidiary, had the world's largest single block of bananas, approximately 12,000 hectares (about 30,000 acres). In the port, one can see the sun rise over the Pacific Ocean, since the Isthmus of Panama runs from west to east, and the Pacific Ocean is on the south coast of the Isthmus.

Fritz had many challenges but always found innovative solutions to problems. There were always drainage problems during torrential rains from May through December. When the dry season began in January, irrigation had to be ready. Sometimes, during hot, dry, dusty months, shipments of green bananas looked off-color and became known as "tired fruit." I was sent to California twice as a troubleshooter to assess this problem.

Fritz believed in educating his monthly employees in different departments so they would better understand the business. Participants were selected from engineering, accounting, production, legal, management and agriculture. This led to the creation of what became known as "The Banana College." Fritz charged me with organizing and running the curriculum that included soils, planting, agronomy, nutrition, diseases, insects, nematodes, field practices, harvesting, packing, quality and other related subjects. The sessions were taught in the classroom and in the field. An exam was given at the end of the training, followed by a ceremony to present a diploma to each graduate *bananero*. That called for a celebration afterwards.

Problems with labor were chronic and frequently caused great tensions and economic losses for the Chiriqui Land Company and the workers. One case in point was a union strike in 1986 because of a list of grievances. When the strike was settled after 38 days, the labor force resumed working.

During the loading of a ship that first night, the powerful and belligerent stevedores stopped their work after only an hour on the job. The reason was a power struggle going on between the wharf workers and the labor union leaders—not the Company. Personalities, politics and personal leadership ambitions were clashing like clan rivals. The Company failed to negotiate with the intransigent stevedores and decided to release the ship from the dock and send it around Punta Burica to Golfito in Costa Rica.

Cameron Forsyth, manager of production and any other tasks at hand, was a rugged executive who had experience as a former British Army lieutenant keeping Catholics and Protestants in Northern Ireland from killing each other. This background gave him the confidence to walk along the wharf with an assistant, Nelson Serrano, and go through the gang of angry strikers. The two of them quickly climbed the ladder to the control cabin where Nelson knew how to raise the banana-box loading gantry out of the ship so that the vessel was free to sail to Costa Rica.

Cameron and Nelson found the angry mob waiting for them at the bottom of the ladder; the stevedores grabbed

Cameron and beat him up, inflicting cuts and bruises. After the struggle, the mob tried to throw him overboard as the ship was moving away from the wharf. The stevedore leader then called off the attack. Nelson received severe head injuries and was hospitalized for a long time. Fritz was outraged that his right- and left-hand man and his assistant had been brutally assaulted. He told the Company lawyers to take legal action against the union.

Meanwhile, on the other side of the border in Costa Rica, unemployed former stevedores successfully loaded the ship with bananas from Panama that had been trucked to Golfito. This port had been completely shut down, along with all the banana plantations and the railroad in 1984 after a militant communist-led labor strike that lasted three months. The violence and economic losses caused Chiquita Brands International to close this enterprise permanently. This total closure would foretell the same fate for Puerto Armuelles when Cameron Forsyth was the general manager and had to close down that giant banana operation in 2008. The closure was caused by an intransigent labor leader, Jose Morris, who refused to compromise with the banana company. It was a terrible fate for thousands of workers, families and the economy of Panama. Traditionally, most banana shipments went from Puerto Armuelles to Wilmington, California, the port for Los Angeles. Nonetheless, in 1987, shipping and marketing took an interesting turn. Ten kilograms of green fruit were packed in a vacuum bag called a "Banavac" that greatly extended the life of the bananas. This made it possible to ship and store fruit for six weeks without ripening. Cardboard boxes were printed in the Japanese language. After a six-day voyage, the ship reached California where half of the banana boxes, the ones that had English markings, were unloaded. The empty space was filled with California citrus. It took three weeks to cross the Pacific Ocean to Japan where the bananas and citrus were unloaded. Despite the long voyage, the Japanese market was very profitable.

In August 1987, I was transferred as director of banana research and technical services to La Lima, Honduras. Once again, Fritz gave us a memorable send-off at the Golf Club.

FRITZ IN ASIA

By 1989, the United Fruit Company, and later United Brands International, had changed its name to Chiquita Brands International based in Cincinnati and was controlled by the Carl Lindner family. The Company set up an Asian Division because relations were declining with Antonio "Tony" Floirendo, their big grower and supplier in the Philippines. The United Fruit Company had set him up in banana production and technical assistance in Mindanao in the 1960s and 1970s but later, strategies changed and disagreements developed. This caused Chiquita to start looking for other sources of bananas to supply Japan and possibly South Korea and some of China in the future.

In 1986, Tony Floirendo went through a major crisis and setback when his close friend and business associate, president and long-time dictator Ferdinand Marcos and his wife, Imelda, were deposed and exiled to Hawaii. It was no coincidence that the mansion in Honolulu where they lived was owned by Antonio Floirendo.

The day before Marcos was deposed in 1986, the Floirendo family suddenly left the Philippines and flew in their private jet to the independent Islamic sultanate Brunei. Antonio Floirendo's son was known as "Tony Boy" and was a friend of the sultan of Brunei. They had played polo together and now the Sultan hosted them as political refugees—in a friendly way. Soon, Tony Floirendo exiled himself to London where he had investments, including race horses, because he liked to gamble. Meanwhile, the "Banana King" put his son Tony Boy in charge of the empire in the Philippines. Business as usual was no longer the same after the people elected Corazon "Corey" Aquino as their new president.

While Floirendo was in exile, the general manager of Chiquita in the Philippines, Buck Kaiser, became friends with the new president's brother and efforts were made to acquire control of banana lands that were traditionally cultivated by Tony Floirendo. At the same time, another major financial default happened. Chiquita and Floirendo had a profit-sharing contract to split the profits made in the Japanese market. However, while Tony was in exile in London, Chiquita did not deposit Tony's half of the profits into his account as agreed. The sum amounted to many millions of dollars. These two deeds by Chiquita infuriated the banana tycoon and relations became even worse.

In 1988, Floirendo "arranged" a way to safely return from exile to the Philippines. The cost of his return home was never revealed. He had to adjust and do business under the new government of President Corazon Aquino—and her brother.

In 1989, Drs. W.G.C. Forsyth, Francisco Sierra, Nadar Vakili and Cameron Forsyth explored prospective areas along the Mekong River in Viet Nam. They found good soils. However, because of numerous old villages and ancient memorials to the dead, such areas were not recommended. Areas in Thailand were also rejected.

Mike Utley replaced Cameron Forsyth as general manager in the Philippines in 1989 when Cameron was transferred to Changuinola, Panama. As the top official for Chiquita in the Philippines, Mike was a friend of Tony Floirendo. He also enjoyed scuba diving with Tony Boy at the Floirendo resort known as The Pearl Farm on Samal Island. Tony Boy had married the former Miss Universe, Margarita "Margie" Roxas Moran who was managing the luxury resort.

Mike made a trip to Nha Trang, Viet Nam, to represent the Floirendo family. Tony Boy made all of the arrangements because there were no diplomatic relations between the United States and Viet Nam in 1990. The vice president of the Asian Project for Chiquita was Micky Rotolo, who had to give Utley permission to make this mystical and secretive exploration in Viet Nam. At the last minute, Tony Boy backed out and did not go. Mike was on his own. When he arrived, he was given a royal reception by Vietnamese promoters who desperately wanted foreign investments. Mike was only interested in finding good silty loam soils for bananas and port infrastructures. He was impressed with what he saw but no further action was taken. Chiquita was not interested in more partnerships with Tony Floirendo.

EXPLORATIONS

These were the complex situations in the Western Pacific and Southeast Asia in 1990 when Chiquita selected Fritz Stargardter to explore opportunities there. He hit the tarmac running, taking Alvaro Espinosa from Panama and Jose Cuevas from Honduras—both good banana and soils men. They assessed the possibilities in Sulawesi, Indonesia, and other such places.

After the Asian budget meeting in Singapore, Bob Kistinger, Micky Rotolo and Mike Utley went with Fritz Stargardter and his men to explore new possibilities, including old rubber plantations in Sumatra. However, competitors like Dole and Del Monte were already ahead of Chiquita with their own plans.

The coup de grace for all banana growers and shippers was the discovery of the Race Four strain of *Fusarium oxysporum*, the devastating fungal disease in Sulawesi, Indonesia. Investors and planters quickly retreated from areas infected with this terminal disease and moved elsewhere. Fritz Stargardter concentrated his efforts was in Cairns, a major banana-growing region in Australia. He liked the area and spent much of his time there. Fritz bought several banana farms for Chiquita to market fruit in Australia and to export. Consequently, conditions, banana production and labor traditions were different from Central American plantations. A major problem was high overhead and production costs—something that financial analysts reported in red ink.

Sometimes Fritz was too busy to stay within his budget and was asked to explain some of his expenses at budget meetings. One particular area of concern was the cost of so many communications around the world. He had to explain the need for phone calls that cost $1,200 each.

Suddenly, Chiquita lost interest in the Asian project and it was cancelled. Fritz had other brief assignments in Spain, Venezuela and the Canary Islands. Afterwards, he reluctantly retired to Colombia in 1993.

It had been a long road of ventures and adventures for Fritz, the boy from Strehlen in Silesia, Prussia, to his final resting place in South America in 2008. He left a legacy that few people could believe or understand. Rafaela followed him in 2015. They are buried in Jardines de Paz in Santa Marta, Colombia.

His son, Francisco, summarized the Stargardter heritage best by saying that the children were raised as good Catholics, but they respected their Jewish heritage.

HOMAGE

In an homage to the Stargardter sense of humor and good story-telling, Sandra Margarita Stargardter, one of Fritz's two daughters, wrote the following experience while visiting her ancestor's graves in Guatemala:

> I had never been to Guatemala before but heard many stories from my dad and Uncle Hans about their life there. In 1994, Rolando Bissot Reynolds, my husband, who is a medical doctor, was invited to speak at a surgeons' conference in Guatemala City. We went with all expenses paid to a plush hotel—and our second honeymoon! I spoke to Pipo (Fritz) and asked him about the family landmarks he remembered: Where their home was, where he went to school and where was the resting place of our Jewish ancestors. We were successful in finding the neighborhood where they lived, but I'm not sure if we found the right house. We couldn't find the Jewish school but, on Saturday, we set off to find the resting place for Rosa Bocka [sic], my great-grandmother, and my grandfather, Alfred Stargardter. We were driven to the Cementerio Central and searched for the Jewish Cemetery. Of course, it was closed on the Sabbath.
>
> Rolando said, *"A lo que vinimos"* (On with what we came for). Flowers in hand, we climbed over the two-meter gate to get to the graves. After a bit of searching, we found Rosa's grave. I swept the pile of rocks from the tomb to put the beautiful flowers on the slab. We did the same at Alfred's grave. After that ritual, there was a problem of getting out of the cemetery.

We realized the cemetery was actually below ground level. Instead of two meters to climb up, it was now about three meters above the surface to get out. We screamed for help but got no answer. I asked Rolando to airlift me, and he did. I pulled him up from above, *muertos de la risa* (dying from laughter).

Well, it was quite an adventure. We told Pipo about it on our return to Panama. He explained that all Jewish sacred places are closed for the Sabbath and that we were trespassing! My father also explained that it is a tradition for Jewish people to leave pebbles on graves instead of flowers because stones are not perishable. We had committed a double no-no! Later, I understood this tradition when I watched *Schindler's List.*

My father, a brave and quiet man, made sure his kids fit in wherever they went and respected all creeds and races. But he did not share his Jewishness with us. It was the only thing he kept to himself.

Shalom don Fritz.

Friedrich "Fritz" Stargardter Bücka
1 Nov. 1931—29 Junio 2008

✡

Rafaela De Andreis de Stargardter
3 Abril 1936—26 Octubre 2015

†

Alfred Stargardter and Rosa Bücka were buried
in the Guatemala City Jewish Cemetery.
Alfred in 1952 and Rosa in 1956.
Tomas Stargardter collection.

Florida Southern Student Invents Weird Contraption

Waterski contraption.
The Southern, p. 6, May 16, 1952.

Fritz wearing a Guatamalan shirt, circa 1948.
Tomas Stargardter collection.

Esther Williams (left) rehearsing at Cypress Gardens for the movie Easy to Love, 1953. Fritz is 5th to the right of Esther Williams. Photograph purchased from Alamy.com by Clyde Stephens.

Wolfgang Schultz and Karl Friedrich Koch (2nd and 3rd from left) and Fritz Stargardter (end of table were Germans. Berend Faber (1st on right) was Dutch and spoke German, but not at the table or in public. Santa Marta dining hall, 1956. Wolfgang Schultz collection.

Some of the places where Fritz lived and worked.
Source: Windows map. Used with permission from Microsoft.

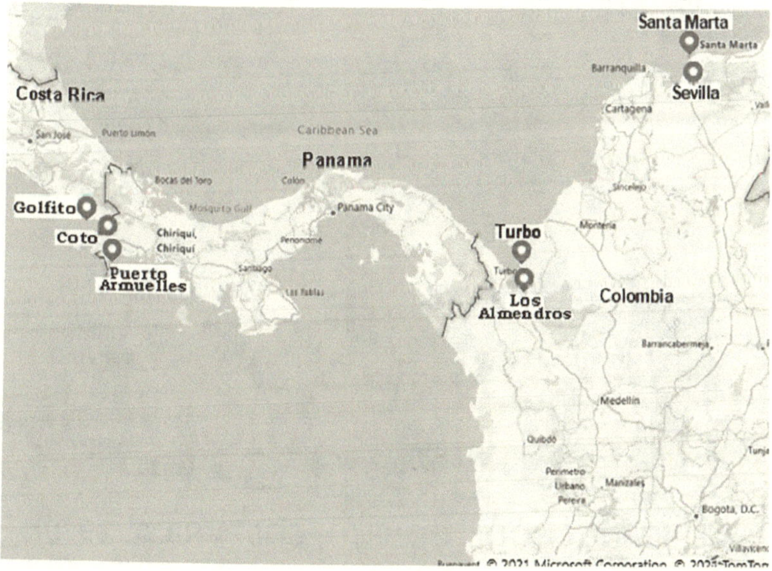

Fritz and Rafaela

Mount Adams School, circa 1979. My wife Phyllis (A), our daughter Linda (B) and our son Carl (C). Two of Fritz Stargardter's children: Sandra (D) and Tommy (E). Clyde Stephens collection.

Puerto Armuelles, Panama. The region was a virgin rainforest up to the beach before 1927. Cameron Forsyth collection.

Recent view of Golfito, Costa Rica. The Stargardter house is hidden from view at the base of the mountain. The Stephens house (white) is on the right, rear. Former United Fruit Company golf course on both sides of the airport grew back to jungle after closure in 1984. Jaime Villalobos collection.

Clyde Stephens, Cameron Forsyth and Fritz Stargardter. The 10-kilogram "Banavac" took a month to reach Japan via California.

"Banana College" Alfredo Stargardter (2nd row, 3rd from left), Clyde Stephens (2nd row, bald). Clyde Stephens collection.

Clyde and Phyllis Stephens farewell party from Puerto Armuelles to Honduras, 1987. Fritz Stargardter (left) and Scott Dinsmore (right). Scott Dinsmore collection.

Notes

Book 1

Notes: 1. Bananas and the Twin Towers

1. Construction of the World Trade Center's North Tower began in 1968 and the South Tower in 1969. The original World Trade Center was a large complex of 6 buildings mostly finished by 1973. A seventh building was completed in 1987. Sadly, 60 workers were killed in construction accidents while the World Trade Center was being built.

Notes: 2. Three Trips to Cuba

1. When Jim and I decided to go to Cuba, we did not tell our parents, as one would expect then and now. We just did it. I suppose that there was a certain trust with my parents because they were used to me not returning home at night. This was because I often remained in town in order to participate in school activities, or I hitchhiked back and forth to our farm. The end result was that my parents did not make a case of the trip to Cuba. In fact, I think they might have considered it an educational opportunity, but we will never know. They did not call the police or a tracking squad to find us. Life went on and all turned out fine.

2. A living fence post is a limb cut from a live tree and put into a hole in the ground. The limb sprouts roots and leaves and new shoots and becomes a healthy post for attaching wire fencing. The advantage is that the posts do not rot, and termites do not feed on them.

Notes: 3. Clyde the Guide

1. Readers can find numerous accounts on the internet and in books about this shocking and dramatic suicide. He smashed the window of his office and jumped from the 44th floor of the Pan American Building in New York City. The best insider account on Eli Black was written by Thomas McCann in his book, *An American Company: The Tragedy of United Fruit*. New York: Crown Publishers, 1976.

2. Eduardo Palavicini, Alcides Mora, Daniel Perez and Eduardo Vega.

Notes: 4. Saving Locomotive No. 84

1. William "Bill" Kenneth Saravanja was born in 1947 in Golfito, Costa Rica, and grew up in Costa Rica and Honduras. His Yugoslav father, Johnny, was a builder for the UFCo. Bill's mother, Alva Bailey, was born in Guatemala of British parents. After graduating from Mississippi Southern in Architecture in 1973, Bill worked in the Company engineering department in Costa Rica and Panama. He lives in Lady Lake, Florida, with his wife, Cathy Gabourel, originally from Utila Island, Honduras.

Notes: 5. Jose Antonio Velasquez: Primitive Artist

Jose Antonio Velasquez died February 14, 1983, and María Raquel Maradiaga Velasquez followed him July 12, 1988. Both are buried in El Cementerio de Suyapa in Tegucigalpa, Honduras.

1. Rosengarten, Frederic Jr. Wilson Popenoe: *Agricultural Explorer, Educator and Friend of Latin America*. National Tropical Botanical Garden. Lawai, Kauai, Hawaii, 1991.

2. Yuscarán, Guillermo. *Gringos in Honduras*. Nuevo Sol Publicaciones, Tegucigalpa, 1995.

3. Malo, Simón E. *El Zamorano: Meeting the Challenge of Tropical America*. Simbad Books, Kansas,1999.

4. *Unifruitco* magazine of United Fruit Co. U.F. Report, No. 1, 1954, cover and pg. 22-25.

5. Yuscarán, Guillermo. Velasquez: *The Man and His Art*, Nuevo Sol Publications, Tegucigalpa, 1994.

6. The author (CSS) found contradictory descriptions of Valentín Velasquez, Jose Antonio Velasquez's father. References to Valentín Velasquez in the biography by Yuscarán always portrayed him as a good father and husband who was active in the village of Caridad. For a New Years event in 1917, he was selected as the actor for the leading role of a play depicting the life of Francisco Morazán, a revered patriot in Central America. When Jose Antonio asked the director of the play why his father had been chosen, the answer was, "Valentín is a man of character." (The play was *La Muerte de Francisco Morazán*).

Valentín Velasquez was musical and played the guitar, accordion, violin and trumpet and was the leader of an eight-piece band that played for private and public events in Caridad. He encouraged his children to play music. There was always music in their home. The father was also the accountant and secretary for the village. And just like generations before him, he and his sons worked the land as farmers. When he was murdered by political party rivals, Jose Antonio was traumatized forever and never wanted to talk about losing his father because of the grief.

During a visit, Wilson Popenoe asked his friend Jose Antonio about his father. He declined to answer and said that he did not want to talk about his father. That response misled Wilson Popenoe to believe that Antonio's father was a rogue of immoral character and a womanizer who neglected the family. These unsavory and disreputable comments were later quoted by Simón Mala et al. In my opinion, the truth was just the opposite. It was an unfortunate misunderstanding.

Valentín Velasquez was best described in the biography by Guillermo Yuscarán. Tulio Enriquez, Antonio's youngest son and constant companion stated, "I saw my father suffer. I saw the tears in his in eyes when he spoke of his father."

7. López R, J. Evaristo and Longino Becerra. *Honduras: 40 Pintores*. Litografía López, Tegucigalpa, 1989.

8. Carr, Archie. *High Jungles and Low*. University Press of Florida, Gainesville, 1953.

9. The artist frequently painted a couple of black, skinny dogs and a priest in his street scenes. In our painting, he added a self portrait of himself playing the violin, accompanied by a guitar player and two women dancing. He said this was a celebration of a festive holiday. The bridge is the main entrance to the village.

Notes: 6. The Yard

1. Arthur Gordon Kidd was born on July 18, 1920, in Austin, Texas. He graduated from the University of Texas at Austin and was employed by the United Fruit Company in New York City in 1946. He became director of publications. After retirement, he was an avid senior swimmer and medaled in senior games in Houston in 2011. He died at age 100.

Notes: 9. Banana People under a Dictatorship

1. https://en.wikipedia.org/wiki/Rafael_Trujillo

2.https://www.imdb.com/title/tt0066808/faq?ref_=tt_faq_1#fq0008294

3.https://americandiplomacy.web.unc.edu/2009/06/tropical-ziongeneral-trujillo-fdr-and-the-jews-of-sosua/

4. https://gradschool.psu.edu/sites/GradSchool/McNairJournals/ mcnair_jrnl2007/files/cordis.pdf.

5.https://www.ibtimes.com/parsley-massacre-genocide-still-haunts-haitidominican-relations-846773

6.https://content.time.com/time/subscriber/article/0,33009,869601,00.html

7. www.seacloud.com/en/sea-cloud-history.

8. https://en.wikipedia.org/wiki/Fair_of_Peace_and_Fraternity_ of_the_Free_World.

9. *Trujillo, mi padre: en mis memorias, Angelita.* María de los Angeles Trujillo de Domínguez. Ediciones Unicaribe.

10. De Lara Viñas, Fernando. *Reminiscencias de un bananero.* Santo Domingo: Amigo del Hogar, 1995.

11. Ibid.

12. Stephens, Clyde S. *Bananeros in Central America.* Fort Myers: Press Print Co., 1989. Pg. 31.

13.*Unifruitco*, magazine of United Fruit Co., March, 1951, pg. 9.

14.https://www.facebook.com/historiadominicanaengraficas/posts/ 1133196223545526/.

15. Turnbull, Mildred Maxwell Hamer. *Turnbull, Mildred Maxwell Hamer, Her Life, Times, and People (1912-2019).* Written by Walter Hamer. Sarasota: 2022.

16.https://www.britannica.com/biography/Henry-Christophe#ref941016.

17.https://dominicantoday.com/dr/local/2021/05/30/what-is-knownabout-the-plan-that-oended-trujillo/.

18. www.bbc.com/news/world-latin-america-13560512.

Notes: 10. A Crooked Path to Citizenship

1. Walter Hamer grew up in the United Fruit Company in Latin America, as did his parents. They were pioneers in new banana plantations that were usually in the most remote parts of the host country. He married Sue Trafton who was also born in a Company hospital in Puerto Armuelles that had just been built in southwestern Panama. Her parents worked in the same banana company. The couple lives in Sarasota, Florida, and stays connected with other banana people.

Notes: 11. When Truth Ends and Fiction Begins

1. *My Macondo*, Dal Weldon, dir., (Channel Four, London, 1990), in British Film Institute, London.

2. Posada Carbó, Eduardo (May 1998). "Fiction as History: The bananeras and Gabriel García Márquez's *One Hundred Years of Solitude*". *Journal of Latin American Studies*. 30 (2): 395–414.

3. Bourgois, Philippe I. *Ethnicity at Work: Divided Labor on a Central American Banana Plantation*. Baltimore: Johns Hopkins University Press. 1989.

4. Stacy May and Galo Plaza, *The United Fruit Company in Latin America*. Washington D. C. National Planning Association.1958.

Notes: 12. From Bananas to Cacao, Abaca, Rubber and Back to Bananas

1. Much of this information came from interviews in 1987 with *don* Teófilo de Jesús Alvarado Segovia who retired to Boquete, Panama. His career began with the United Fruit Company in 1928 in the new banana division of Puerto Armuelles, Panama. In July 1943, he and the family were transferred to Changuinola across the Continental Divide of Panama to Farm 8 where he became the superintendent of abaca.

2. "Cacao." Merriam-Webster.com Dictionary,. https:// www.merriamwebster.com /dictionary/cacao. Accessed 6 Sep. 2002.

3. These letters were written by both spouses who wrote about hardships, living on the island, diseases, their banana farm, racism and prejudice, labor, employees, weather, debts, ships and shipping, gossip, bad banana prices, the banana boom in Bocas del Toro and travel.

4. Series II. Paul Osterhout, 1881-1939 | ArchivesSpace Public ... archives.library.rice.edu/ .../2/archival_objects/210887. http://www.lib. utexas.edu/taro/ricewrc/00014/rice-00014.html#a0. Angelique Stephens made numerous trips to the archives at Woodson Research Center, Fondren Library, at Rice University.

5. Read more about Aaron King on page 190 of *Banana People*, Dollar Bill Books, 2002, by Clyde S. Stephens.

6. Names of two indigenous tribes in Bocas del Toro changed before the 21st century. Guaymi are now called Ngäbe and the Buglé remain Buglé. The Teribes are now Nasos. Indigenous people from the San Blas archipelago and their shore were called San Blas, then Kunas and now Gunas. The islands are referred to as Guna Yala.

Notes: 13. Abaca in Honduras

1. Stephens, Clyde S. 1989. *Bananeros in Central America—True Stories of the Tropics*, pp. 33-80. Press Printing Co., Ft. Myers, FL.

2. *Unifruitco*, pp. 8 - 10, March-April, 1948.

3. Rosengarten, Frederick. 1991. *Wilson Popenoe, Agricultural Explorer and Friend of Latin America*, p. 103. National Tropical Botanical Garden, Hawaii.

4. Malo, Simón E. 1999. *El Zamorano, Meeting the Challenge of Tropical America*, p. 155. Simbad Books, Manhattan, Kansas.

5. *Unifruitco* magazine, pp. 12 - 13, April 1949, feature Wimmer's panoramic, double page aerial view of the new banana port of Puerto Libertador on the north coast of the Dominican Republic. The caption identified him as the Fruit Dispatch photographer and company traveler, probably based at Pier 3, North River, New York City. He was the son of Hans Wimmer Daner, a photographer who lived many years in Costa Rica where he died in 1949. Two other sons also worked for the United Fruit Company.

6. Wilson, Charles Morrow. 1968. *Empire in Green and Gold*, the Story of the Banana Trade. Pages 290-92. Greenwood Press, New York.

7. Panama and Costa Rica also used Philippine names for their abaca farms. Guatemala's abaca farm district was known as Los Andes and had several abaca farms. Today, banana farms on former abaca plantations in Guatemala are known as Chinook on the map.

Notes: 14. An Abaca Family in Guatemala

1. Millicent is a 3rd generation Crichton in the tropics, although she was born in New York City in 1944. She attended United Fruit Company schools in Los Andes and Bananera, Guatemala, then Quepos, Costa Rica, and Sevilla, Colombia. High school was at Northfield School for Girls in Northfield, Massachusetts. She graduated from Middlebury College, Vermont, with a degree in mathematics. Millicent has lived in Ann Arbor, Michigan, for 52 years in the same house and enjoys a cottage in Maine with her husband Doug Turner. Her sister Louise and husband Gerry Lopez have a cottage nearby.

2. As told to the author of this book (CSS).

3. Stephens, Clyde S. 1989. *Bananeros in Central America—True Stories of the Tropics*, pp. 85 and 197. Press Printing Co., Ft. Myers, FL. All photographs are from the Millicent Crichton Empedocles collection.

Notes: 15. The End of Abaca in Guatemala

1. Mario Mena Stricker was born in Retalhuleu, Guatemala, in 1941. His father worked for 40 years in irrigation for United Fruit Company in Tiquisate and Bananera. Mario attended the American School in Tiquisate and then studied engineering at Texas A&M. Afterwards, he worked for United Fruit Company in Changuinola and Puerto Armuelles, Panama.

He married an American school teacher, Cathy Marqua, who taught in banana company schools. Four of their children were born in the Tropics, and his son was born in Texas.

Mario advanced to management, serving 55 years in the banana business as follows: 34 years with United Fruit Company/Chiquita, 13 years with Del Monte and eight years with the Nicaragua Banana Corporation.

He was a banana producer and consultant. These positions took him to Panama, Guatemala, Philippines, Mozambique, Angola,

Ethiopia, Ivory Coast, Cameroon, Cuba, Chile, Mexico and Nicaragua. He lives in Guatemala City with his wife. Mario pilots his own airplane and two of his sons are also pilots.

2. As revealed to the author of this book (CSS).

Notes: 16. Life on an Abaca Farm in Costa Rica

1. Harold L. Brent was born in the United Fruit Company port of Tela, Honduras, in 1939. His mother was from Guatemala and his father was Robert Oliver Brent, who was the superintendent of construction for the Company. The family lived in Progreso, Departamento de Yoro, until 1942, when his father was transferred to the Caribbean side of Costa Rica to build an abaca plant in Bataan. Harold went to Holy Cross High School in New Orleans, graduated from Texas A&M with a degree in Civil Engineering, then became an officer and paratrooper in the U.S. Army. He served in Viet Nam followed by a career in civil engineering until his retirement in Houston.

In 2014, two sons of Robert Oliver Brent, Sammy and Bobby, published some interesting background data about their father. In his book, Activated Angels (Got to Go): Notes from a Working Man, 2014, Sam Brent wrote that his father left his grandfather's ranch near Omak, Washington, in the late 1920s when he was about 17. After joining the US Marines, he was sent to Nicaragua to fight Sandinista rebels. He had a different name then.

On November 26, 2014, Bobby Brent sent details about his father in an e-mail to numerous friends. A website described battles in Nicaragua and the names of US Marines in those conflicts. A Patrol Report in 1928 listed Clyde H. O'Bryant, who later went AWOL, found his way to Honduras and changed his name to Robert Oliver Brent. He worked for Tela Railroad Company of the United Fruit Company until the 1960s. See: http://www.sandinorebellion.com/PCDocs/1928a/PC280409-Putnam.html.

2. Billy Saravanja was born in 1947 in Golfito, Costa Rica, and grew up there and in Honduras. His father, Johnny, a Croatian Yugoslavian, was a builder for the United Fruit Company. Bill's mother, Alva Bailey, was born in Guatemala of British parents. Her father also worked for the United Fruit Company. After graduating from Mississippi Southern in architecture, Bill worked for the Company in Costa Rica and Panama He lives in Lady Lake, Florida, with his wife, Cathy Gabourel, originally from Utila Island, Honduras.

Notes: 17. The United Fruit Company in Cuba

1. *Blog de genealogía e historia de Holguín y Banes* - por WILLIAM NAVARRETE (*desde* París, Francia).

2. *Unifruitco*, Golden Jubilee 1899-1949, June-July, 1949, pg. 34. Special Edition.

3. Published in *Bananeros in Central America*, by Clyde S. Stephens, 1989, p. 95.

4. In 1976, the province of Oriente was redrawn and the name Oriente was deleted. Banes and Preston are now in the province of Holquín.

5. Fidel Castro changed the name from Preston to Guatemala. It is now a display of relics of the past.

6. Oscar Zanetti and Alejandro García. *The United Fruit Company in Cuba*. 2020. Duke University Press, Durham. Page 268. https:// doi.org/10.1515/9781478004561-058

7. May 22, 2006, edition of *Forbes*, "Kings, Queens and Dictators." Forbes alleges Castro has a personal fortune of $900 million. https:// www.cbsnews.com/news/castro-i-am-not-rich/.

8. https://en.wikipedia.org/wiki/Fulgencio_Batista.

Book 2

Notes: 1. Sea Level Canal Study

1. Unfortunately, our house burned down in February 1966, because of a faulty old kerosene stove.

2. The Chocó are now known as Emberá-Wounaan.

3. After several days at Boca Lara, a piragua took me down the river and across the bay to La Palma. Pacific tides that went up and down three and four meters left mud flats along the slummy shore of the town. This was a contrast for me, because I was accustomed to half meter tides at Bocas del Toro on the Caribbean side. Lodging overnight was in a minimal *pensión*. The next morning, I flew to Panama City in a small airplane, then continued on to the Changuinola banana plantations and landed at Base Line near our home.

4. Eventually, the existing canal was expanded with a third set of locks. The inauguration occurred on June 26, 2016.

Notes: 2. Science versus Tradition

1. Corporate executives thought staying with the Gros Michel variety would help maintain the Company's monopoly over the trade. United Fruit Company could afford the losses from Panama disease; its competitors could not. If disease-resistant plants were found, competitors might successfully edge their way into the business. See Van Diepen, William T., *Saving the Company*, in Stephens, Clyde S., *Banana People*, Dollar Bill Books, 2002.

2. Malo, Simón E. *El Zamorano, Meeting the Challenge of Tropical America*. 1999. Pages 154-155. Sinbad Books.

3. Ibid.

Notes: 3. Canals in Bocas del Toro

1. Araya, Giselle Marín (2006). "Proyectos para construir una ruta transístmica en la zona limítrofe de Costa Rica y Panamá en el siglo XIX." *Revista de Historia de América*. 131-165. 10.2307/20140166. Download at: (PDF) PROYECTOSCANALEROS | Cecilia Marin Araya Academia.eduwww.academia.edu/34847202/PROYECTOSCANALEROS.

Notes: 5. Hospital Point: A Lost Story of Medical History

1. In the 1880s, when the French attempted to build a sea level Panama Canal, France provided excellent doctors and good hospitals. Nevertheless, over 20,000 victims died because it was not yet known that mosquitos were the carriers of yellow fever and malaria. At that time, Panama was not considered part of Central America.

2.https://www.thefreelibrary.com/Taming+mankind%27s+ancient+foes%3a+despite+inherent+dangers+and.a0135888144.

3. It probably helped that United Fruit Company had banana and sugarcane operations in the northeast part of Cuba at Banes and Preston and quickly learned first hand about the overwhelming success of controlling yellow fever in Havana. This knowledge spread to the corporate level in Boston that was then successfully applied in Bocas del Toro. This is a personal belief of the author.

4. It is ironic that the incredible successes in Cuba and Bocas del Toro did not convince Major General George W. Davis, the first governor of the Canal Zone, and other members of the seven-man Isthmian Canal Commission. They did not take seriously Gorgas' "theory" that mosquitoes transmitted yellow fever and malaria. Rear Admiral Commissioner John Walker said the notion of mosquito transmission was "balderdash." President Theodor Roosevelt solved the problem by disorganizing the Commission and naming Dr. William Crawford Gorgas to head yellow fever and malaria control in the Panama Canal Zone.

5. Dr. Russell H. Wilson was the son of John "Jamaica" Wilson (1832 to 1895) who was an entrepreneur in the tropics in the 1800s until his death in 1895. He had banana plantations, banana ships and hotels in Bocas del Toro and Bluefields, Nicaragua. His son became a medical doctor in Ohio and worked at Hospital Point. Dr. Wilson married a local woman and they remained in the tropics. The head matron, Cora Lawrence, was the doctor's aunt.

6. Reid, Carlos. *Memorias de un Criollo Bocatoreño*. Edited by Dr. Stanley Heckadon-Moreno. Panamá: Asociación Panameña de Antropología, 1980.

7. In 1994, Dr. Elizabeth Swain of Harvard University obtained nearly 300 pages of United Fruit Company annual reports of the Medical Department and mailed them to me.

8. Scoullar, William T. *The Blue Book of Panama*. The Latin America Publicity Bureau, Inc., 1916-17. This publication presents many notable employees of the United Fruit Company in Bocas del Toro, plus private business people in the province.

9. Details in this section came from interviews at Hospital Point with Enos Flemming in 1970-75 and Bruce Brown in 1991-95. The author was taken to the colored cemetery site by three men who either witnessed burials or helped with the burials.

Bruce Brown was with his half-brother, Martin Sparks, during one of my interviews. Martin was born in 1920 in the Nancy's Cay village where they lived after the hospital closed.

Brown began the first interview by saying that he never knew his father, who was an Englishman and a steward on a ship that loaded bananas in Bocas del Toro. This was the version written in the baptismal record—because it looked good in the eyes of the church and for social reasons.

A year before Bruce Brown died in the year 2000, he corrected the record by telling me that his father was a German, Egers Dorffer, from Hamburg. Dorffer worked in the store of Herbert Leer, another German, who was also an executive of the United Fruit Company in Bocas del Toro. Leer and his wife took Rebecca Swartz on a German ship that sailed from Bocas del Toro to Hamburg to help as a nanny to care for the two Leer children. Dorffer accompanied the Leers on that trip. Thus, Bruce should have been a Dorffer, not a Brown. Ironically, his mother had a German name, Swartz, but she was a local Afro-Antillean with German Jewish origins from Corn Island, Nicaragua.

Notes: 6. With C. H. Ellis through Central America and Panama, 1906

1. Ed Loyd served during 2007-2015 as Director of Corporate Communications and Investor Relations at Chiquita's headquarters in Cincinnati, Ohio; and later Charlotte, North Carolina, until Chiquita's more than a century of history as a public company ended in a hostile takeover. Previously, he had been press secretary of the U.S. Department of Agriculture and today is the head of public relations for Fifth Third Bank. He continues to collect and write about Chiquita's history, especially on Dr. Wilson Popenoe's work at the Lancetilla Botanical Garden, Zamorano University and Casa Popenoe in Antigua, Guatemala, where Ed and his wife, Melissa Neffa, were married in 2017.

Notes: 8. Natural Entertainment on Hospital Point

1. Dr. David Roubik of the Smithsonian Tropical Research Institute in Panama, dedicated years unraveling this biological puzzle. He was generous and patient in helping me put this story together in layman terms. Dr. Roubik and his colleagues continue to study tropical bees. Their scientific publications are numerous and can be found on the internet.

Notes: 10. Black Physicians in Early 20th Century Bocas del Toro

1. Ariel René Pérez Price was born in Panama City in 1975. He studied Law and Political Science at the University of Panama and obtained a Master of Laws from the University of Utrecht, Netherlands. He was an Erasmus Mundus fellow from the European Commission which led him to pursue postgraduate studies in Economic Analysis of Law at the University of Vienna and the University of Hamburg. He is an amateur historian of the Caribbean and Bocas del Toro, a cultural activist and author of two historical novels:

Memorias de Bocas Town: la vida y tiempo de Jose Antonio Price (2017) and *Pionero: la historia de Luis Russell* (2022). Ariel is a grandson of Dr. Jose Antonio Price. He started from zero to find his roots in Bocas del Toro, resulting in his historical novels.

2. Philippe I. Burgois, *Ethnicity at Work. Divided Labor on a Central America Banana Plantation* (Baltimore and London: The Johns Hopkins University Press, 1989).

3. Ibid., p. 85.

4. The Criollos were black and mulattos, mainly the descendants of the British settlers and/or Americans with strong links with the British Caribbean and their slaves, that arrived at Bocas del Toro during the beginning of the 19th century. They were mostly members of the Methodist Church and amongst them the following original families were identified: Brown, Shepperd, Knapp, Humpries and Peterson. Other members of this social group were freedmen and slaves on the run that arrived during the same time period at the archipelago from Providencia and San Andres, seeking a living from the hawksbill sea turtle fishing industry. Some of these family names were Ellis, Warren, Howard, Machore, Matheus, Reid and others. Donaldo Velasco, *Recuerdos historicos de la Bahia de Almirante* (Bogota: Imprenta de vapor de Salhma Hermanos, 1898), P.11; *A Creole from Bocas del Toro: The Story of Carlos Reid*, ed. Stanley Heckadon (Panama: Editorial Exedra, 2011), P. 19, 37; "Sinopsis historica de Bocas del Toro" in Revista Lotería, 2da. Época, vol. V, No. 56 by Armando Aizpurúa, 1960, P. 90.

5. Philippe I. Burgois, *Ethnicity at Work. Divided Labor on a Central America Banana Plantation* (Baltimore and London: The Johns Hopkins University Press, 1989), P. 45-65.

6. Celestino Araúz Monfante, *Bocas del Toro y el Caribe Occidental: periferia y marginalidad* (Panamá: Editorial Mariano Arosemena, 2007), p. 179.

7. "Censo demografico de la provincia de Bocas del Toro de

1920" (Panama, Imprenta Nacional, 1924), p. 1

8. *The Panama Canal Records*, Volume XI, August 22, 1917 August 14, 1918 (Canal Zone: The Panama Canal Press, 1919), p. 36.

9. Clyde Stephens, *History of Hospital Point, Pioneer Medical Center, 1899-1920* (Florida: Leesburg Printing Co., 1997), p. 32.

10. Ibid, p. 33; The Gleaner (Jamaica), May 7, 1908, p. 4.

11. Clyde Stephens, *History of Hospital Point, Pioneer Medical Center, 1899-1920* (Florida: Leesburg Printing Co., 1997), p. 40.

12. Ibid., p. 40, 41.

13. Ibid., p. 35, 37.

4. Ibid., p. 41.

15. *Gaceta Oficial 3150* (Panama), August 2, 1919.

16. Clyde Stephens, *History of Hospital Point, Pioneer Medical Center, 1899-1920* (Florida: Leesburg Printing Co., 1997), p. 44.

17. *A Creole from Bocas del Toro: The Story of Carlos Reid*, ed. Stanley Heckadon (Panama: Editorial Exedra, 2011), p. 19, 37.

18. "Archivo Iglesia Nuestra Señora del Carmen, Defunciones 1912" (Bocas del Toro), p. 89.

19. Alfredo Vaz reported that his Jamaican grandfather was a medical doctor and had a Jewish heritage.

20. *The Gleaner*. (Jamaica), June 12, 1917, p. 13.

21. *The Star & Herald* (Panama), April 20, 1924, p. 2.

22. *The Gleaner* (Jamaica), May 5, 1924, p. 3.

23. *The Gleaner* (Jamaica), July 19, 1915, p. 4.

24. "Catalogue of The Officers and Students of Howard University 1898-1899" (Washington, D.C. : Howard University, 1899), p. 62.

25. Interview conducted by the author with Jose Thomas, grandson of Dr. Joseph E. Thomas, in Bocas del Toro on January 7, 2014.

26. *The Gleaner* (Jamaica), July 19, 1915, p. 4.

27. *The Panama Tribune* (Panama), March 24, 1929, p. 15.

28. *The Gleaner* (Jamaica), June 12, 1917, p. 13.

29. "Tuskegee Historical Timeline" (Tuskegee University website) https://www.tuskegee.edu/about-us/centers-of-excellence/bioethics-center/historical-timeline Accessed August 11, 2022.

30. *The Gleaner* (Jamaica), June 12, 1917, p. 13.

31. Reavis L. Mitchell, Meharry Medical College, (Tennessee Encyclopedia, online) https://tennesseeencyclopedia.net/entries/me-harry-medical-college/ Accessed September, 15, 2022.

32. *The Central American Express* (Bocas del Toro), July 21, 1917, p. 4.

33. "Osterhout Family papers, 1836-1941", (MS 355, Woodson Research Center, Fondren Library, Rice University) http://archives.library.rice.edu/repositories/2/resources/980 Accessed August 11, 2022.

34. William T. Scoullar, *El libro azul de Panamá/ The Blue Book of Panama*, (Panama: Imprenta Nacional, 1916), p. 308.

35. "Archivo Iglesia Nuestra Señora del Carmen, Bautismos 1896" (Bocas del Toro), p. 218.

36. *The Gleaner* (Jamaica), April, 15, 1915, p. 2; Ariel Pérez Price, *Memorias de Bocas Town, La Vida y Tiempo de Jose Antonio Price*, (Panama: Miss Hilda Press, 2019), p.46.

37. *Annual Catalogue of Officers and Students, 1907-1908* (North Carolina: Shaw University), p. 50, 58, 60.

38. Todd L. Savitt, *Race & Medicine in Nineteeth and Early Twentieth Century America*, (Kent: The Kent State University Press, 2007), p. 346.

39. Ibid., p. 212-221.

40. *The Star & Herald* (Panama), April 16, 1920, p. 5.

41. *El Progreso* (Bocas del Toro) July 26, 1914, p. ; William T. Scoullar, *El libro azul de Panamá/ The Blue Book of Panama*, (Panama: Imprenta Nacional, 1916), p. 308.

42. *La Critica* (Panamá), December 2, 2003.

43. *The Star & Herald* (Panamá)16 January,1921.

44. *The Star & Herald* (Panama) March 4, 1920, p. 5

45. *Gaceta Oficial 4378* (Panama), April 4, 1924, p. 14309.

46. *Gaceta Oficial 4415* (Panama), May 30, 1924, p.

47. The task of accurately determining whether Dr. Price is the first Afro-Panamanian physician to acquire such qualifications is complex. The reason being the distortion of the ethnic ancestry of a large number of Panama's historic figures, making Afro-Panamanians invisible in the context of history. Our research, for example, identified some candidates: Jose Domingo Espinar (1791-1895), a graduate of the Universidad de San Marcos de Lima around 1840; Mateo Iturralde (1791-1865), graduated from the University of Quito, and Ciro Urriola (1863-1922) with studies at the University of Bogotá, where he graduated in 1888, continuing his education in Europe before returning to the country and becoming President of the Republic in 1918. Scattered sources classify these individuals as Afro-descendants under the denomination of "mulattos" or "criollos" and in other cases the educational centers from which they graduated have been identified as universities for "castas" or "mulattoes." Much research is needed to determine how many others escape this analysis. The truth is that the documentary evidence shown here aims to corroborate, for the time being, that Jose Antonio Price was the first Afro-Panamanian to obtain his degree as a doctor and surgeon within the republican era that began in 1903. "Dr. Jose Antonio Price: pionero de la medicina republicana en Bocas del Toro", Ariel Pérez Price, (Panama: Prisma Tecnológico, 13(1)), p.102. https://doi.org/10.33412/pri.v13.1.3471

48. Document delivered to this author by Julio Cotes Surgeon on August 9, 2022.

Notes: 11. A Boy from Bocas

1. Ibu Alvarado was born on April 21, 1945, and raised in the

United Fruit Company, Bocas Division. He attended the American School of the banana company. At age thirteen, he left for Holy Cross High School, a Catholic boarding school in New Orleans. His college degree was in agricultural engineering from Louisiana Tech University. This was followed by postgraduate studies in hydrology and hydraulic resources in the United Kingdom, a year in Haverfordwest, Wales, where a company was building a dam, and a year in Manchester, England, where a tunnel was being built under a river. Ibu returned to Changuinola in 1971 to manage the family's banana farm. After one year, he left Bocas del Toro and was employed by the Panama Canal Company as a field hydrologist, remaining there until retirement as environmental division manager in 2005. This was followed by a career in flying until 2009. He now flies a 2-place, 1951 Cessna 140A. Ibu lives in Boquete, where he retired with his wife Pat, whom he met when she came to Almirante to teach...you guessed it, at the Chiriqui Land Company school in 1969. They were married in South Wales in 1970. Ibu wrote a book about his adventures in flying: *9,000 Hours and Counting: A Pilot's Log*.

2. Told by *don* Teófilo de Jesus Alvarado Segovia to his son Luis Alonso Alvarado Kinkead who compiled and edited the memoir, *Conversando con Papá: AnécdoTAS*, 2007. Cecropia Press.

3. People remember his airline that flew from Panama City to David to Puerto Armuelles. On page 34 of the 1948 March-April issue of *Unifruitco*, Gelabert is identified as "ex-United Fruit Company pilot" who visited Almirante and Bocas. He was the aviation inspector for the government and received the Bocas airfield for the government. There was a note that he had been the pilot for Reginald Hamer, the general manager of Costa Rican operations. (Note: Reginald Hamer died in a LACSA airline crash in Costa Rica in 1946.) Later, Gelabert died in a plane crash in Veraguas, Panama. The former Albrook Air Force Base in Panama City is now called Marcos A. Gelabert International Airport.

Notes: 12. From Italy to the Western Hemisphere

1. Dr. Paul Cyprian Messina was born June 28, 1943, in San Cipriano Picentino, Italy, during the heavy bombardment during World War Two by the Americans and British in preparation for their landing at Salermo. His parents, Dr. Nicola Ambrogio Messina and Margherita Padovani Messina, migrated to Guatemala in 1950. In 1952, the family moved to Almirante, Panama, where Paul's father was a medical doctor in the United Fruit Company hospital. Paul and his two sisters enjoyed their early years in this banana port on the Caribbean coast of Panama. Paul enjoys opera, something he first heard in Italy as a child. He organizes and participates in international symposiums in the United States and Europe and visits Italy several times a year. He is fluent in Italian, Spanish and English.

2. *Cesare Mormile Addresses the People Rebelling against the Decrees of the Inquisition.* (https://en.wikipedia.org/wiki/Vincenzo_Marinelli).

3. Sadot Venegas, based in San Jose, Costa Rica, conferred with Dr. Messina and the medical department regarding mosquito control in all areas of the plantations and in houses. They checked on water and sanitation issues, and were especially vigilant of malaria and yellow fever, diseases that had been controlled since the early 1900s.

Notes: 13. Sharks

1. John George Munch was born on September 24, 1944, in Limon, Costa Rica. His parents were Suzanne Repta Munch and George Del Munch who was the General Manager of the United Fruit Company in Bocas del Toro, Panama. John lived in Almirante from 1951 to 1959 and finished the first eight grades in the Almirante American School. By 1977, he received a Bachelor of Arts degree after studying at several universities in California. In recent years, John owned Le Cuvier vineyards and winery in California. He resides in Templeton, California.

2. Erin Dillon wrote this when she was a graduate student at UC Santa Barbara. She now holds a doctorate and continues as a fellow at the Smithsonian Tropical Research Institute.

3. As the author of this book, I found some notes that relate to catching, marketing and eating sharks in Bocas del Toro and Chiriqui on the Pacific side of Panama.

Notes: 14. Peace Corps Volunteers

1. Joseph Looby was born in 1942 in Darby, Pennsylvania. He is a graduate of the Wharton School where he earned a graduate degree in Education. He served as a Peace Corps Volunteer in Panama in 1964-1966 and later became a college recruiter and fundraiser for the Corps. He was President and Board Chairman of the Chamber of Commerce of West Philadelphia and a Director of the Greater Philadelphia Chamber of Commerce. He served in several administrative positions at the University of Pennsylvania and later co-founded a microcomputer company that specialized in developing and promoting workstations for the blind and visually-impaired. Joe loves Panama and has maintained lifelong friends there. He married Katrina Hellebush, a graduate of Mt. Holyoke College and a Peace Corps Volunteer in the Philippines in 1964-1966. They have three children who were also bitten by the volunteer bug.

2. More details on the lives of the Englers can be found on pages 144153 in Banana People, by Clyde S. Stephens, 2002. A free copy can be downloaded at: https://collections.ushmm.org/search/catalog/bib271388 or https://archive.org/details/bib271388_001_001/

3. Dr. Engler was giving prenatal care to Phyllis Stephens during the pregnancy of her first child, Linda Anne Stephens. Linda was born three months after Dr. Engler died.

Notes: 17. Bluefields — Panama: Going Back in Time

1. Wikipedia states that "Outside of the Comarca [administrative region], balsería has a negative reputation in Panama; it is officially

outlawed by the government. The general attitude is that balsería is a drunken, violent mess. According to the Ngäbe, while alcohol and violence are present during balsería, it is primarily a cultural event, a unique sport, and a chance to demonstrate pride in their heritage." https://en.m.wikipedia.org/wiki/Ng%C3%A4be#Balseria.

Notes: 18. Striking it Rich at an Oil Well

1. Stephens, Clyde S. Bananeros in Central America: True Stories of the Tropics. Press Printing, Fort Myers FL. 1989. P. 186.

2. Caption by Stewart D. Redwood.

3. Redwood, Stewart D., "The History of Oil and Gas Exploration in Panamá." Tales of Old Panama and Castilla de Oro, Vol. 3, No.3, Dec. 2022. Edited and Published by Dr. Stewart Redwood, P.O Box 0832-0757, World Trade Center, Panama, Republic of Panama. stewart@sredwood.com

4. Heckadon-Moreno, Stanley. A Creole from Bocas del Toro: The Story of Carlos Reid. Panama: Exedra Books, 2011.

5. Alvarado, Teofilo. Interview in Boquete, Panama. 3 Feb. 1987.

6. https://en.wikipedia.org/wiki/Luis_Russell

7. Courtesy of Ariel René Pérez Price, Catherine Russell Collection. And with permission from Catherine Russell, the daughter of Luis Carl Russel of jazz fame. She and her mother were both well-known American jazz singers. Catherine Russell is still a star performer.

8. Redwood (Ibid.)

9. Also, on our return to Changuinola, we discovered that Linda Anne Stephens had accompanied us through Colombia, Ecuador and Peru. At the time, we did not know this. She was born 27 April 1965 in the Almirante banana company hospital. Consider her a bonus in winning that lucky lottery ticket number 0113.

Notes: 19. The Zipper and La Solucion

1. Recently, I took my visiting family on the bus to Boca del Dra-

go at the far end of Isla Colon. Suddenly, the rainforest disappeared along the un-repaired road, and hundreds of small concrete block houses appeared on a bulldozed clay clearing. The bus driver told me that this is where officials will relocate people from La Solucion. It is a good idea except that I suspect that many will choose to continue life in the mangrove swamp where they find work in town and nearby islands, where their children go to school and where there are stores and the hospital. Time will tell. By the way, the name of this new housing development is La Solucion.

Notes: 20. James Norman Morris Maas: One of a Kind

In my searches, I found that the internet has a wealth of personal accounts of other inmates, books, general information, drawings and photographs of the Lunghau Civilian Assembly Center. Several references are listed as follows:

1. Betty Barr, who was a young prisoner, published her mother's diary which is called *Ruth's Record: The Diary of an American in JapaneseOccupied Shanghai – 1941-1945.*

2. Desmond Powers wrote *Little Foreign Devil.*

3. *Captives of Empire*, by Greg Leck, is by far the most complete book on Japanese prison camps in China, which is why it costs about a $1,000.

4. *Maas, James Norman Morris* (Oral history). Imperial War Museums (iwm.org.uk). July 21, 2003. Accessed at: https://www.iwm.org.uk/collections/item/object/80024144.

5. Wikipedia provides references at Lunghau Civilian Assembly Centre. The Ballard home in the International Settlement was converted into an upscale restaurant. One account is by Peter Brigg: brigg visit to ballard house in shanghai.

6. Information about Shanghai's elite High School, where the Lunghau Prison stood, can be found at https://en.wikipedia.org/wiki/Lunghua_ Civilian_Assembly_Centre.

Book 3

Notes: 1. Close Encounters with Germans

1. Kirkpatrick, Kathy. *Prisoner of War Camps Across America*, 2012, GenTracer, Salt Lake City.

Notes: 2. German Prisoners of War

1. A version of this story appeared in Tales of Old Panama, April 2022, v.3.1. Dr. Stewart D. Redwood, editor and publisher, PO Box 08320757, World Trade Center, Panama, Republic of Panama.1. Stephens, Clyde S., Banana People, Dollar Bill Books, Ann Arbor, 2002, 133-144.

2. *The Gleaner*, March 27, 1914. This Jamaican newspaper had a correspondent in Bocas del Toro because of the high West Indian population that immigrated there to work for the United Fruit Company banana operations.

3. James, Winifred, *A Woman in the Wilderness*, George H. Doran Company, New York, 1915.

4. Perez Price, Ariel Rene, *Memorias de Bocas Town, La Vida y Tiempo de Jose Antonio Price,* Miss Hilda Press, Panama, 2017.

5. *The Gleaner*, Nov. 22, 1915, page 11.

6. Amazingly, there were seven consuls, vice consuls or consular agents in Bocas del Toro at the time. There had been eight agencies before the German Consul, Carl Friese, was disbarred when the war broke out: Costa Rican Consul, Juan Raphael Mora Escalente; Nicaragua Consul, Solomon Conoan; Great Briton Vice Consul, William H. Ponton; Norway Vice Consul, Hans F. W. Kandler; Portugal Vice Consul, E. C. McFarland; United States of America Consular Agent, Dr. Paul Osterhout; France Consular Agent, E. Cochez. The Panama Canal Record, Volume II, 1917, page 36.

7. The speakers were: W. H. Ponton, British Vice-Consul; Dr. Paul Osterhout, American Consular Agent; John M. Kyes, Manag-

er of United Fruit Company; several priests, reverends and other orators inspired donors in Almirante, Guabito, and Bocas del Toro.

8. White, Herbert L., former judge advocate, Panama Canal Department, April 27, 1918. Accessed at: fold3.com/image/ 56561793.

9. Dr. Stewart Redwood confirmed that the Aspinwall Hotel closed on April 11, and opened the next day as a prison camp. "Taboga in World War One, The German Detention Camp", page 3, The Panama and Canal Zone Junction, October 23, 2015, edited by Luis R. Celerier.

10. David Zemer: Much of his data and references came from his interest in the Canal Zone Philatelist. Letter covers, stamps, postmarks, messages and special notes provided details about the detained Germans and their fates. COPACARTA, "Mail Related to Enemy Aliens Interned in the Canal Zone and Panama during World War I", Vol. 27, No. 4, Pg. 5, June 2010.

11. The Reverend S. Moss Loveridge, a Baptist preacher from England, was in Panama from 1900 to 1919 and kept a meticulous diary. His descendants published the diary in 1995 in Book One and Book Two. The comments about the Germans in Panama is found in Book Two, *Panama Padre*, Chapter 28, "Changing Times" on page 156.

12. Luis R. Celerier reviews the history of the Aspinwall Hotel in his book *Panama & the Canal Zone in Bits and Pieces*, 2015, www.panamahistorybits. com on page 35.

13. The cover letter of Major Butcher reads: "The enclosed is a confidential list of internees sent to the United States on the transport *Kilpatrick* April 19, 1918, also reasons and dates of internment and other data on file in this office." ttps://www.fold3.com/image/60217739. The long list and descriptions of internees sent from Taboga Island to the United States begins in alphabetical order with Arndt, Paul. https://www.fold3.com/image/60217746.

14. On the first day, April 12, 1917, 12 enemy aliens on Butcher's list were interned, followed by 17 more on the second day. By the

end of April, another 12 were brought to "the German Camp," as it was known. A year later, on April 18, 1918, the last ones on Major Butcher's list were an intern's wife and her three young children whose father had arrived earlier.

15. The record states that he was re-interned at Taboga on April 11, 1917, probably an error. Another statement noted that he was re-interned April 12, 1917, the first day others were interned. Haack asked for a release to buy pearls in the Pearl Islands but the authorities were suspicious.

16. Troetsch, Dimitri, *Genes Alemanes en Chiriqui y Bocas del Toro*, Baru Publicidad, Doleguita, Chiriqui, Panama, 2016, pg. 37-46. With my permission, the chapter on "German Genes in Bocas del Toro" in *Banana People* was translated to Spanish and published in the Troetsch book as "Genes Alemanes en Bocas del Toro" on pages 29-36.

17. Major Edmund Butcher, 29th U.S. Infantry Provost-Marshal detail, April 18, 1918. https://www.fold3.com/image/ 60217858.

18. Dr. Stanley Heckadon-Moreno interviewed Allis Hartmann, the granddaughter of Alois Hartmann, who stated that her grandfather was born June 20, 1891, in Moravia. After a voyage to Ecuador, he came to Panama City in 1912 and walked to David, then continued on to Bugaba where he met the Troetsch family and married Susana Troetsch. Pages 217-219, *Alexander Wetmore y las Aves de Panama*, 2020, Panamericana Formas e Impresos, S. A., Colombia.

19. Hellen Margaret Lutz, 28, was interned voluntarily as a spouse and given permission to come and go. Permission was withdrawn when she was caught smuggling German letters in and out of Taboga.

20. Dr. Stanley Heckadon-Moreno told me that Alois Hartmann was not arrested as an enemy alien in Panama during World War II. Personal communication.

21. The ship's manifest list declared that Hartmann was: Age 27, Bohemian, Austria, Moravian, with residence at Volcan, Pan-

ama. This may explain why the name of his place of birth can't be found on current maps because German names were changed to Czech names when these regions, boundaries and languages in Central Europe were in a convulsive state of flux after wars. He declared that he had $20 on him, and had been in New York in 1907 and 1911.

22. https://www.fold3.com/image/ 60217812. Despite what Major Butcher thought, Dimitri Troetsch reported how the Hartmanns and Troetsch families bonded and presented them in a respectable manner.

23. David Zemer's publication in COPACARTA noted that Jaeger had a wife and five children in Panama City.

24. American Consulate, San Jose, Costa Rica, to the Secretary of State in Washington D.C. May 24, 1918. https://www.fold3.com/image/60833969.

25. Price letter to the U.S. Secretary of State, September 30, 1918. Accessed at https://www.fold3. com/image/56440866,/image/56440879,/image/56440891 and /image56440904.

26. https://www.foitimes.com/internment/Eiserloh.htm.

27. John B. Murphy, intelligence report, Aug. 20, 1918. FBI Case Files. P. 13.https://www.fold3.com/image/2860365.

28. One group of early prominent German merchants was not found on Enemy Trade Lists, internment or deportation records. They were: I. W. Beckman, Adolph and Otto Dolder, J. W. Kroesman, Rudolph and Ernst Braden, and Albert Knoefel. They may have returned to Germany before the roundup of enemy aliens, or some possibly had retired or died. Elsa Knoefel, who was born in Bocas Town, was the daughter of the German merchant, Albert Knoefel. At age 87, in 2000, she told me that her father went to Germany for the Great War and that she never saw him again. His fate was unknown.

29. Strangely, Paul Haack was not added to the black list in 1918 although Julius D. Dreher, of the American Consul in Colon,

noted that Haack was in charge of Carl Friese & Company and was a known German propagandist.

30. From the American Consular Agent in Bocas del Toro, dated December 19, 1918: "The name should read: Antonia C. de Rylance; not Antonia Ryland. I have her signature before me on a Costa Rican passport which she surrendered to me today upon her arrival from San Jose to see Mr. Friese. She told me that she was an equal partner with Friese and Kandler in the business here."

31. Letter from William Jennings Price to the Secretary of State, Dec. 7, 1918. Accessed at: https://www.fold3.com/image/56410486, Page 6.

32. Friedman, Max Paul, *Nazis & Good Neighbors*. Cambridge University Press, New York. 2003. Page 20.

33. The day before departure, on April 18, a cover letter confirms the status of those passengers: "The enclosed is a confidential list of internees sent to the United States on the transport *Kilpatrick* April 19, 1918; also reasons and dates of internment and other data on file in this office. Edmund Butcher, Major, 29th Infantry, Provost-Marshal."

34. Philadelphia is located across the Delaware River in the state of Pennsylvania but the Philadelphia Immigration Station was in Gloucester City, New Jersey, on the East bank of the river.

35. Frida Bartel and her son Otto were held in this Detention Center in Glouchester, New Jersey, until their departure to Panama, according to Jacob L Wasserman in his Senior Essay, "Internal Affairs: Untold Case Studies of World War I German Internment," 2016, Yale University. They were on the USAT *Kilpatrick* that sailed from New Orleans to Panama on June 15, 1919.

36. A well-connected and devious Irish entrepreneur, Billy Thompson, sold his Victorian mansion on five acres of waterfront to the government for $100,000an inflated price. The luxury home on South King Street was converted by the immigration service to

an administration building by 1912. A large white building was constructed nearby to process inspections, detentions, hearings and deportations. Enemy aliens sent there were en route to internment camps in other parts of the country, although some were detained there for months.

37. John B. Murphy, intelligence report, Aug. 20, 1918. https://www.fold3.com/ image/2860365.

38. https://library.sewanee.edu/c.php?g=118671&p=773217. Jessie Ball duPont Library, 178 Georgia Avenue, Sewanee, TN 37383.

39. Jenifer Chrisman, February 21, 2017: https://gordon.armymwr.com/fyi/ culturemil/internment-camp.

40. https://en.wikipedia.org/wiki/Fort_Oglethorpe_(prisoner-of-war_camp)

41. Paintings by prisoners are shown at: http://cookiehouse.net/enemy-alieninternment-camp-fort-oglethorpe-ga-1918-1920/40.

42. Erich Posselt, "Prisoner of War No. 3598 [Fort Oglethorpe]," American Mercury, v. 11, no. 43 (July 1927) 313-323.

43. Murray, Smith; Robbins, Mickey (July 19, 2015). "Murray Robbins: WW I POWs at Fort Oglethorpe". timesfreepress.com. Times Free Press.

44. *Orgelsdorfer Eulenspiegel and the German Internee Experience at Fort Oglethorpe, 1917-19*; by Jeanne Glaubitz Cross and Ann K. D. Myers. The Georgia Historical Quarterly Vol. XCVI, No. 2 Summer 2012, page 233. This account is one of the best in the literature.

45. On February 4, 1919, the American Minister in Panama, William Jennings Price, wrote a letter to Dr. Ernesto T. Lefevre, Secretary for Foreign Affairs for Panama, stating that "several days ago" a group of enemy aliens were returned to Panama from the United States. They were being held in the Canal Zone and were awaiting transfer to the Panama authorities for repatriation to their homes. https://www.fold3.com/image/60217679.

46. https://www.fold3.com/image/1/60217708.

47. The American Minister states that Eduardo Lange had sailed from New Orleans on January 5, 1919 and arrived in Cristobal about January 11. That explains why his name was not on the list with his brother Ernest Lange who came later. Another name not on the list of men who returned to Panama was Peter Ulhmann, age 18, a mechanic from Almirante. He was the only enemy alien not accounted for out of the original 19 men deported from Bocas del Toro on the USAT *Kilpatrick* on July 27, 1918.

48. https://www.fold3.com/image/1/56849254 is the first page from: Aliens deported from Panama. Page 2 is at: https://www.fold3.com/image/1/56849265

49. This letter from William Jennings Price to the Secretary of State in Washington D.C. was dated April 11, 1919, and noted that some Panama Germans had already been repatriated back to their homes in Panama. He attaches a list of the prisoners sent from Taboga Island the year before and includes details about each prisoner. His letter ponders about Panama German prisoners being deported from the United States directly back to Germany instead of returning to Panama. The Panama Minister LeFevre seems to agree with Price. Page 1 is located at https:// www.fold3.com/image/60217653 Page 2 is https:// www.fold3.com/ image/60217662 page 3 is https://www.fold3.com/ image/60217679.

50. Maria Marta Kandler, a granddaughter in San Jose, Costa Rica, provided their marriage certificate and Kandler family photographs to me.

51. https://www.fold3.com/image/2860384.

52. The *Star & Herald* on August 14, 1919. Courtesy Ariel Rene Perez Price.

53. More information about von Wedel is found in the next chapter.

54. Clyde S. Stephens, *Outline of History of the Province of Bocas del Toro, Panama* (bilingual), 2008, SPS Publications, Eustis Florida.

55. Clyde S. Stephens, *Bananeros in Central America*, 1989, Press Printing Company, Fort Myers, Florida. See pages 104-154 where Mark Trafton writes details of how the Puerto Armuelles banana division was developed. Trafton arrived there in 1928.

56. *Unifruitco*, Volume 5, Number 5, December 1929, pages 260-271. A detailed story of the development of the Puerto Armuelles Division. Boston.

57. Ted Zimmerman, a German who became a naturalized American citizen and retired to Bocas del Toro, told me that he and Carlos Lange used to drink beer in Bocas. They spoke German together. Carlos Lange told Zimmerman that his father sent him to study in Germany in the 1930s from Bocas del Toro and was in the German military during the war. Carlos Lange's mother was a Guaymi Indian from the Cricamola River.

58. *The Unknown History of Japanese Internment in Panama*, by Greg Robinson and Maxime Minne, 26 April 2018. http://www.discovernikkei.org/en/ journal/2018/4/26/japanese-internment-panama. European Jews, German and Italian prisoners are also discussed.

59. Friedman, Max Paul, *Nazis & Good Neighbors*. Cambridge University Press, New York. 2003. Page 148

60. Yoshitaro Amano, Canal Zone Resident and Prisoner #203, by Esther Newman, editor@DiscoverNikkei.org! These accounts reveal the circumstances of Germans and Italians.

61. Video: "Camp of the Innocents," by Mira Kohl et al. 2017, Tulane University.

62. Stanley Heckadon-Moreno, *A Creole from Bocas del Toro: The Story of Carlos Reid*, Editorial Exedra, Panama, 2011, page 211.

63. Stephens, Clyde S., *Banana People*, Dollar Bill Printing, Ann Arbor, MI, 2002, 141.

64. Previously misspelled as Hertz. A ship's manifest shows that in 1938, Ewald, Cristel and Irmgard traveled from Jamaica to

England aboard the SS *Ariguani*. At that time, Ewald was listed as a mechanic. This voyage suggests to me that Hortz was already working for United Fruit Company in Almirante in 1936. The clue is that Europeans who worked for United Fruit got a vacation every two years and traveled on Company ships like the *Ariguani* of Elders and Fyffes, which was controlled by United Fruit.

65. Ruby Vargas was listed as a fourth Hortz child on the manifest of the USAT *Florida*. Cleveland Nelson never mentioned her when I interviewed him in 1996, and his was the only first-person account available to me. It's always possible she was from an earlier marriage, might have been taken in as a homeless child, as often happens in Latin America, or perhaps she was adopted. Also, Irmgard was previously misspelled as Hemgot, and Ingribud was misspelled as Ingabud. When Cleveland wrote his story, he wrote those names in a phonetic vernacular....thus Hemgot and Ingabud.

66. Names of students in the 1940 Fall Class of the American School of the United Fruit Company in Almirante, Bocas del Toro, Panama. Front row, left to right: Jimmy Paterson, Margarita Bogantes, Irmgard Hortz, Andrew Paterson, Diane Dudley, Aline Toupin. Back row, left to right: Yvette Michaud, Marvin Marsh, Giselle Michaud, Mike Marsh, Irene Toupin, teacher Mattie Grant.

67. National Archives, Hull to Attorney General Francis Biddle, Nov. 9, 1942. 740.0115/4570, RG59.

68. Report on Alien Enemy Internment Camp, Stringtown Oklahoma, Sept. 2123, 1942. Accessed at: https://gaic.info/wp-content/uploads/2017/05/21-23-Sept -1942-report.-Stringtown.pdf.

69. The older daughter, Ana, married an American during the time of their imprisonment in Crystal City, Texas, or somewhere during their imprisonment in the USA. She remained in the USA the rest of her life although the Probst families keep in touch. Ana was not shown on the ship manifest list of enemy aliens when the family was repatriated to Germany in 1944 because she had married

the American. There were 11 children and the two parents, a family of 13 that went to Germany without Ana.

70. Foitimes.com/Memoirs.htm.

71. https://gaic.info/gripsholm-event-held-february-14-15-2019-in-n-y-c-and-new-jersey/.

72. https://warfarehistorynetwork.com/2020/12/11/the-blue-division-and-thespaniards-who-fought-against-stalin/

73. In the 1980s, Ligia Thomas de Paget and her husband Alfredo Paget, bought the Kandler house in Bocas del Toro. During restoration, they found sensitive documents hidden in the attic above the second floor. These papers were copied by Larry Mahan who passed them to me. Travel applications from Germany to France, New York and Panama, for Richard and Edmund Kandler left a paper trail of the challenge of returning to their home in Bocas del Toro.

74. Wasserman, Jacob L., "Internal Affairs: Untold Case Studies of World War I German Internment" (2016). MSSA Kaplan Prize for Use of MSSA Collections.

Notes: 3 The Kandler Family

1. The references in this chapter are essentially the same as those that are found in the endnotes of the previous chapter: "German Prisoners of War from Bocas del Toro." Other details are based on personal interviews and exchanges between Maria Marta Kandler and me.

2. "He agreed with his father, probably reluctantly, and so it happened that Richard K. in March 1934 boarded with other young Germans and their mothers the vessel *Tacoma* of the Norddeitscher Lloyd towards Bremen, Germany." Translation provided by Wolfgang Schulz.

3. Personal conversation with Walter Hamer, who related his conversation with Samuel Maña.

4. Britannica, The Editors of Encyclopaedia "Iron Cross." *En-*

cyclopedia Britannica, 3 March 2013, https://www.britannica.com/topic/IronCross. Accessed 7 July 2022.

Notes: 4 Alwyn Hasso von Wedel

1. Olson, Storrs and Stephens, Clyde: *Alwyn von Wedel (1873-1957): Bird and Plant Collector on the Caribbean Coast of Panama.* Archives of natural history 45.2 (2018): 317-334.

2. *Epocas* is a monthly historical magazine of *La Prensa*, a Panama newspaper. Dr. Stanley Heckadon-Moreno contributed eight monthly stories on Hasso von Wedel (January-August 2019) based on the publication of Olson and Stephens (2018). Dr. Heckadon-Moreno enhanced the biography with numerous historical accounts of the times.

3. Stephens, Clyde, et al.: *Banana People*, First published, 2002. 133-143.

4. Olson and Stephens: p. 317.

5. *Star & Herald*, July 6, 1918. Courtesy: Ariel Perez.

6. Fold3.com/images: FBI Case Files Old German Files, 1909-21 Various (#254409) Page 37.

Notes: 5 The Probst Family

1. In 2023, I visited the plaque along Road 43 near the bridge that crosses the Cotito River. The memorial plaque states that the common grave for the ten victims was 200 meters from that point. In fact, the mass burial is about one kilometer south of the plaque in a coffee farm. The grave is neatly groomed with plants and has a fence around it. It is contiguous to the entrance of the house of the owner of the coffee farm.

2. Friedman, Max Paul, *Nazis & Good Neighbors*. Cambridge University Press, New York. 2003.

3. Several names in this chapter weren't always spelled the

same: Gretchen's name is spelled with a double t. On ship manifests, the spelling typically is Gretchen; Hedgwig is often referred to as Hedy or Heddy (Jeanette Probst says the spelling was typically Heddy); Richard (German) is sometimes spelled Ricardo (Spanish); Josef is sometimes spelled Joseph; Fred is Alfredo; and Wilhelm is sometimes spelled William.

Notes: 6 The Stargardter Family

1. Max Stargardter's grandson, Andres Bustos, wrote a privately published book in Spanish: *MAX, Prusiano y Alemán, pero Judío.* (MAX, Prussian and German, but Jew). In chapter 2, he states the following: "Stargardter proviene del idioma polaco, donde star es una derivación de stari que significa viejo y gard es ciudad o castillo." (The name Stargardter comes from the Polish language, where star is derived from stari that means old and gard means city or castle. Bustos provided me with an electronic copy.

2. Davies, Norman and Moorhouse, Roger. *Microcosm, Portrait of a Central European City.* Jonathan Cape, London, 2002.

3. Davies and Moorhouse, pp. 369-370.

4. Stargardter, Jessica. *Hans Stargardter, My Experiences in Hitler's Germany* (updated in 1997).

5. Stargardter, Jessica, ibid.

6. Personal telephone interview with Hans Stargardter, 2021.

7. Moses, Fritz. *Strehlen: Memories of a Small Silesian Town and Her Jewish Citizens.* (German edition, Temmen. 1995).

8. Bustos, Andres. *MAX, Prusiano y Alemán, pero Judío.* (MAX, Prussian and German, but Jew). This is a private publication in Spanish.

9. The following explanation is from https://en.wikipedia.org. wiki/ Kinder-transport: "The Kindertransport (German for "children's transport") was an organised rescue effort that took place during the nine months prior to the outbreak of the Second World

War. The United Kingdom took in nearly 10,000 predominantly Jewish children from Nazi Germany, Austria, Czechoslovakia, Poland and the Free City of Danzig. The children were placed in British foster homes, hostels, schools and farms. Often they were the only members of their families who survived the Holocaust."

10. Personal communication with Michael Heymann.

Appendix I

The Banana Family: The Early Years in Montecristi

The headquarters of the Grenada Company in the beginning years was Montecristi. Old abandoned offices, stores and houses in the remote, rundown town were renovated for offices and the banana company staff and their families from 1940 to 1947. Bachelors lived in the Hotel Noroeste and El Pájaro Lindo.

Company employees included the manager and his assistant manager, engineers, surveyors, accountants, agronomists, materials and supplies people, research, and heavy equipment operators. Additionally, there were the labor and employment offices, transportation and other departments. A Company medical doctor and medical clinic dealt with health problems. Most of the agriculture staff was located at La Cruz by the river to be closer to the banana farms. Jeeps and pickup trucks and were provided as work vehicles and required a repair shop to keep them in good condition in the dusty, salty air.

With the development of the new agricultural enterprise, Montecristi went from being a quiet, old dilapidated town to a busy municipality with a big Company payroll. This bonanza started to decline in 1947 when employees began moving from Montecristi to the new banana port of Manzanillo as infrastructure, houses and offices were completed there.

One of the founding dynamos of the Grenada Company who organized and managed these early operations was Warren Breck, originally a soils chemist. His wife, Toni, was from Honduras and their daughter, Shirley, was born in Montecristi in the 1940s. Dominicans respected and adored this man to the extent that he was honored by the *Ayuntamiento Municipal de San Fernando de Montecristi* (the Municipality of Montecristi) as an adopted son. This great honor had always been reserved for Dominicans, not foreigners like Warren Breck.

Besides Warren Breck, other early officials in Montecristi were Felix Aycock, David Cloward, Alfred Butler, Wilson Popenoe, Collin Shaw, plus numerous Dominicans at all job levels. Some of the Dominican names were Jimenes, Espaillat, Santelises, Hernández, Regalado, Ginebra, Tavaras, Marchena, Lister, Borrell, Helú, Abréu, Ricardo, de la Maza, Joubert, Bobea, de Jesús Viñas, Viñas and so many more. There was a cordial and respected relationship between all nationalities who were building this challenging new enterprise in the desert.

One of the Dominicans from Moca, who arrived in Montecristi in 1944 in the back of a dusty truck looking for work, was Fernando de Lara Viñas. After a distinguished career with the Grenada Company, he wrote his memoirs in 1995. His book begins in 1937 when Dr. Dunlap came to survey soils. Chapters continue with historical details of the times until the Grenada Company shut down in 1966. Much of the text in this chapter is taken from his book, *Reminiscencias de un bananero* (1995).

Fernando de Lara Viñas worked in surveying, agricultural operations, the research department, landscaping and planting trees in Manzanillo, inspecting and rejecting fruit and supervising stevedores on the wharf in Manzanillo. He was a Jack-of-all-trades. His descriptions of Montecristi, Manzanillo and the farms are unique, as are his subtle but unmistakable comments about the brutal dictator of the country.

The New Company Town: Puerto Libertador (the same place as Manzanillo and later Pepillo Salcedo)

To review this major Company investment, high profile executives from the corporate offices of the United Fruit Company visited the

Grenada Company operations in November 1949. They were: Samuel Zemurray, who controlled the Company; Walter Turnbull; Bill Turnbull; Joseph Montgomery; Robert LaFollette, Jr.; and Hartley Rowe. After touring the port and banana farms, they helped to inaugurate the new Manzanillo Golf Club built mostly by volunteer employees and shareholders. Bill Cook was the first president of the club.

New department offices were built as follows: Management and Administration, Engineering, Materials and Supplies, Accounting, Medical and a hospital, Legal, Personnel, Fruit Purchasing, Electric, Radio Communications, Heavy Equipment Railroad, Marine Shipping and the Banana Puree Plant managed by Dr Otto Reif, from Vienna. He had barely escaped from the Nazis two decades earlier.

Don Fernando reminisces about moving to a new house on the hill, Number 4017, just after he was married in 1949. His children attended the American School and they were members of the golf club. Their children loved the pool. There was a large Company store for household shopping, and it was open to the public. The Company had a good hospital with competent doctors, like Dr. Jose Ernesto Antoni, Dr. Mario Cino and Dr. Raphael Viñas. They were assisted by registered nurses. The hospital was available to all classes of employees. At that time, there were almost no non-Company people in the port except for military and government employees.

Sports leagues, especially baseball, were extremely popular and Fernando de Lara Viñas was an organizer of these games. Ball clubs competed with Company teams in Manzanillo, the banana farms, and big city teams in other parts of the country. According to *don* Fernando, Puerto Libertador was an impressive place to work and raise a family. But one thing was certain—people never criticized the government. If they did, they and their family might pay the ultimate price.

Open Door Policy

The Dominican Republic had an "open door" policy and sometimes encouraged white immigrants and refugees to settle in the country. Japanese immigrants were given land to grow vegetables to sell in the country. They demonstrated successful techniques of agriculture and fishing in hopes that Dominicans would learn from them. The same applied to 800 Jewish refugees from Europe, via Shanghai, who were given land at Sosua to grow crops, pastures, beef cattle and dairy farming for milk and cheese. Some of the land and houses at Sosua were part of the old United Fruit Company town that was abandoned in the early 1900s. People of color were not invited, especially Haitians.

Japanese colonies were of special interest to us banana people because we bought their products. I visited the small, isolated farm commune that was located just south of Manzanillo. The land and modest houses had been given to them by the Dominican government as an incentive to immigrate from Japan. The area was arid but had irrigation water. Salt in the soil was a problem. Once a week, Japanese farmers drove a small pickup truck with fresh vegetables to Manzanillo to sell. Their Spanish was minimal because of the isolation of the colony near the Haitian border. It helped that Japanese children went to a Spanish school provided by the government.

From my bachelor's room on the hill, I could see bright lights at night on a small Japanese boat that was catching squids. The lights attracted squids into nets and this seafood was sold in the port. I ate in the Company dining room for fifty U.S. dollars a month, and sometimes fresh squid was on the table. Mrs. Shelby Thompson was in charge and knew how to prepare seafood because she was from the Bay Islands of Honduras.

Warren Breck, a Legend

Warren was one of the most energetic people that I have ever known. In 1961, I rode with him on a trip up the valley of the Yaque del Norte River to take care of some kind of business. The driver was Enrique, a very short man who sat on an elevated cushion in order to see the road. We rode in a station wagon that the managers used. The Company vehicle was recognized by the general public.

When the car stopped at police check points, or at little villages along the dusty, gravel road, people would come rushing to the car to greet Mr. Breck. It was common for someone to put a demitasse of strong coffee in his hand. After a few sips, he would hand the cup back, and Enrique would drive to the next stop where the same ritual was repeated. Everybody knew his name and fame because of his huge impact on the region.

His greeting was the same for young and old alike, "*¿Como estás tú, joven?*" (How are you, young man? or young woman) Frequently, little old wrinkled ladies chomping on a cigar would come rushing to the car to greet Mr. Breck. He rolled down his window for everyone. Depending on the person, he would sometimes put a few coins in their hands as a friendly gesture. Even children along the road would yell, "Mr. Breck!"

We were in a desert, but when the window was rolled down, swarms of house flies came into the car. Fly populations were high because of chronic deposits of goat, burro and human manure in the villages. Flies were part of life in those villages. It was no secret that Warren Breck was a reformed alcoholic. He personally told me that he would start the day in a canvas-top Jeep to examine soils, drainage and irrigation. By noon, his bottle of Dewar's White Label Scotch would be half empty. At the end of the day, the bottle would be empty. Luckily, he had a doctor who told him that he was going to die if he did not stop drinking. He made the decision to stop, and he never drank again. At dinner parties, he looked kind of bored, but actually he was always thinking of business. He was one of the first to leave the party.

Later, Warren Breck became the president of the Company with his office in New York City. In the 1970s, Paul and Seymore Milstein honored him at the Time Life Building for his 40 years of service with the Company.

Appendix II

The Trujillo Family

Book Review: María de los Ángeles del Sagrado Corazón de Jesús Trujillo Martínez de Domínguez

Numerous books, articles and documents are available to the public to learn about the Trujillo dynasty. But one book is unique in that it was written by Trujillo's daughter, known as Angelita. Her book, *Trujillo, mi Padre, en mis memorias*, 2009, has 460 pages, including 214 pages with multiple photographs. The content is written as a personal tome describing her life and the lives of numerous Trujillos. Angelita's inside knowledge and many memories are detailed. In this book, she makes no apologies for the greed, nepotism, corruption, terror, fear, tortures and atrocities caused by her father, her brother Ramfis and their supporters for 31 years. The life of luxury and privilege is presented in a matter-of-fact way. By default, the book is informative in that there are no regrets. No remorse is shown except for her father's assassination, the overthrow of the Trujillo dynasty and their ultimate expulsion from the country.

Ramfis Trujillo: Son of the Dictator

His real name was Leonidas Rhadamés Trujillo Martínez, but he acquired the nickname Ramfis, after the high priest in Giuseppe Verde's opera, *Aida*. Ramfis was born in 1929 and died in Spain in

1969 at age 40. He had been in exile under the protection of Generalissimo Francisco Franco, dictator of Spain.

Ramfis became a full colonel in the Dominican Army at the age of four with full pay and all the benefits. He married in the early 1950s and had six children. Later, his father sent him to the United States Army Command and General Staff College in Fort Leavenworth, Kansas. He did not graduate for lack of attendance. This was because he became a renowned playboy in Hollywood, entertaining Hollywood stars like Kim Novak, Zsa Zsa Gabor and various other women. Ramfis was accompanied by another famous international philanderer, Porfirio Rubirosa, his brother-in-law. The two were hot news items in the press, living *le bon vivant.*

When the dictator saw that his son had spent about a million dollars on expensive gifts for movie stars and the high life, he ordered Ramfis to come home, along with the Trujillo yacht, the *Angelita*, which he was using to entertain celebrities. One possible reason for this stern order was that American tax payers and some members of congress in Washington, D.C., took note of how American foreign aid to the Dominican Republic was being spent. Reports of gross misappropriation of American tax money made scandalous news, and this made the dictator very uncomfortable—he liked foreign aid.

When Ramfis returned to the Dominican Republic, his wife Octavia filed for divorce. His unruly behavior, including gang rapes of young women and frivolously ordering murders, forced his father to send him to a sanatorium in Belgium. Ramfis Trujillo apparently suffered from psychological problems. He received electroshock treatments in Belgium as early as 1958; there were also stays in mental hospitals after that. After these treatments, he moved to Paris to resume life as a *bon vivant.*

After his father's assassination, Ramfis returned to the Dominican Republic to continue the dictatorship, with the help of some loyal supporters. One of these supporters was his infamous head of military intelligence, the *Servicio de Inteligencia Militar (SIM)*, Johnny Abbes García, who personally took part in the vengeful slaughters that were occurring. Johnny Abbes had lots of experience managing international spy and murder networks for the Trujillos in other Latin American countries and the United States. His profession was torture and murder. After the Era of Trujillo, President Balaguer shuttled Johnny Abbes out of the country in order to shut down SIM. Later, Abbes became useful to the Haitian dictator, François "Papa Doc" Duvalie. According to journalist Bernard Diedrich, Abbes and his wife had come under suspicion for plotting against Duvalier. Johnny Abbas García disappeared in Haiti in June 1967 and was never heard from again.

La Rebelión de los Pilotos (The Rebellion of the Pilots) was a military uprising carried out by six members of the Dominican Military Aviation on November 19, 1961. Their heroic action forced the entire Trujillo family to leave the country. This allowed president Joaquín Balaguer to start making reforms and try to stabilize the country.

The Exodus of the Trujillo Family

The exodus of the Trujillo family was a drama for the world to witness. Ramfis exhumed the casket with the body of his assassinated father and loaded it aboard the famous Trujillo yacht, the *Angelita*. The yacht was loaded with huge amounts of cash and assets and accommodated some of the Trujillo family and their prized belongings. The plan was to acquire safe haven in Cannes, France, and rebury the dictator in Paris.

One reference reported that the *Angelita* went to Guadeloupe, French West Indies, where the Trujillo family flew on Air France to Paris, presumably with the casket and body of the dictator, but this last point was not made clear. Ramfis's plan was for the yacht to continue on to France with many assets aboard. Meanwhile, elements of the military back in the Dominican Republic were disturbed that the Trujillos had escaped with assets that belonged to the country. Other references claimed the yacht with the family and body of the dictator arrived in the Canary Islands of Spain but was turned back to the Dominican Republic. The reports are confusing.

At some point and after many intrigues, President Joaquin Balaguer ordered the captain and crew of the yacht to return to the Dominican Republic. They complied. When the vessel arrived, the government confiscated the yacht and the many millions in assets found on board. The name of the yacht was changed to *Patria* and it was put up for sale. On the 21st of November, 1961, the name of the capital, Ciudad Trujillo, was changed back to Santo Domingo as it had been called since 1496.

Burials

Rafael Leonidas Trujillo was buried in Paris. Later, he was exhumed for the second time and moved to Madrid. Ramfis, the international Hollywood playboy and mass murderer of his people, died after his sports car crashed near Madrid in 1969. He was buried in Madrid but was later exhumed and interred with his father in another cemetery in Madrid. Only four years earlier, the dictator's former son-in-law and Ramfis's former brother-in-law and playboy partner, Porfirio Rubirosa, died when he crashed his sports car in Paris. He was buried near Paris.

Appendix III

People Not Listed as Detainees

In some reports, certain spouses and children were listed for deportation to the United States. However, a clarification was issued later by Infantry Captain A. K. Bernap: "Mrs. John Hagenbarth (Wilhelmina), Mrs. Ana Maass, Benedictos Maass, Jr. (son), Elma von Karger and Charlotte Walter were not sent to the United States. I believe they are at the Colon Quarantine Station." This report was dated February 8, 1919.

One German who was not accounted for in the deportation was Herbert Leer, a former executive for the United Fruit Company, a consular agent in 1905 and a well-known merchant and investor in Bocas del Toro and Chiriqui. His wife was Lilia Bonilla, from Costa Rica, and it seems probable that Leer escaped from being arrested by living with his in-laws in the neighboring country during the war.

Another prominent Prussian in Bocas del Toro that escaped deportation was William Gerchow.

Appendix IV

Departure to U.S. on April 19, 1918[75]

Condensed list of people aboard the USAT *Kilpatrick* that departed from Cristobal, Panama, on April 19, 1918, and arrived at New York Harbor for processing at Ellis Island on April 28 and 29:

Hermann Albert Deistler, 31, engineer for the Panama Canal. He was interned because his house was too close to the Panama Canal.

Venedia Ruth Deistler, 34, his Jamaican wife, was interned voluntarily because of destitute condition.

Eugen Josef Lutz, 37, teacher of mathematics at Panama National Institute. Also a lawyer in Panama City. Wrote anti-American articles in newspapers.

Hellen Margaret Lutz, 28, was interned voluntarily as a spouse and given permission to come and go. Permission was withdrawn when she was caught smuggling German letters in and out of Taboga.

Wilhelm Probst (sometimes listed as William Probst), 23, born in Meseritz, Posen, Germany, farmer in Llano Bonito, Chiriqui. Visited David town frequently and met with Max Neumann and other Germans.

Juanita Probst, 19, born in Boquete, Chiriqui, Panama. Volunteered for internment—destitute condition.

Wilhelm, Probst. Jr. (also sometimes listed as William Probst, Jr.), 8 months of age, born in Llano Bonito, Chiriqui, Panama.

Clementa Selar, 36, born in Colombia.

Anatulia Selar, 10, born in Panama City.

Matilda Selar, 13 months, born in Panama City.

Bartolo Selar, 37, born in Austria, speaks Croatian. Panama Canal foreman.

(On the ship's manifest: All 4 are listed as Hungarian nationality, and race as Croatian.)

Bernard Sgypritt, 34, German, iron worker, operated a launch service between Panama City and Rio Bayano.

Eladia Sgypritt, 24, interned April 18, 1918, voluntarily. Lilian Maria Sgypritt, 4.

Alfonso Sgypritt, 2.

Raul Antonio Sgypritt, 4 months.

Gustav E. Horst Zilian, 36, born in Kiel, Germany, engineer. Arrived in San Francisco, California, in 1903 and processed naturalized papers. Arrived in Panama in 1908 and worked as a machinist in the Panama Canal Zone until 1917. Spent time with dangerous Germans.

Zenobia Zilian, 31, born in Paraiso, Panama.

Fritz Zilian, 6, born Panama City.

Gustavo Zilian, 5, born Panama City.

Paul Zilian, 2, born in Panama City.

Consuela Zilian, 4 months, born in Panama City.

Only one of these enemy aliens was from Bocas del Toro—Hans Kandler.

Appendix V

Departure to U.S. on July 27, 1918[76]

Condensed list of 19 Enemy Aliens from Bocas del Toro, Panama, on the USAT *Kilpatrick* that sailed form Colon, Panama, on July 27, 1918 and arrived in New Orleans on August 8. The list was copied from the ship manifest:

Bernhardt von Karger, 58, surveyor, German, Almirante, contact: Wife Elma von Karger, Colon.

Benedictus Maass, 54, planter, German, Bocas del Toro, contact: Wife, Ana Rebecca Maass, Bocas del Toro.

Peter Uhlmann, 18, mechanic, German, Almirante, contact: Mrs. J. O. Posey, mother, Almirante.

Richard W Knoefel, 25, clerk, German, Bocas del Toro, contact: Captain Gerchow, Almirante, friend.

Fred Filter, 61, planter, German, Bocas del Toro, contact: Carmelita Filter, wife, Bocas del Toro.

Carl Walter, 54, hotel keeper, German, Bocas del Toro, contact: Charlotte Walter, wife, Colon.

John Emil Hagenbarth, 47, bookkeeper, Austria, Bocas del Toro, contact: None.

Hermann Grabowski, 44, merchant, German (the word Hebrew was penciled in under race), Bocas del Toro, contact: Gabriela Grabowski, wife, Bocas del Toro.

Otto Gebert, 39, agent, German, Bocas del Toro, contact: Policarpa Lopez de Gebert, wife, Bocas del Toro.

William Broderson, 46, merchant, German, Bocas del Toro, contact: Jose Maria Sanchez, friend, Bocas del Toro.

Albert Leer, 41, merchant, German, Bocas del Toro, contact: None.

Erich Rassow, 31, farmer, German, Bocas del Toro, contact: Belarma Selles de Rassow, wife, Bocas del Toro.

William von der Heyden, 26, farmer, German, Bocas del Toro, contact: Flora von der Heyden, Bocas del Toro.

Hasso von Wedel, 43, clerk S&T, German, Bocas del Toro, contact: Dr. C. A. Vaz, friend, Bocas del Toro.

Paul Haak, 29, merchant, German, Bocas del Toro, contact: Adelina Richardson, friend, Bocas del Toro, 7th Street.

Ernst Lange, 50, farmer, German, Bocas del Toro, contact: None.

Edward Lange, 46, farmer, German, Bocas del Toro, contact: Raphael Arciniagas, friend, Bocas del Toro.

Alfred Schieck, 33, merchant, German, Bocas del Toro, contact: None.

John Applegrun, 48, laborer, from Panama, German, Bocas del Toro, contact: Gabriela Alxons, wife, Bocas del Toro.

William von der Heyden, 26, farmer, German, Bocas del Toro, contact: Flora von der Heyden, Bocas del Toro.

Appendix VI

Puerto Armuelles

In 1927, the United Fruit Company's Bocas del Toro manager, Henry Stirling Blair, was transferred across the mountain to the Pacific side in Progreso. This was an isolated and abandoned facility of the former Panama Sugar Company that went bankrupt. Progreso was close to the Costa Rican border and near the Chiriqui Viejo River. Blair's mission was to develop new banana plantations and to build a new open seaport and Company town. To do so, he selected his best engineers, agriculturalists, railroad and bridge builders, dragline operators, mechanics, water and sanitation experts, public health and disease control officials, constructionists, accountants, overseers and general staff and moved them to Progreso in Chiriqui.

Starting with no infrastructure, Henry Blair and his team built a long open sea wharf, railroads, a water system, a three-story modern hospital, electric plant, office buildings, machine shops, schools and well-stocked stores. Housing for thousands of laborers, professional employees and management was constructed at the new port of Puerto Armuelles, where future banana plantations would be developed.

When the infrastructure was in place, banana corms were planted in holes and the forest was felled on the top. An airport was built to get Company staff in and out of the new port. Recreational facilities were eventually added, including a club house, golf course and a pool. Puerto Armuelles was an autonomous Company town built from scratch and located between the Pacific coast and a lush virgin rainforest.

This massive investment replaced the operations in Bocas del Toro that had declined to the point of no return—a dead banana division.

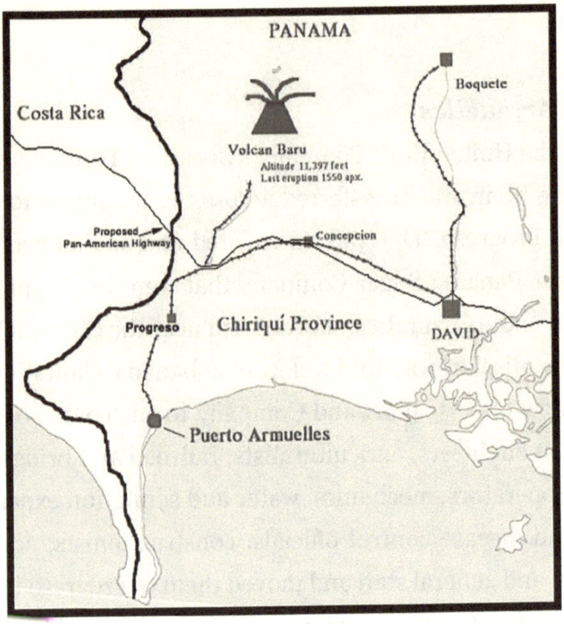

Puerto Armuelles.
Public domain. https:// en.m.wikipedia.org/wiki/ File: ClosePuertoChiriqui. JPG#mw-jump-to-license.

Appendix VII

Remembering Friends

My old banana friend, Karl Friedrich Koch, was 16 when the war was over in 1945. In 1961, Fred and I were employees of the United Fruit Company in Puerto Libertador, later re-named Manzanillo, in the Dominican Republic. After drinking a few bottles of El Presidente beer (named for the brutal dictator Trujillo at that time), Fred loosened up with some experiences as a teenager in Germany. His father and uncle were Nazi officers. As a boy, Fred remembers going on the train to the Ukraine with his father to acquire horses for the Third Reich. His father told him to select a horse of his choice for himself. The teenager rode back to Germany in the same railcar with his own horse.

Fred told me that he had a rifle in his hands from the time he was 11. Like millions of others, he was in the Hitler Youth, *Hitlerjugend*. At age 16, he was east of Berlin when the Germans were crushed between the Russian Front and the American, British and French armies coming from the west. The teenage soldiers were led by young lieutenants. One day, these scared kids realized that their officers had disappeared. Fred threw down his rifle, and started walking in the direction of his family home in Minden, 356 kilometers to the west. It took him weeks to walk through fields, back roads and the rubble of towns and villages. He slept in abandoned barns along the way. Food and water were scarce but he eventually arrived in Minden and found his home.

Later, Fred became the general manager of United Fruit Company's Chiriqui Land Company in Puerto Armuelles. The banana company then moved him north across the mountains to Changuinola in Bocas del Toro. In the late 1980s, he was transferred to Honduras at La Lima to manage the Tela Railroad Company, a subsidiary of Chiquita Brands International. He retired in San Pedro Sula, Honduras.

Another friend in the banana company was Bernie Firchow. When he, Fred Koch and I socialized in the Dominican Republic, I was astonished by their war stories—after a few bottles of El Presidente. Bernie's father was the German consul in San Jose, Costa Rica, and his mother was Costa Rican. Their children were born there. In 1941, the entire family suddenly moved to Germany for the duration of the war. When Bernie and Fred exchanged stories about their whereabouts at the end of the war in 1945, they discovered that they had been only a few kilometers from each other during the destruction and chaos of Berlin.

The Firchow family was separated during the war. Bernie and his father were in Berlin after the surrender on May 7, 1945, and began walking. Trains and railroads had been destroyed. They walked for weeks, eating and sleeping wherever they could. Eventually, the family reunited and returned to Costa Rica after a long struggle, thanks to the help of the United Nations Relief and Rehabilitation Administration (UNRRA) in Munich. Bernie graduated from Harvard University and was employed by the United Fruit Company soon after graduation. He became an executive for the banana company.

Robert "Bob" Bruce told me stories of the destitute masses of displaced persons roaming around Germany and the rest of Europe after World War II. As a U.S. Consular Service official, he was stationed in Munich interviewing hundreds of refugees, helping them fill out repatriation applications and assisting them with food, housing and other basic needs. Bob acted as a liaison with the UNRRA that helped hundreds of thousands of out-of-place people in need.

This assignment enabled him to become fluent in German. I have wondered if the Bernie Firchow family or Richard and Edmund Kandler might have passed through his office in Munich. What is extremely coincidental is that Bob Bruce knew the Kandler family in Bocas del Toro when he was an office manager for the United Fruit Company in Almirante, Panama, in the 1950s. He told me that he spoke German with the Kandlers even though they all spoke English and Spanish.

About this Book

Two previous books that I wrote, *Bananeros in Central America* (1989) and *Banana People* (2002), presented numerous personal accounts by employees of the long-gone United Fruit Company. Some of those *bananeros* began their careers in the 1920s when the banana industry was ripe with incredible experiences in the American Tropics.

Banana Man consists of three books, available individually or complete in this special edition, which includes stories by banana people who followed the footsteps of those early pioneers. Some are second-generation "banana kids" who were born, raised and educated in banana plantations and Company towns. It also records the appalling effects of two world wars on German immigrants living in Panama and other Latin American countries. Entire families were uprooted and put in concentration camps in the United States. Many were repatriated to Germany and exchanged for Jews and Americans being held by Nazi Germany.

One Jewish banana family, whose father and relatives barely escaped the Nazi Holocaust fled to Guatemala. Another family from war-torn Italy immigrated to Guatemala, then Panama. While studying in Germany, the two sons of an old German family in Panama were drafted into Hitler's military.

One chapter describes how banana people lived and worked under a dictatorship in the Dominican Republic. Another chapter depicts the last years and days under the dictatorship of Fidel Castro.

The lack of family planning and over-population are identified as the common evils that cause hopelessness and eternal poverty in places like Haiti and some parts of Panama.

There are chapters about the production of abaca, a strategic war material; rubber tapping and harvesting; finding an abandoned oil well on an island in Panama, retracing many footsteps in jungles that once had thriving banana plantations.

The abandoned Tropical Radio and Telegraph tower that transformed communications in Latin America, tropical medicine and public health, all are discussed. Natural history is presented in various ways. The Peace Corps story in Almirante is told by a volunteer. Sharks are reviewed from several perspectives.

There is something for everyone in these books.

Acknowledgments

In 2020, when I was 86, I decided to write another sequel to *Bananeros in Central America* and *Banana People*. Chiquita, now Brazilian-owned, somehow found this old *bananero* and asked me to help orient some of their key executives from Europe and the United States. We gathered in March 2020 in Changuinola, Panama, where I had lived for 11 years, between 1960 and 1971. The vibrant young group made me feel like a rock star after presenting them with PowerPoint lectures and participating in a field trip. I told them stories about pioneer banana people and the challenges they overcame in uncharted corners of the Tropics. The participants' appreciative response inspired me to think about another book.

A few months later, Francisco Stargardter wrote me from London asking for United Fruit Company historical information for his son in Brazil. Others from around the world frequently asked questions and showed interest in what I had in my files and library. It was always a pleasure to provide what I could.

My library has many scientific publications and books on bananas, banana people and banana companies. Monthly *Unifruitco* magazines, that date from 1926 and erratically end in the early 1970s, are historical and fascinating. The library also has about 3,000 early photographs of the first plantations, railroads, banana canals, ports, ships, and banana company towns in Bocas del Toro. In the late 1960s, I salvaged and restored hundreds of old photographs that had been hopelessly discarded by the manager's office. From these old relics, I have developed several PowerPoint programs that are presented to the public in Spanish and English.

An album bought with some of my children's inheritance has over 200 early postcards that illustrate the evolution of the banana industry in Bocas del Toro. Commercial banana plantations started there in the 1880s and by the early 1900s, postcards were in vogue. Senders wrote brief descriptions about their lives and the times where several banana companies were exporting this exotic fruit in high demand in the United States. This was during the era of the large, prosperous German colony in Bocas del Toro and many postcards were written in German and mailed to *das Vaterland* (the Fatherland). These postcards tell a special story about the region over a century ago.

Writing this book was a challenge because I have been migrating back and forth to Panama. Some of the time, I was in Tavares, Florida, where my main library is located. Other times, I wrote in my solar-powered retirement house on Hospital Point located on Isla Solarte in Bocas del Toro. Fortunately, a Wi-Fi signal reaches my laptop just a few steps away from a giant shady mango tree by the Caribbean shore. From this remote tropical island, I communicated with numerous people in so many places around the world. They provided me with missing details that I needed for these chapters. Despite the help from so many people, I know there will be mistakes for which I take responsibility. There was a minor dilemma in writing the incredible story about the Probst family. Two elderly siblings disagreed about which port they departed from in Europe as displaced persons after World War II. They were still children at the end of the war. Amazingly, I interviewed four of the 12 Panamanian Probst siblings who had been repatriated to Nazi Germany in an exchange for Americans and Jews held by the Germans. A fifth member, Alfred Probst, was unable to talk because of poor health.

The unethical capture of the Probst family in their own country of Panama and then their exchange to Hitler's Germany should be a lesson learned by all nations during wartime.

Sometimes, history is boring simply because certain facts must be recorded for clarity or for posterity's sake. I am guilty of that in some places, but I hope the reader will sort out the subject matter of greatest interest and enjoy it. It is my hope that used copies will be deposited in libraries so others can learn about these chronicles.

In recognition, I will begin with the person who made this book possible. Dr. Ken Huff, my editor and friend, has patiently polished off so many rough edges in my manuscript. Ken was a journalist for *Time* magazine and knows how to deal with untrained writers like me. He also helped with research and formatted the text and photographs for publication. Ken, I am forever grateful, more than you will ever know.

Dr. Stanley Heckadon-Moreno wrote the foreword and embellished it with so many comments that I don't deserve. He also found and helped to interview two members of the Probst family. Stanley has made so many contributions in so many ways. Thanks, Stanley.

Luis "Ibu" Alvarado spent much time locating and interviewing the Probst family in Panama. With the help of his brothers, he also wrote about the careers of his parents and himself in Chiriqui and Bocas del Toro. Patricia Alvarado gave valuable advice about the manuscript and publishing. The album of the Almirante School was special. Thanks to all of you.

Frank Amador Selles explained the Selles, Probst and Rassow family relationships, plus making contact with some of them for me to interview. *Un buen amigo y gracias, Frank.*

Dr. Jorge Selles discussed his brother-in-law Otto Probst and the Selles family. *Gracias, Doctor.*

Special thanks to Dr. Siegelinde Probst, Otto and Josef Probst in Panama City and Gretchen Probst Bentley of North Carolina. You shared your incredible story that the world needed to hear. Your survival is an example of determination in time of war. I am grateful for your contributions.

Dr. Jeanette Probst Selles dedicated much time relaying interviews from her father, Otto Probst. *Gracias, Doctora.*

Tom McCann provided sensitive corporate information about the United Fruit Company, especially about operations on the wharves and offices in New York and Boston. No one else knew it like you did. I am so grateful, Tom.

Rene Ariel Perez Price is now the leading historian in Bocas del Toro and magically finds lost documents hidden in old archives. Ariel, thanks for your many contributions, too.

Emily LaRusch is a dedicated genealogist of her Snyder family of Bocas del Toro and provided new facts about this old banana family. Thanks, Emily.

Mario Mena always responded with photos and authoritative answers about abaca, bananas, Guatemala and Aaron King questions. Thanks, Mario.

Gary Bratcher always helped with questions about Costa Rica and Colombia, for which I am grateful.

Millicent Crichton Empedocles was generous with numerous photos taken by her father of abaca in Guatemala and operations in the Dominican Republic. And thanks for writing your family story.

Ulises Pallais sent data about his grandfather, Hermann Grabowski, who later moved to Nicaragua. Thanks.

Dimitri Troetsch Wilcox gave information about his Panama German family and other Germans in his book. Thanks, Dimitri.

Richard Yudin always came to the rescue about banana railroads and banana operations. Thanks, Richard.

Dr. George Wilson wrote a section about him and Claudia traveling from New York to Honduras on a banana ship. Thanks, George.

Sandie, Hallie and Linda Baker provided hundreds of pictures and data on the Wilson and Lawrence families in Bocas del Toro in the late 1800s. Thank you all for an incredible labor of love.

Tito and Pepo Thomas provided historic data about Jose Thomas and other historic figures in Bocas del Toro. *Gracias, amigos.*

Vicente Pascual always gave me data on postcards and just about anything historical about Panama. He has shown me some of his colossal collection of historical treasures. Thanks for sharing, Vicente.

Bill Saravanja always had answers based on experience or what he learned from his Yugoslav father. Thanks, Bill.

Bernal Saborio always responded with quick and accurate answers on so many questions, especially about Costa Rica. I am ever so grateful, Bernal.

Harold Brent sent the story of his family on an abaca farm in Costa Rica. Thanks, Harold.

Wolfgang Schulz, always helped as a German translator. *Wunderbar, Herr Wolf!*

Thanks to my brilliant granddaughter Ashley Stephens for digitally converting numerous *Unifruitco* magazines so I could access them better.

Maria Kandler of Costa Rica provided details about the Kandler family and sent photographs. *Gracias, Maria.*

Stewart Redwood provided data about oil wells in Bocas del Toro. His electronic journal, *Tales Of Old Panama And Castilla Del Oro*, Vol. 3, No. 13, March 2022, published my "German Prisoners of War from Bocas del Toro." Stewart, thanks for your publications, always par excellence. Also, for references about Germans in Panama.

Mr. Federico Pardo steered me in the right direction to find the old oil well lost in the jungle. Thank you.

The two Stargardter chapters were made possible by the following: The five children—Alfredo, Francisco, Ana Elisa, Sandra and Tomas; Hans Stargardter, his daughter Susan Stargardter Jaafar; and his granddaughter Jessica Stargardter. Michael Heymann aligned the family globally. Andres Bustos Stargardter of Chile authored the book on his grandfather Max Stargardter and answered many questions. Mike Utley, Cameron Forsyth and Scott Dinsmore provided information while working with Fritz Stargardter. What a team effort! I thank you all.

Fernando de Lara Viñas's book, *Reminiscencias de un bananero*, was a valuable source of data used in the story of banana people in the Dominican Republic. I am deeply grateful for his many contributions to Dominican banana history.

Leovigildo Julian Lorenzo educated me about his indigenous people in so many ways. (Note: There is no word in the Ngäbere language to say thanks.)

Thanks to Dr. David Roubik for unraveling the mysteries of bee and orchid interactions—an incredible science.

Thanks to Dr. Erin Dillon for her expertise on sharks and palaeoecological studies in Bocas del Toro.

David Zemer provided philatelic records of German prisoners and passenger lists; Brad Wilde swapped data on Bocas del Toro postcards and Federico Brid availed old Bocas del Toro shipping records since the late 1800s. All were my guests at Hospital Point in 2019. Thanks also to Bob Karrer for his many historic contributions.

Thanks to good neighbor Colin Hill, my computer guru and map maker.

Simon James Lang told me family history about his godfather James Mass. Thank you, Simon.

Thank you, Ed Loyd, for your in-depth knowledge of United Fruit and Chiquita.

Thanks to my linguistic friend, Horace Thompson, who translated passages to Jamaican.

Eduardo Melendez answered and clarified so many questions. Thanks, *amigo*.

Joe Looby wrote a neglected account about the Peace Corps in Almirante. Thanks, Joe.

Thanks to Esperanza Judith Vargas for the history of her family, *Madame* and *Monsieur* Léger.

Thanks to Dr. Paul Messina for his family story from Italy to Panama.

John Munch told us what some of the kids did in Almirante. Thanks, John.

Very special thanks to Angelique Stephens, a student in 2014, for making numerous trips to the archives at Woodson Research Center, Fondren Library, at Rice University. She patiently took pictures of the many Osterhout letters with her cell phone and e-mailed them to me. A special thanks to Jose "Chepe" Oquendo for solving the mystery of the burial of Hasso von Wedel in Changuinola.

Elsie Howard gave many hours of interviews on Hasso von Wedel for which I am so grateful.

Charlotte Elton graciously provided a reference on abaca: *Guide to Summit Park*.

Thanks to Luis Melara for the reference about the Gabriel Garcia Marquez interview.

A heartfelt thank you to Teo Alvarado for taking part in the final edit of my manuscript.

Patricia Alvarado, editor and publisher, patiently and skillfully brought this book to fruition. I am so grateful, Pat.

About the Author
Clyde S. Stephens

Clyde S. Stephens was born in 1934 and is a fifth-generation Floridian. After graduating from the University of Florida in 1959, he worked for 32 years as an entomologist with the United Fruit Company, now Chiquita Brands International. His positions in banana research and technical services were in Central America and the Dominican Republic. In 1963, he met and married Phyllis Steffen, a Company school teacher in Honduras. Their two children, Linda and Carl, were born in Panama.

Clyde's career took him to banana-producing countries in Central and South America, the Caribbean, the Philippines and Taiwan. After retiring in 1991 to a solar-powered house on an island in Bocas del Toro, Panama, he became a banana consultant. As a volunteer with Winrock International, Stephens has advised hundreds of small banana farmers in Nepal, Bangladesh, India, Guinea and Mexico. He led friends and club members on eight nature, birding and banana tours in Central America. Clyde lives part of the year in Tavares, Florida.

Books by Clyde S. Stephens

Banana Man, Peeling and Revealing,
    Book 1
    Book 2
    Book 3

Banana People

Bananeros in Central America: True Stories of the Tropics

Bosquejo histórico de la provincia de Bocas del Toro, Panamá / Outline of History of the Province of Bocas del Toro, Panama

Bosquejo histórico del cultivo del banano en la provincia de Bocas del Toro (1880-1980)

La historia de Punta Hospital

History of Hospital Point

Stephens Ancestors and Pioneer Relatives

Thank you for buying this special edition of ***Banana Man***. I hope you enjoyed reading it as much as I enjoyed writing it. If you can, please visit wherever you bought the book, leave a review and recommend it to your friends. You can write to me at bananabyte@aol.com.

C. S. S.

Visit us!

Cecropia Press
cecropiapress.com

Made in United States
Troutdale, OR
01/11/2026

45495797R00444